Environmental Effects of Natural Processes and Human Activities on the Water Environment in Watershed

Environmental Effects of Natural Processes and Human Activities on the Water Environment in Watershed

Guest Editors

Guilin Han

Xiaolong Liu

MDPI

Basel • Beijing • Wuhan • Barcelona • Belgrade • Novi Sad • Cluj • Manchester

Guest Editors

Guilin Han
Institute of Earth Sciences
China University of
Geosciences (Beijing)
Beijing
China

Xiaolong Liu
Tianjin Key Laboratory of
Water Resources and Environment
Tianjin Normal Universty
Tianjin
China

Editorial Office
MDPI AG
Grosspeteranlage 5
4052 Basel, Switzerland

This is a reprint of the Special Issue, published open access by the journal *Water* (ISSN 2073-4441), freely accessible at: www.mdpi.com/journal/water/special_issues/watershed_water.

For citation purposes, cite each article independently as indicated on the article page online and using the guide below:

Lastname, A.A.; Lastname, B.B. Article Title. *Journal Name* **Year**, *Volume Number*, Page Range.

ISBN 978-3-7258-2878-4 (Hbk)
ISBN 978-3-7258-2877-7 (PDF)
https://doi.org/10.3390/books978-3-7258-2877-7

Contents

About the Editors

Guilin Han

Guilin Han (Prof.) is currently a professor at China University of Geosciences (Beijing). She received her Ph.D. in environmental geochemistry at the Institute of Geochemistry, Chinese Academy of Sciences, in Guiyang. She studies the surficial environment geochemistry in China. Her major contributions include developing methods for purifying K, K-Ca-Sr from a geological matrix for precise isotope analysis, measuring the Ca and Sr isotopic composition of rainwater in different ecosystems in China, and watershed weathering and global C cycles based on the Sr, C isotope. She has published many papers during her career in journals such as *Earth and Planetary Science Letters, Geochimica et Cosmochimica Acta, Water Research, Chemical Geology, Atmospheric Environment, Sciences of Total Environment, Journal of Hydrology, Journal of Analytical Atomic Spectrometry*, and *Agriculture Ecosystems and Environment*, among others.

Xiaolong Liu

Xiaolong Liu (Prof.) is a research professor at the Tianjin Key Laboratory of Water Resources and Environment, Tianjin Normal University. He received his Ph.D. in environmental geochemistry at the Institute of Geochemistry, Chinese Academy of Sciences, in Guiyang, and a B.S. in geochemistry from China University of Geosciences (Beijing). His research is mainly related to aquatic carbon and nitrogen biogeochemical cycles. He has great interests in the research on stable isotopic composition and its application in studying the carbon/nitrogen transformation in ecological systems and greenhouse gases, including N_2O, CO_2, and CH_4, and their impacts on the natural environment. He is the author or co-author of about 70 publications in international journals and 4 book chapters. He teaches topics related to global change and environmental geochemistry.

Preface

This Special Issue brings together a diverse collection of research and perspectives, offering insights into how natural processes and human activities interact within watersheds to shape water environments. Moreover, it explores various methodologies and measurements with the dual goals of providing implications for future water management as well as exploring potential alternatives. The papers included in this Special Issue draw from numbers of perspectives, including heavy metal source identification, water salinization and alkalization, water environment deterioration, pollutant assessment methods, the influences of fluvial geomorphological processes, and hydraulic characteristics of water environments. They also emphasize the importance of integrative approaches that consider both the ecological and social dimensions of watershed management. The contributions of this Special Issue highlight different facets of the environmental effects of natural processes and human activities on the water environment of different aquatic ecosystems, including rivers, streams, reservoirs, lakes, and groundwaters connected to surface water on a watershed scale.

Guilin Han and Xiaolong Liu
Guest Editors

water

Editorial

Environmental Effects of Natural Processes and Human Activities on the Water Environment in Watershed

Guilin Han [1,*] **and Xiaolong Liu** [2]

[1] Institute of Earth Sciences, China University of Geosciences (Beijing), Beijing 100083, China
[2] Tianjin Key Laboratory of Water Resources and Environment, Tianjin Normal Universty, Tianjin 300387, China; xiaolong.liu@tjnu.edu.cn
* Correspondence: hanguilin@cugb.edu.cn

1. Introduction

The Special Issue focused on the water environment changes within watersheds due to its critical role in sustaining ecosystems, supporting biodiversity, and providing essential resources for human life. The water environment in a watershed is also highly sensitive to both natural processes and human activities, which can significantly alter its balance [1,2]. As an essential contributor to global biodiversity and ecological productivity, watershed aquatic ecosystems provide a wide range of services for all life on Earth, including water for drinking, irrigation, power generation and industry, biological habitat, food, and recreation places. However, the direct and/or indirect threats or changes in watershed water environments and hydrology driven by human activities and natural processes have significantly increased [3,4]. Globally, the environmental problems caused by water and hydrology changes in watersheds have been a matter of concern for more than a century [5,6]. In addition to water quality deterioration, a series of water environment changes, including eutrophication, pollutant exposure and migration, and nutrient biogeochemical cycling, have become on-going research hotspots since these are essential to current water protection and management policy makers [7,8].

In this Special Issue, we assess the fate and transformation of nutrients, chemical ions, heavy metals, trace elements, and other pollutants at different watershed scales and discuss the influence of anthropogenic and natural processes. Generally, the papers in this Special Issue focus on the following: (1) revealing the watershed-scale transformation and migration of nutrients and heavy metals and how they are influenced by human activities and natural processes; (2) modeling and assessing the pollution risks and sources of solutes, including nutrients, ions, and heavy metals; (3) evaluating the environmental effects of dam construction, power generation operation, and protocols on the water environment in watersheds.

check for updates

Citation: Han, G.; Liu, X. Environmental Effects of Natural Processes and Human Activities on the Water Environment in Watershed. *Water* **2024**, *16*, 2929. https://doi.org/10.3390/w16202929

Received: 30 September 2024
Accepted: 11 October 2024
Published: 15 October 2024

2. Contributions to the Special Issue

The papers in this Special Issue addressed specific topics such as the evolution of the water environment in river basins, the sources of water solutes, the tracking and assessment of pollution sources, and the effects of changing hydrological regimes on the watershed environment. The papers described and analyzed water issues and new ideas in several typical watersheds, including rivers in the Tibetan Plateau, the karst region, the black soil region of northeast China, and rivers in the UK. Beside that, some papers also emphasized the important role of reservoirs for water environment changes in a watershed, especially in the largest reservoir in the world, the Three Gorges Reservoir. In this editorial, we discuss the contributions of these papers in this Special Issue in light of the pollutants types, characteristics of the source apportioning, and environmental effects.

2.1. Nutrient Accumulations and Water Environment Deterioration

The surface water and groundwater in the watershed are subjected to significant external pollutants, such as agricultural fertilizers, soil erosion, surface runoff, urban sewage, and industrial wastewater, leading to a rapid accumulation of nutrients and pollutants [7,9]. Various environmental issues, including threats to drinking water safety, agricultural irrigation efficacy, and hypoxic conditions in water bodies, ensued [10]. Several of the pieces in this issue, including those by Wang et al. (contribution 1), Zhang et al. (contribution 2), and Li et al. (contribution 3), specifically consider the effects of the anthropogenic activities on the riverine nutrient accumulation, while the study conducted in the Pearl River in south China clarifies the nutrient level in the river water was promoted by the fluvial geomorphological processes and its driving forces.

In their article on "Pollutants Source Assessment and Load Calculation in Baiyangdian Lake Using Multi-Model Statistical Analysis", Wang et al.'s (contribution 1) study focuses on identifying the primary sources of pollution in Baiyangdian Lake, North China, and calculating their contribution to the water quality degradation. The study found that agricultural and rural non-point source pollution are the main contributors to riverine nitrogen, phosphorus, and organic pollutants in Baiyangdian Lake, accounting for more than 50% of the total load. By employing the models like the export coefficient model (ECM), the revised universal soil loss equation (RUSLE), and PLOAD, the study quantifies the nitrogen, phosphorus, and chemical oxygen demand (COD) entering the lake from various sources such as agriculture, domestic sewage discharge, soil erosion, and runoff. The authors suggested that better management of agricultural pollution, including livestock waste treatment, reducing fertilizer use, and controlling soil erosion, would be helpful for maintaining good water quality. This study provides a comprehensive approach to pollutant source identification and load estimation, aiding in the future management and restoration efforts for Baiyangdian Lake.

In "Hydrochemical Characteristics, Water Quality, and Evolution of Groundwater in Northeast China", Zhang et al. (contribution 2) found that the water quality of the groundwater in Northeast China, specifically in the Sanjiang Plain, was good and suitable for drinking. The majority of the samples analyzed in the study were categorized as high-quality water sources, accounting for approximately 93.8% of the total samples. The water quality index (EWQI) values were lower than the World Health Organization's drinking water standards. The study also found that the concentration of NO_3^- in water was the main influencing factor for water quality. The presence of low-quality water samples was primarily associated with human activities in densely populated cities. Additionally, the chemical composition of groundwater is mainly influenced by the dissolution of carbonate rock and silicate rock, as well as the alternative adsorption of cations and human activities. Overall, the study provides a hydrochemical basis for understanding the interaction between groundwater and the environment in the Sanjiang Plain and can serve as a reference for groundwater management in cold areas globally.

In "Evolution Characteristics of Water Quality in Plain Reservoirs and Its Relationship with the Economic Development Response: A Case Study of Daheiting Reservoir in Northern China", Li et al. (contribution 3) concluded that the water quality changes in Daheiting Reservoir are closely related to upstream water, economic development, and cage fish culture. The study found that the concentration of total phosphorus (TP) and ammonia nitrogen (NH_3-N) in the reservoir is significantly correlated with the upstream Panjiakou Reservoir and the primary industry, particularly cage fish farming. During the second stage of the reservoir, the exogenous input of nutrients from upstream water and fish culture was the main factor affecting the TP and NH_3-N concentration. The study also observed that the water quality improved due to the slowdown in economic development and strengthened environmental protection efforts. Overall, this study provides insights into the evolution of water quality in Daheiting Reservoir and its relationship with economic development, and the findings contribute to the understanding of the impact of economic development on water resources and the need for environmental protection measures.

In the article on "Sustaining the Pearl River: A Critical Review of Changes in Fluvial Geomorphological Processes and the Driving Forces in the Pearl River Basin", Ou et al. (contribution 4) concluded that the fluvial geomorphological processes in the Pearl River Basin have undergone significant changes in recent decades, seriously affecting the river's sustainable development. These changes are attributed to anthropogenic impacts, such as the construction of reservoirs and water pollution from industrial and domestic wastewater. The article recommends enacting comprehensive regulations for the protection and utilization of water resources in the Pearl River Basin to strengthen and standardize ecological protection efforts. The contributions from this article include the examination of the impacts of human activities and climate change on fluvial geomorphological processes, as well as the analysis of influencing factors such as anthropogenic impacts and land cover change.

2.2. *The Source Identification and Pollution Risk Assessment of Heavy Metals*

Heavy metal contamination in watershed waters has consistently been a significant research hotspot of water environment in aquatic systems. Heavy metals are generally featured with diverse origins, enduring toxicity, resistance to biological degradation, and ongoing health risks to living organisms [11,12]. Consequently, the assessment of heavy metal pollution, particularly its occurrence, source identification, and contamination in aquatic and terrestrial environments, holds significant environmental importance and practical necessity for the protection of watershed water quality and the development of water management policies [13,14]. Several papers in this Special Issue studied heavy metal contamination and source identification at the watershed scale.

In "Geochemistry and Sources Apportionment of Major Ions and Dissolved Heavy Metals in a Small Watershed on the Tibetan Plateau", Xing et al. (contribution 5) studied water quality in the Duilong Qu (DLQ) watershed on the Tibetan Plateau, focusing on the origins of heavy metals. The concentrations of heavy metals, such as arsenic (As), lead (Pb), and chromium (Cr), are higher than the global average but still below World Health Organization (WHO) standards. Arsenic concentrations are significantly impacted by geothermal activity, especially during the low-flow season when dilution is weaker. Using the PCA-APCS-MLR model, the study identifies that most heavy metals, including arsenic, chromium, manganese, and iron, originate from geothermal water. Meanwhile, copper and nickel are linked to mining activities, and lead and zinc stem from natural sources. Increased precipitation and runoff due to climate change have led to reduced arsenic concentrations in DLQ over the years, highlighting the dilution effects of increased river flows during the high-flow season. This study emphasizes the joint effects of natural processes (geothermal activity, rock weathering) and human activities (mining) on water quality from the perspective of heavy metals in the Tibetan Plateau's small watersheds and suggests the need for careful monitoring to prevent heavy metal contamination.

In the paper "Distribution Characteristics and Genesis of Iron and Manganese Ions in Groundwater of Eastern Sanjiang Plain, China", Wang et al. (contribution 6) concluded that the iron and manganese in groundwater exist in the form of Fe^{2+} and Mn^{2+}, respectively. The primary sources of iron and manganese are the minerals in the original geological environment of the study area. Surface runoff and soil organic matter contribute to the reduction of groundwater, leading to excessive iron and manganese content. High mineralization and evaporation concentrations are conducive to increased iron content. The study provides insights into the sources and factors influencing the high levels of iron and manganese in groundwater, which is important for understanding water safety and addressing potential health risks. Additionally, the study highlights the need for further research on treatment methods and strategies for high-iron and high-manganese groundwater, as well as the relationship between agricultural activities and iron and manganese content.

The paper titled "Spatial Distribution and Sources of Rare Earth Elements in Urban River Water: The Indicators of Anthropogenic Inputs" (contribution 7) explores the distributed characteristics of rare earth elements (REEs) in urban river systems, evaluates their sources, and further emphasizes the contributions of anthropogenic input. The study reveals that rare Earth elements display uneven spatial distribution in the urban river water, with higher concentrations observed near industrial and urbanized areas. This suggests significant human impact on the distribution of these elements. The research identifies that anthropogenic activities, particularly industrial discharges, urban runoff, and wastewater effluents, are major contributors to the presence of REEs in river water. These elements are often used in technology, manufacturing, and urban infrastructure, leading to their elevated concentrations in urban waterways. The presence and concentration of REEs, particularly heavy REEs like dysprosium (Dy) and neodymium (Nd), serve as indicators of human-induced pollution. Their correlation with other pollutants, such as heavy metals, reinforces their role as markers of industrial and urban pollution. The paper raises concerns about the potential ecological and health risks posed by elevated REE concentrations, given their persistence in the environment and possible bioaccumulation in aquatic ecosystems. The authors recommend increased monitoring of REEs in urban rivers as part of broader water quality management strategies. Understanding their distribution can help mitigate the impacts of urbanization and industrial activities on water environments. The study underscores the importance of recognizing REEs as critical indicators of anthropogenic pollution in urban water systems.

2.3. The Origin and Transportation of Major Ions and Solutes

Major ions and solutes in river water are influenced and dominated by numerous sources and geochemical processes. The fluctuations in major ion chemistry of surface water and groundwater facilitate the identification of geochemical processes that govern water quality [15]. The weathering of minerals within rocks is a significant process that regulates the concentration of dissolved ions [16]. Salinity in river water increases with rising groundwater levels due to the dissolution of additional mineral salts found in soils and weathered products [17]. The impact of evaporation and irrigation return flows on the major ion chemistry of river water has been highlighted globally [18]. However, it is still inadequate to understand how the natural processes and human activities alter the chemical composition of water. In this Special Issue, three papers focused on the variation and origins of major ions in river water.

Xing et al. (contribution 5) found that the water chemistry in a small watershed in the Tibetan Plateau is mainly controlled by the weathering of carbonate rocks, with the dominant ions being calcium (Ca^{2+}) and bicarbonate (HCO_3^-). High sulfate (SO_4^{2-}) and chloride (Cl^-) concentrations were observed in this small watershed, illustrating that major ions in the river water were affected by geothermal springs. This showed the importance of natural processes in this region. The paper titled "Hydrochemical Characteristics and Risk Assessment of Tongzi River, Guizhou Province, Southwest China" (contribution 8) investigates the water quality, hydrochemical features, and potential risks of the Tongzi River and identified that the water chemistry of the Tongzi River is dominated by calcium (Ca^{2+}) and bicarbonates (HCO_3^-), suggesting that the river's water composition is primarily controlled by the weathering of carbonate rocks in the region. The research shows that the concentrations of major ions are within the permissible limits for drinking water based on China's national standards.

Jiang et al. (contribution 9) investigated long-term trends of river salinization and alkalinization in UK river water. The study found that UK rivers have generally become more alkaline but less salty over the past 20 years. This is contrary to the initial hypothesis of increasing salinization but aligns with the global trend of rising river alkalinity. The decrease in salinity was largely influenced by reduced road salting and other anthropogenic factors, while the increase in pH was strongly linked to agricultural activities, land use changes, and urbanization. The study also highlights regional and seasonal variations in both conductivity and pH, particularly showing greater changes in rivers passing through areas with higher anthropogenic disturbances, such as urban and agricultural lands. Moreover, the findings underscore the importance of understanding these trends to better manage freshwater resources and mitigate the risks posed by river alkalinization and salinization.

This research offers valuable insights into the environmental impacts of anthropogenic activities on UK rivers and highlights the importance of monitoring freshwater ecosystems.

2.4. The Effect of Hydraulic Characteristics on Water Environments in Large Reservoirs

In the paper "The Impact of the Three Gorges Reservoir Operations on Hydraulic Characteristics in the Backwater Region: A Comprehensive 2D Modeling Study", Xu et al. (contribution 10) applied a comprehensive 2D modeling approach to analyze the impact of Three Gorges Reservoir (TGR) operations on hydraulic characteristics in the backwater region. The researchers designed a hydrodynamic and water quality model and implemented 15 operational scenarios, considering flood season, drawdown, and impoundment periods. They used an advanced dynamic storage capacity method to improve the precision of water level simulations. The study also analyzed the sensitive, sub-sensitive, and insensitive areas during different reservoir operation periods and quantified the delay of temperature waves. The findings in this work provide critical insights into the complex sensitivities of the TGR to varied operational modes, aiding in the development of eutrophication and water resources control strategies. Additionally, the modeling application offers different operational scenarios and insights for ecological management strategies in large dam systems globally, informing future water resource management and policymaking.

3. Conclusions Remarks

The articles in this Special Issue examine the spatiotemporal variations, origins, and migratory and transformation features of chemicals in surface water and groundwater of the watershed, including major ions, nutrients, heavy metals, and other pollutants. The topics and understandings in this Special Issue could be very helpful for comprehending and identifying the prevailing trends and developmental trajectories of the aquatic system's water environment changes. The markedly divergent research findings from various regions demonstrate temporal and spatial variations in the influence of natural processes and human activities on the sources and variations of nutrients and pollutants in water. Future study should persist in exploring the research interest highlighted in this Special Issue, such as (1) the watershed-scale transformation and migration mechanisms of nutrients and other pollutants influenced by human activities and natural progress; (2) mechanisms of different kinds of water pollution and deterioration in watershed, including eutrophication, hypoxia, and greenhouse gas emissions; (3) quantitatively and qualitatively identify the contribution of potential sources of nutrients and pollutants; and (4) assessing the environmental effects of human activities coupling with natural processes on biogeochemical cycling and transformations of nutrients and pollutants in watershed.

Author Contributions: Conceptualization, G.H.; methodology, G.H. and X.L.; investigation, G.H. and X.L.; data curation, G.H. and X.L.; writing—original draft preparation, G.H.; writing—review and editing, G.H.; supervision, G.H. All authors have read and agreed to the published version of the manuscript.

Conflicts of Interest: The authors declare no conflicts of interest.

List of Contributions

1. Wang, G.; Lv, C.; Gu, C.; Yu, Y.; Yang, Z.; Zhang, Z.; Tang, C. Pollutants Source Assessment and Load Calculation in Baiyangdian Lake Using Multi-Model Statistical Analysis. *Water* **2022**, *14*, 3386.
2. Zhang, T.; Wang, P.; He, J.; Liu, D.; Wang, M.; Wang, M.; Xia, S. Hydrochemical Characteristics, Water Quality, and Evolution of Groundwater in Northeast China. *Water* **2023**, *15*, 2669.
3. Li, B.; Chen, K.; Liu, X.; Liu, C.; Wang, S. Evolution Characteristics of Water Quality in Plain Reservoirs and Its Relationship with the Economic Development Response: A Case Study of Daheiting Reservoir in Northern China. *Water* **2023**, *15*, 3229.

4. Ou, H.; Cai, S.; Fan, W.; Qiu, J.; Mu, X.; Zhou, T.; Yang, X.; Picco, L. Sustaining the Pearl River: A Critical Review of Changes in Fluvial Geomorphological Processes and the Driving Forces in the Pearl River Basin. *Water* **2024**, *16*, 1001.
5. Xing, W.; Wei, L.; Ma, W.; Li, J.; Liu, X.; Hu, J.; Wang, X. Geochemistry and Sources Apportionment of Major Ions and Dissolved Heavy Metals in a Small Watershed on the Tibetan Plateau. *Water* **2022**, *14*, 3856.
6. Wang, M.; Wang, M.; Yang, L.; Yang, T.; Li, J.; Chen, Y. Distribution Characteristics and Genesis of Iron and Manganese Ions in Groundwater of Eastern Sanjiang Plain, China. *Water* **2023**, *15*, 2068.
7. Gao, X.; Han, G.; Liu, J.; Zhang, S. Spatial Distribution and Sources of Rare Earth Elements in Urban River Water: The Indicators of Anthropogenic Inputs. *Water* **2023**, *15*, 654.
8. Lü, J.; An, Y. Hydrochemical Characteristics and Risk Assessment of Tongzi River, Guizhou Province, Southwest China. *Water* **2023**, *15*, 802.
9. Jiang, S.; Wu, X.; Du, S.; Wang, Q.; Han, D. Are UK Rivers Getting Saltier and More Alkaline? *Water* **2022**, *14*, 2813.
10. Xu, Y.; Yu, S.; Liu, D.; Ma, J.; Chuo, M. The Impact of the Three Gorges Reservoir Operations on Hydraulic Characteristics in the Backwater Region: A Comprehensive 2D Modeling Study. *Water* **2024**, *16*, 2045.

References

1. Akhtar, N.; Syakir Ishak, M.I.; Bhawani, S.A.; Umar, K. Various natural and anthropogenic factors responsible for water quality degradation: A review. *Water* **2021**, *13*, 2660. [CrossRef]
2. Goudie, A.S. *Human Impact on the Natural Environment*; John Wiley & Sons: New York, NY, USA, 2018.
3. Duan, W.; Takara, K. *Impacts of Climate and Human Activities on Water Resources and Quality*; Springer: Singapore, 2020.
4. Khatri, N.; Tyagi, S. Influences of natural and anthropogenic factors on surface and groundwater quality in rural and urban areas. *Front. Life Sci.* **2015**, *8*, 23–39. [CrossRef]
5. Poulopoulos, S.G.; Inglezakis, V.J. *Environment and Development: Basic Principles, Human Activities, and Environmental Implications*; Elsevier: Amsterdam, The Netherlands, 2016.
6. Barnett, T.P.; Pierce, D.W.; Hidalgo, H.G.; Bonfils, C.; Santer, B.D.; Das, T.; Bala, G.; Wood, A.W.; Nozawa, T.; Mirin, A.A. Human-induced changes in the hydrology of the western United States. *Science* **2008**, *319*, 1080–1083. [CrossRef] [PubMed]
7. Liu, X.-L.; Han, G.; Zeng, J.; Liu, M.; Li, X.-Q.; Boeckx, P. Identifying the sources of nitrate contamination using a combined dual isotope, chemical and Bayesian model approach in a tropical agricultural river: Case study in the Mun River, Thailand. *Sci. Total Environ.* **2021**, *760*, 143938. [CrossRef] [PubMed]
8. Zeng, J.; Han, G.; Yang, K. Assessment and sources of heavy metals in suspended particulate matter in a tropical catchment, northeast Thailand. *J. Clean. Prod.* **2020**, *265*, 121898. [CrossRef]
9. Wang, F. Impact of a large sub-tropical reservoir on the cycling of nutrients in a river. *Water Res.* **2020**, *186*, 116363. [CrossRef] [PubMed]
10. Bond, T.; Huang, J.; Templeton, M.R.; Graham, N. Occurrence and control of nitrogenous disinfection by-products in drinking water—A review. *Water Res.* **2011**, *45*, 4341–4354. [CrossRef] [PubMed]
11. Ouyang, W.; Wang, Y.; Lin, C.; He, M.; Hao, F.; Liu, H.; Zhu, W. Heavy metal loss from agricultural watershed to aquatic system: A scientometrics review. *Sci. Total Environ.* **2018**, *637*, 208–220. [CrossRef] [PubMed]
12. Moore, J.W.; Ramamoorthy, S. *Heavy Metals in Natural Waters: Applied Monitoring and Impact Assessment*; Springer Science & Business Media: Berlin/Heidelberg, Germany, 2012.
13. Vareda, J.P.; Valente, A.J.; Durães, L. Assessment of heavy metal pollution from anthropogenic activities and remediation strategies: A review. *J. Environ. Manag.* **2019**, *246*, 101–118. [CrossRef] [PubMed]
14. Ali, M.M.; Ali, M.L.; Islam, M.S.; Rahman, M.Z. Preliminary assessment of heavy metals in water and sediment of Karnaphuli River, Bangladesh. *Environ. Nanotechnol. Monit. Manag.* **2016**, *5*, 27–35. [CrossRef]
15. Ivanova, I.; Savichev, O.; Trifonov, N.; Kolubaeva, Y.V.; Volkova, N. Major-ion chemistry and quality of water in rivers of Northern West Siberia. *Water* **2021**, *13*, 3107. [CrossRef]
16. Tsering, T.; Wahed, M.S.A.; Iftekhar, S.; Sillanpää, M. Major ion chemistry of the Teesta River in Sikkim Himalaya, India: Chemical weathering and assessment of water quality. *J. Hydrol. Reg. Stud.* **2019**, *24*, 100612. [CrossRef]
17. Berger, E.; Frör, O.; Schäfer, R.B. Salinity impacts on river ecosystem processes: A critical mini-review. *Philos. Trans. R. Soc. B* **2019**, *374*, 20180010. [CrossRef] [PubMed]
18. Misaghi, F.; Delgosha, F.; Razzaghmanesh, M.; Myers, B. Introducing a water quality index for assessing water for irrigation purposes: A case study of the Ghezel Ozan River. *Sci. Total Environ.* **2017**, *589*, 107–116. [CrossRef] [PubMed]

water

Article

Spatial Distribution and Sources of Rare Earth Elements in Urban River Water: The Indicators of Anthropogenic Inputs

Xi Gao, Guilin Han *, Jinke Liu and Shitong Zhang

Institute of Earth Sciences, China University of Geosciences, Beijing 100083, China
* Correspondence: hanguilin@cugb.edu.cn; Tel.: +86-10-8232-3536

Abstract: Rare earth element (REE) concentrations in river water reflect both local geological background and the intensity of anthropogenic activities. The increasing demand for high-tech applications is accompanied by an increase in REEs in water environments, especially in urban regions. Therefore, the dissolved REEs in a typical urban river (Yongding River) were analyzed to reveal the influence of anthropogenic inputs. It was found that the sum of dissolved REE (ΣREE) concentrations in river water were 6.27~86.26 ng L^{-1} (mean 31.05 ng L^{-1}). The spatial distribution of REEs is significantly affected by both natural processes and human activities. To eliminate the effect of upstream land use on downstream measurements, this study established a 500 m buffer zone alongside the river and identified the anthropogenic origin of REEs through correlations among land use proportions within the buffer zone, the population density, and REEs. The Post-Archean Australian Shale (PAAS)-normalized REEs revealed pronouncedly positive Gd anomalies in the range 0.30~20.16 in all river samples. Spearman correlation analysis indicated that the proportion of cultivated land was positively correlated with most of the REEs. A three-dimensional tracer system was established via Gd_{anth}, NO_3^-/Na^+ and Cl^-/Na^+ to characterize the impact of sewage treatment plants, hospitals and agricultural activities on river water. The results indicate that with the acceleration of urbanization, abnormal REEs can be considered as a sensitive indicator to assess the influence of anthropogenic activities on water ecosystems.

Keywords: rare earth elements; Yongding River; agricultural activities; urban sewage; land use; anthropogenic Gd

Citation: Gao, X.; Han, G.; Liu, J.; Zhang, S. Spatial Distribution and Sources of Rare Earth Elements in Urban River Water: The Indicators of Anthropogenic Inputs. *Water* **2023**, *15*, 654. https://doi.org/10.3390/w15040654

Academic Editors: Junhong Bai and César Andrade

Received: 3 January 2023
Revised: 28 January 2023
Accepted: 6 February 2023
Published: 8 February 2023

1. Introduction

Rare earth elements (REEs) generally comprise fifteen Lanthanide elements from Lanthanum (La) to Lutetium (Lu) [1] and Scandium (Sc) and Yttrium (Y) [2–4]. This group of elements exhibit similar chemical properties and geochemical behaviors [3,5,6]; however, Cerium (Ce) and Europium (Eu) are two particular elements which can stably exist under certain environmental conditions at the valence of +4 and +2, respectively [2,7–9]. REEs enter the water, soil and sediments mostly via denudation (a geological process eroding the land's surface via moving water, wind, ice, and waves), transportation, weathering, and deposition of hydrocarbon-source rocks [10–12]. In addition, REEs are also regarded as key technology elements that are widely applied in electric vehicle engines, catalytic converters, agriculture, biomedicine, and other modern industrial fields [13–15]. Therefore, the worldwide demand for REEs has grown rapidly in the past decade [16,17], resulting in increasing amounts of REEs being released to the environment [18,19].

The REEs released into the environment are ecologically toxic. They can further accumulate in the food chain and be absorbed by the human body through food intake. Therefore, long-term exposure/ingestion of low-dose REEs may have adverse effects on the metabolic processes of the human body [20]. For example, excessive REEs consumption could cause abnormal biochemical indications in the blood and lead to the elevation of cholesterol [21]. In addition, recent works have revealed that some Gd chelates decomposed in vivo are enriched in tissues, bones and the human brain [22,23]. Song et al. (2017) reported positive anomalies of Gd associated with anthropogenic origins and slightly positive anomalies of La and Sm from point source pollution in the Han River, South Korea [24]. It is likely that excessive human-derived REEs alter the original REEs distribution pattern in aquatic ecosystems, leading to harmful consequences [25,26]. For instance, Gd can bioaccumulate in freshwater bivalve tissues and trigger a brief biochemical reaction [27]. In addition, Gadoteric acid is a medical MRI contrast agent. Although it does not cause lethal effects at an exposure level of around 200 mg/L, Gd may inhibit the growth zebrafish cell lines at levels as low as 20 ng/L [28]. Bispo et al. (2021) observed the REE concentrations of topsoils in Brazil and found that the divergence of REEs content in cultivated soil resulted from the variation of land use and agricultural inputs such as phosphate fertilizer [29]. The variability in soil REE concentration is influenced by parent material, land use, pedogenic processes and biomes [29]. In addition, cultivated soils could potentially be polluted by REE via the input of phosphate fertilizers. Phosphate rock is the raw material for the manufacture of phosphate fertilizers and are widely applied as a source of phosphorus [30]. Phosphates naturally contain REEs, and thus may be of great concern for environmental contamination [31]. Liu et al. (2022) also reported the degree of REE anomalies in 13 artificial lakes and the relationships between artificial REE concentrations in lakes and the surrounding environment. They accurately characterized the impact of sewage treatment plants, hospitals, factories and agriculture on lakes [32]. Therefore, a detailed investigation on anthropogenic-derived REEs is rather urgent to protect the human health of local residents.

Tianjin and Beijing are two famous mega cities in north China with a total of more than 30 million residents [33]. By 2020, the permanent population had reached 18.33 million, including an urban population of 14.81 million (accounting for 80.80%) and a rural population of 3.65 million (19.91%). However, the study area is a typical low-flow urban river with poor capacity for self-purification [34]. Although recently restoration measures have been undertaken in river ecosystems by the Chinese government [35], the water quality of the study region is still far from satisfactory due to inadequate investigation of urban water pollution and a tremendous quantity of point/non-point pollution sources [36]. Therefore, the mixed toxicity of various REEs and the water quality in this typical urban river are still unclear at present and require more attention. This study collected 17 water samples in the Yongding River and analyzed the geochemical characteristics of dissolved REEs. The main aims of the study are to (1) determine the variation in and spatial distribution of REE content in the Yongding River, (2) identify the influence of population, land use, hospitals and other factors on the abnormal concentration of REEs, and (3) track the pollution sources in this typical urban river via REEs and hydro-chemical indicators.

2. Materials and Methods

2.1. Study Area

The Yongding River is 747 km in length with an area of 47,016 km^2 and drains through Inner Mongolia and Hebei Province before finally entering Bohai Bay (Figure 1) [37]. The Sanggan River and the Yang River are referred to as the headwaters [38]. The basin is in a semi-humid/semi-arid transition zone and has a temperate continental monsoon climate with an average temperature of 12.4 °C and precipitation of 513 mm annually [39]. The main land use types include woodland, cropland, grassland, urban land, wetland and unused land, of which grassland and farmland occupy 34.33% and 42.86% of the basin area, respectively [37]. The mountainous upstream areas are rich in mineral resources and are also an important energy base, renewable energy demonstration area in China as well as an important area for production of food and vegetables. The central part of the basin has the densest population and highest urbanization rate. This area will become a cluster of high-quality service industries, technology and modern manufacturing in the future. Furthermore, the New Coastal District of Tianjin city is also regarded as a development base of manufacturing, modern service industries and scientific and technological innovation, which will become the northern shipping logistics center and an important economic development zone in the future.

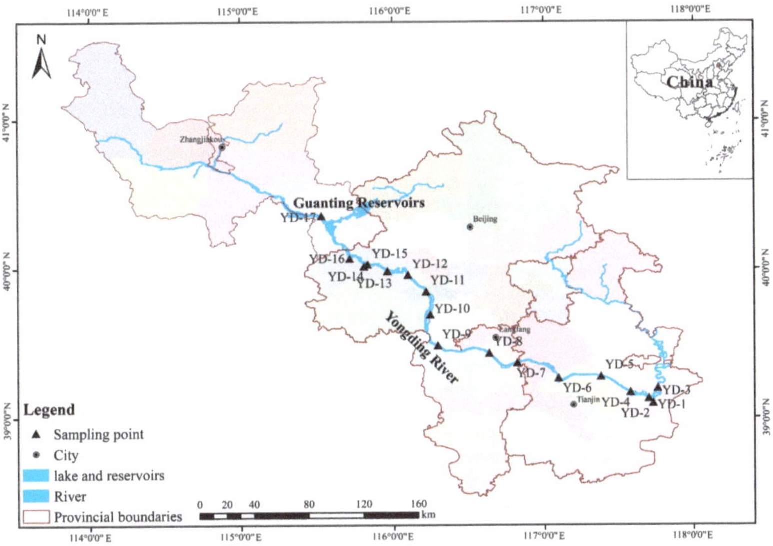

Figure 1. Sampling sites on the Yongding River.

2.2. Sampling and Analysis

A systematic sample collection process was conducted at the depth of 0~50 cm in the Yongding River in August 2021, among which 16 sampling points were located in the mainstream with one from the tributary. In total, 17 river sites were sampled and filtered via a 0.22 μm cellulose membrane. The samples were then acidified via ultra-purified HNO_3 to pH < 2 and sealed in pre-cleaned polyethylene bottle in a refrigerator for later measurement. The samples were then prepared for analysis at the Surficial Environment and Hydrological Geochemistry Laboratory (SEHGL, class 1000). First they were concentrated through evaporation at 60 °C for 10–30 min and then re-dissolved in distilled 2% HNO_3), together with 2 blank samples and 1 reference sample (SLRS-6, certified natural water) [40]. The REE concentrations of the Yongding River samples were analyzed using Agilent 7850 ICP-MS at the Key Lab of the Groundwater Circulation and Environmental Evolution (GCEE). The 5-point calibration of standards including the blank was conducted using the multi-element solution GSB 04-1789-2004 (100 mg/L) through a gravimetrically quantitative dissolution process. The concentration ranges of calibrations were prepared in distilled 2% HNO_3 as follows: 10 μg/L, 1 μg/L, 100 ng L^{-1}, and 10 ng L^{-1}. The instrumental drift was corrected by the automatic addition of internal standard solution (ISTD), and the quality control was performed via the calibration standard of 1 ug/L (medium concentration). The diluent 2% distilled HNO_3 was regarded as the sample of the continuing calibration blank (CCB), which periodically repeated during the whole sequence. The accuracy of repeated analysis (10 times) ranged from 0.2% (Ce) to 15.1% (Lu) of the standard materials, with an average of 5.3%, and the relative standard deviation of the REE concentrations was <6% on average for the individual analyses.

The REE contents were normalized to the Post-Archean Australian Shale (PAAS) standard in order to remove the influence of the Oddo–Harkins rule that results in a typical zigzag shape of the lanthanides' abundance in logarithmic plots of content versus atomic number [41]. δCe, δEu, δGd are the Ce, Eu, and Gd anomalies, respectively. δCe, δEu, δGd, and Gd_{anth} can thus be calculated through the following formulas [42–44]:

$$\delta Ce = Ce_N/Ce^* = Ce_{N}/(La_N \times Nd_N)^{1/2} \quad (1)$$

$$\delta Eu = Eu_N/Eu^* = Eu_{N}/(Sm_N \times Gd_N)^{1/2} \quad (2)$$

$$\delta Gd = Gd_N/Gd^*_N = Gd_{N}/(0.33Sm_N + 0.67Tb_N) \quad (3)$$

$$Gd_{anth} = Gd_{measured} - Gd^*_N \times Gd_{PAAS} \quad (4)$$

Here, the subscript of N and * are the normalized value relative to PAAS and the geologic background value, respectively. Gd_{anth}, $Gd_{measured}$, and Gd_{PAAS} are sthe Gd concentrations originating from anthropogenic sources, the measured Gd value, and the Gd concentration in PAAS, respectively.

3. Results and Discussion

3.1. The Content of REEs

The contents of all rare earth elements and the total REE (∑REE) concentrations in the Yongding River samples are presented in Tables 1 and 2. It is clear that the ∑REE content was generally low, with a range of 6.27~86.26 ng L^{-1} (average 31.05 ng L^{-1}), which is far lower than the global average of 532 ng L^{-1} and other rivers such as the Yangtze River and the Connecticut River [1,37,45]. However, compared with other urban rivers, the ∑REE concentration in the Yongding River was higher than in the urban river in Guiyang city and lower than concentrations the Han River, the Pearl River and the Rhine River [24,46–48].

Table 1. Dissolved REE contents in the Yongding River (in ng/L).

Site	La	Ce	Pr	Nd	Sm	Eu	Gd	Tb	Dy	Ho	Er	Tm	Yb	Lu
YD-1	1.53	3.77	0.90	4.94	1.08	0.75	5.15	0.30	3.08	0.72	2.49	0.48	4.58	0.90
YD-2	5.79	5.67	1.78	8.81	1.90	0.83	11.84	0.42	4.01	1.04	3.32	0.53	4.01	0.83
YD-3	2.61	3.42	0.84	4.02	0.96	0.63	4.44	0.21	2.07	0.39	1.71	0.39	3.54	0.81
YD-4	15.79	14.50	3.35	15.71	3.71	1.74	8.50	0.65	6.94	1.65	5.18	0.88	6.38	1.29
YD-5	5.43	5.46	4.87	16.19	0.71	0.50	12.45	0.15	1.65	0.44	1.74	0.29	2.98	0.71
YD-6	5.85	8.20	1.51	7.25	2.35	0.77	7.46	0.45	4.28	0.92	2.91	0.50	4.81	1.19
YD-7	4.41	4.71	1.43	6.50	1.16	0.60	4.29	0.27	2.15	0.54	1.73	0.30	1.67	0.30
YD-8	2.24	3.35	0.57	2.60	0.66	0.78	7.81	0.12	0.87	0.24	0.81	0.15	1.05	0.24
YD-9	1.28	3.81	0.66	2.71	0.74	0.36	3.40	0.15	1.58	0.36	1.34	0.21	1.49	0.27
YD-10	0.87	0.99	0.33	1.53	0.45	0.54	0.96	0.09	0.81	0.24	0.78	0.12	0.75	0.09
YD-11	3.98	5.93	1.05	3.92	0.39	0.30	20.51	0.06	0.51	0.15	0.45	0.09	0.57	0.15
YD-12	0.99	1.61	0.24	0.90	0.15	0.27	1.37	0.03	0.24	0.06	0.15	0.03	0.15	0.09
YD-13	1.35	2.93	0.36	1.44	0.33	0.30	2.18	0.03	0.27	0.03	0.15	0.03	0.15	0.06
YD-14	1.70	2.62	0.36	1.43	0.21	0.27	0.48	0.03	0.30	0.06	0.09	0.03	0.12	0.03
YD-15	1.96	3.75	0.60	3.01	0.74	0.57	4.43	0.15	1.64	0.39	1.64	0.27	1.99	0.48
YD-16	2.07	3.93	0.66	3.63	0.69	0.54	5.22	0.21	2.01	0.48	1.98	0.30	2.49	0.51
YD-17	0.89	2.08	0.65	3.59	0.98	0.53	19.83	0.27	3.53	0.98	3.53	0.65	6.61	1.66

Table 2. Element anomalies of water samples from the Yongding River.

Sample	∑REE	LREE/HREE	La_{SN}/Yb_{SN}	δCe	δGd	δEu	Gd_{anth}	Cl^-	NO_3^-	Na^+	pH	DO
	ng L^{-1}						ng L^{-1}	mg/L	mg/L	mg/L		mg/L
YD-1	30.66	0.29	0.02	0.61	3.42	1.50	4.83	223.90	11.59	178.60	7.72	2.65
YD-2	50.77	0.41	0.11	0.35	5.38	0.83	11.37	226.81	10.50	184.60	7.73	2.95
YD-3	26.01	0.45	0.05	0.47	3.99	1.44	4.20	176.10	10.31	148.80	7.64	2.75
YD-4	86.26	0.63	0.18	0.41	2.34	1.46	7.72	261.42	9.34	212.30	7.75	3.18
YD-5	53.57	0.90	0.13	0.26	15.73	0.80	12.28	211.04	3.96	164.30	7.59	1.83
YD-6	48.45	0.57	0.09	0.56	3.05	0.87	6.93	270.95	3.33	224.90	7.58	2.85
YD-7	30.06	0.62	0.20	0.39	3.06	1.26	3.99	125.22	4.44	96.33	7.79	4.90
YD-8	21.48	0.57	0.16	0.62	11.74	1.62	7.66	46.00	0	35.05	8.67	6.76
YD-9	18.35	0.39	0.06	0.91	4.21	1.06	3.22	169.41	0	177.40	9.04	8.00
YD-10	8.53	0.46	0.09	0.38	1.97	3.87	0.85	79.97	8.14	50.88	8.45	8.10
YD-11	38.05	0.60	0.52	0.67	58.70	0.50	20.43	75.31	5.78	62.16	8.77	9.41
YD-12	6.27	0.94	0.49	0.77	8.49	2.79	1.34	55.50	19.04	55.53	8.24	8.24
YD-13	9.60	1.28	0.66	0.94	10.31	1.66	2.14	80.23	14.38	77.11	8.56	8.69
YD-14	7.72	1.84	1.05	0.75	2.68	4.00	0.44	25.18	24.15	23.77	8.32	11.07
YD-15	21.61	0.37	0.07	0.69	5.50	1.47	4.26	123.62	2.88	120.80	8.63	8.78
YD-16	24.72	0.33	0.06	0.64	5.03	1.34	5.00	142.11	2.80	140.70	8.60	6.79
YD-17	45.77	0.13	0.01	0.52	14.72	0.57	19.54	141.75	5.74	125.40	8.20	7.00

Previous studies have reported that ecological factors, such as microorganisms, aquatic plants, and the area of weathering surfaces, affect the overall mobility of REEs [49]. Therefore, lower REE concentrations in the Yongding River may result from the increase in vascular plants from spring to summer, leading to a dramatic ∑REE decrease [49] since aquatic plants can produce stable REE complexes via unsaturated groups from the surface of a large number of organic colloids [50]. Seasonal redox, hydrological process and suspended load may also affect ∑REE and the relationship among each element. During the sampling period in the summer of 2021, extreme rainfall occurred in Beijing. Therefore, the rainwater would have diluted the river water and reduced the REEs concentration [51–54]. Furthermore, it is reported that a multi-point ecological water replenishment was conducted in the Yongding River in the spring of 2021 to relieve the pressure on the water storage of the Guanting Reservoir and the whole Yongding River. Therefore, the runoff increased to levels 1.3~6.6 times the annual average monthly runoff [55], and this alteration of river runoff caused by supplementary ecological water releases may have been a reason for the lower REE contents [55,56].

3.2. Spatial Variation Patterns of REEs

Figure 2 illustrates the spatial variation of ∑REE contents in the Yongding River. The distribution of ΣREE contents displayed a certain regularity and significant spatial variations. Compared with the upper stream, the concentration of ΣREE was higher in the middle/lower reaches and the estuary. The relatively lower concentrations at the estuary (YD-1) result from salt-induced flocculation which eliminate the colloidal-bond REEs when the river water mixes with seawater. In addition, the influence of estuarine salinity may also deplete the total concentration of dissolved REEs [57].

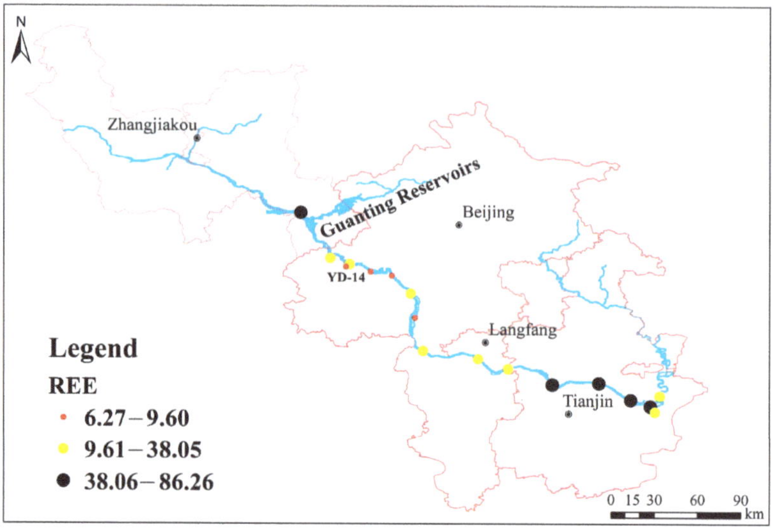

Figure 2. Spatial distribution of the total REEs content in the Yongding River.

The ΣREE distribution in the Yongding River may also be linked to the input of sediment from upper tributaries and the supplementary water released in some reaches. Since the Yongding River flows through urban, industrial and agricultural areas, REEs in the river water may also be related to anthropogenic activities (industrial and medical wastewater, domestic sewage, fertilizer and pesticide usage) in addition to natural processes [24,58]. According to this study, the ΣREE content is relatively high in lower reaches, exactly where the river flows through densely populated urban areas. Therefore, we speculate that the REE concentrations in the downstream reaches of the Yongding River are closely related to urban development and may increase under further intense development of the city. In addition, sites YD-4, YD-5 and YD-6 are adjacent to a large number of coal-fired power plants and thermal-related industry plants. Studies have indicated that large amounts of artificial REEs are contained in the fly ash produced by coal-fired power plants [59], which may fall into the river water via wind and lead to high dissolved REE contents in adjacent river water.

3.3. REE Anomalies

3.3.1. Ce and Eu Anomalies

The δCe of the river water samples from the Yongding River was <1.0 with an average of 0.59, exhibiting a weak negative Ce anomaly (Figures 3 and 4). As shown in Equation (5), Ce^{3+} can be readily oxidized to Ce^{4+} in alkaline environments:

$$Ce^{3+} + 4OH^{-} - e^{-} = CeO_2 + 2H_2O \tag{5}$$

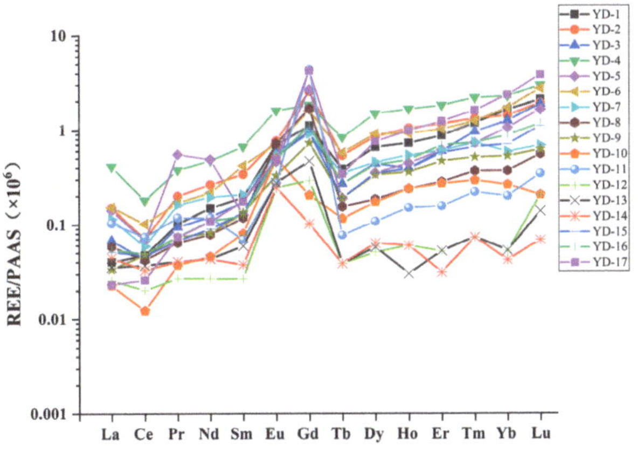

Figure 3. PAAS-normalized REE distribution pattern.

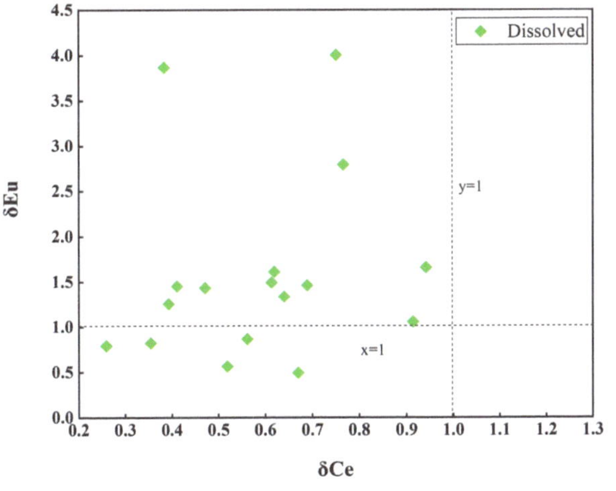

Figure 4. δCe-δEu diagram of REEs in the Yongding River.

Under oxidation conditions, Ce^{3+} forms more stable Ce^{4+} and CeO_2, which is insoluble in water, leading to a negative abnormality of Ce in alkaline water where it absorbs to Fe-Mn oxides and other substances. In this study, δCe exhibited an obvious variation between sites YD-1~YD-2 and YD-9~YD-10 with values in the ranges 0.35~0.61 and 0.38~0.91, respectively. The mixing of river water and seawater near YD-1 and the inflow of tributaries may be the reason of the increasing variability of Ce/Ce* values. Xu and Han (2009) revealed a negative relationship between Ce anomalies (δCe) and pH in river water [60]. However, this phenomenon is not obvious in this work due to slight pH variations in the Yongding River. Meanwhile, more negative δCe was found together with low DO concentrations, indicating a redox potential and geochemical processes such as rock weathering.

As shown in Figure 4, river water in the Yongding River displayed positive Eu anomalies and a weak negative anomaly (δEu = 0.50~4.00), which are lower than those in the Nanming River (δEu = 5.85~16.18) [48] and the Zhujiang River (δEu = 1.23~6.53) [61], but higher than those in the Huainan coal mining area (δEu = 0.23~1.23) [1]. The main mineral assemblages in this area are potassium feldspar and plagioclase. Since Eu could be infiltrated into plagioclase preferentially in the reduced form Eu (II) to replace Ca in reduced magma [62], the Eu enrichment level in plagioclase is higher than that in other

REE-bearing primary minerals [60]. In addition to natural sources, human activities may also make contributions to REE anomalies. Considering that Eu has been widely applied as an activator of barium fluoride in medical X-ray sensitization screens [63], the discharge of medical wastewater is a potential origin of the increase in Eu concentration observed in the Yongding River.

3.3.2. Gd Anomalies and Anthropogenic Gd Contribution

Small positive Gd anomalies (<1.4) are generally common in seawater because of the tetrad effect, which is attributed to the semi-filling of 4f electrons in its orbital shell and is unrelated to normalization [42]. All samples in the Yongding River exhibited pronounced positive Gd anomalies of 1.97~58.70. Therefore, it is unlikely that the abnormal Gd is of natural origin. Gd compounds are commonly applied as contrast agents in magnetic resonance imaging (MRI) in medical examination. Because of the toxicity of Gd^{2+}, the Gd^{2+}-based contrast agents are formed as hydrophilic compounds, which are not only highly stable but also quickly and completely released from the body without being metabolized [64]. However, the patients' excreta are not regarded as medical waste generally, and thus, directly enter urban sewage treatment plants as domestic sewage (including both residents' and hospital domestic sewage). Traditional sewage treatment plants cannot eliminate these compounds and discharge them into water environments, causing serious Gd pollution. In this study, the content of anthropogenic Gd ranged from 0.30 to 20.16 $\mu g\ L^{-1}$ with an average value of 5.29 $\mu g\ L^{-1}$. Compared with the anthropogenic Gd content in other urban rivers located in megacities, our results were lower than the average of 41.35 ng L^{-1} in the Rhine River and 110.67 ng L^{-1} in the Han River [24,46]. However, the potential pollution from nearby hospitals should still be taken into account.

3.4. Relationship between Land Use and Anthropogenic REE

The anthropogenic factors of land use and population density are presented in Table 3 and Figures 5 and 6 and show obvious spatial variations. Factories and hospitals are mainly distributed in the middle reach (YD-9~YD-12) and lower reach (YD-5~YD-6) of the mainstream. It is worth noting that the proportion of urban land and cultivated land increases in the middle reach and reaches a highly stable proportion in the lower reach. The urban land use proportion is highly consistent with the distribution of population density. This finding indicates that the middle/lower reaches of the river are strongly disturbed by human activities.

Table 3. Proportion of land use types in the Yongding River basin.

Buffer Zone	Woodland	Grassland	Cropland	Urban Land	Unused Land	Wetland
1	0.11	0.50	0.27	0.05	0.07	0.00
2	0.17	0.55	0.19	0.03	0.05	0.00
3	0.15	0.53	0.21	0.04	0.06	0.01
4	0.24	0.50	0.17	0.04	0.05	0.00
5	0.14	0.50	0.21	0.08	0.06	0.00
6	0.28	0.37	0.16	0.13	0.06	0.00
7	0.27	0.32	0.17	0.16	0.07	0.00
8	0.26	0.27	0.21	0.17	0.08	0.00
9	0.25	0.23	0.28	0.15	0.08	0.00
10	0.23	0.21	0.35	0.14	0.08	0.00
11	0.20	0.17	0.39	0.15	0.08	0.00
12	0.19	0.17	0.39	0.17	0.08	0.00
13	0.19	0.16	0.39	0.17	0.09	0.00
14	0.18	0.15	0.41	0.16	0.09	0.01
15	0.17	0.14	0.41	0.18	0.09	0.01

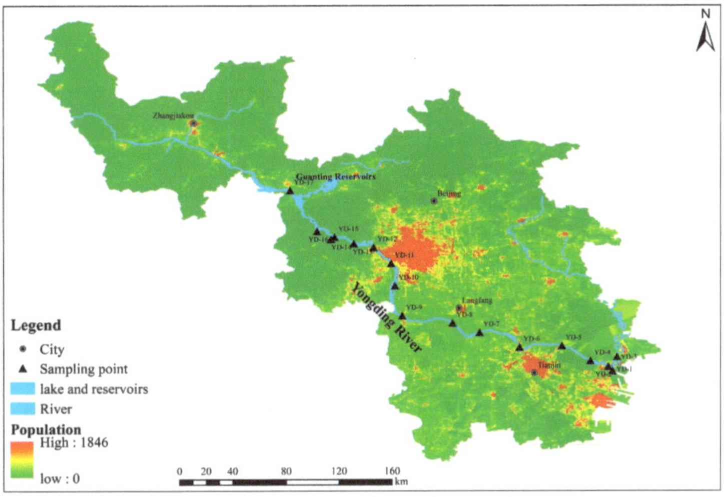

Figure 5. Population density in the Yongding River basin.

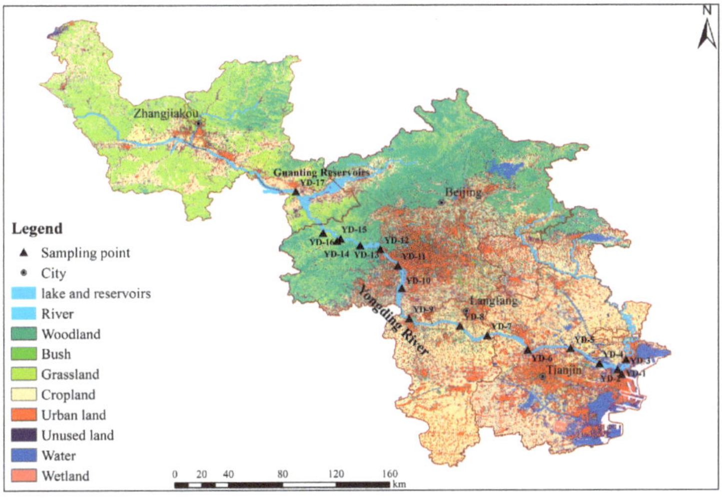

Figure 6. The distribution of land use in the Yongding River basin (data sourced from the website http://landsat.datamirror.csdb.cn, accessed on 20 December 2022).

River buffer zones have become a widely-accepted management tool in identifying diffused pollution. To eliminate the influence of land use in the upstream are on the downstream site, the first 500 m buffer zone was established along the river bank from headwater to the most upstream site, and the second buffer zone was established from headwater to the second upstream site. The remaining buffer zones were successively set up as above. Finally, fifteen buffer zones in total with a width of 500 m were set up along the mainstream via the buffer zone function of ArcGIS, except for YD-3 which is on a tributary of the Yongding River. Furthermore, we applied the land use datasets in 30 m resolution (2020) (http://landsat.datamirror.csdb.cn, accessed on 20 December 2022) and the population density of 2020 (https://www.satpalda.com/landscan, accessed on 20 December 2022) for the analysis of anthropogenic response factors. The land use datasets were summarized as six types: woodland, cropland, grassland, urban area, unused land, and wetland.

The degree of human-derived REEs could be identified and correlated with the proportions of land use. As human activity indicators are continuous numerical variables which do not follow a normal distribution, the Spearman correlation analysis is more suitable in this study. The correlations between land use proportion and artificial REE concentration are presented in Table 4. The results indicate that cultivated land is significantly correlated with Gd_{anth} and REE concentrations, and urban land use is significantly related to the concentration of most REEs. With the extensive use of chemical fertilizer, the agriculture within the basin is relatively developed [65]. Other researchers have also reported that various REE contents in cultivated soil could originate from agricultural sources (i.e., phosphate fertilizers) and land use [29]. Therefore, intensive agricultural activities and phosphate input enriched in REEs may be regarded as the potential sources of anthropogenic REEs in the Yongding River basin.

Table 4. Spearman correlations of anthropogenic Gd and REE concentrations with land use in the Yongding River (** $p < 0.01$; * $p < 0.05$).

	Cropland	Urban Land	Grassland	Woodland
La	0.59 *	0.24	−0.55 *	−0.01
Ce	0.51 *	0.38	−0.60 **	0.12
Pr	0.69 **	0.51 *	−0.74 **	0.05
Nd	0.72 **	0.55 *	−0.78 **	0.03
Sm	0.73 **	0.58 *	−0.74 **	0.04
Eu	0.68 **	0.51 *	−0.71 **	0.04
Gd	0.45 *	0.37	−0.56 *	0.15
Tb	0.74 **	0.56 *	−0.74 **	−0.06
Dy	0.80 **	0.55 *	−0.76 **	−0.13
Ho	0.80 **	0.59 *	−0.76 **	−0.16
Er	0.80 **	0.58 *	−0.77 **	−0.15
Tm	0.79 **	0.56 *	−0.76 **	−0.14
Yb	0.78 **	0.59 *	−0.76 **	−0.15
Lu	0.78 **	0.58 *	−0.75 **	−0.18
∑REE	0.65 **	0.51 *	−0.74 **	0.06
Gd_{anth}	0.44 *	0.33	−0.54 *	0.17

However, agricultural activities are not the only anthropogenic source of Gd and other REEs. Previous studies have revealed that a potential reason for high anthropogenic Gd abnormalities in river water is the application of Gd-based compounds in magnetic resonance imaging (MRI) at hospitals, and the dense population is positively related to the number of MRI [24,43,66]. Samples with abnormally high Gd (YD-5, YD-8, YD-11, YD-13, YD-17) were also observed in the middle and lower reaches with high population density in this study. The sewage treatment plants in this area may receive sewage from hospitals, and the patients' excretions defined as domestic water could also directly enter urban sewage treatment plants and eventually be discharged to the river, leading to high anthropogenic Gd in river water.

Studies have shown that the REE content of topsoil in different land use types varies [67]. For instance, the REE contents of urban soil and agricultural soil are significantly higher than those in forest soil [68]. The pollution sources for these land use types are also different. The main sources of metal pollution in urban areas are traffic emissions and urban sewage. For agricultural soil, the main source is the intensive use of manure and fertilizer, and atmospheric deposition [69]. This is similar to the source of REEs in river water, indicating that the concentrations of rare earth elements in river water are closely related to the land use type.

3.5. Identification of REE Contamination by Co-Indicators

Gd_{anth} and two traditional contamination tracers (NO_3^-/Na^+, Cl^-/Na^+) were selected as co-indicators to investigate the potential traceability of anthropogenic REEs. Traditional sewage treatment plants are designed to treat traditional pollutants, but a series of new pollutants could pass the treatment process and enter the environment through sewage systems. Numerous studies have shown that artificial Gd generally comes from hospital wastewater, and Gd_{anth} can thus be used as a highly sensitive tracer for medical sewage and sewage from treatment plants, since it cannot be effectively removed in conventional processes and would enter the aquatic ecosystem from treatment plants. Some forms of industrial activities and agricultural production discharge water with high NO_3^-/Na^+ and Cl^-/Na^+. In addition, domestic wastewater also exhibits high Cl^-/Na^+ [70]. Therefore, we take the NO_3^-/Na^+ and Cl^-/Na^+ ratios in river water as an indicator of the agricultural and industrial impacts [71].

Three distribution patterns of the river water samples were determined based on three co-indicators (Figure 7). Cluster 1 is located in the mainstream and has low concentrations of Gd_{anth} with high NO_3^-/Na^+ and Cl^-/Na^+ ratios. The mainstream area includes two megacities, Beijing and Tianjin, with developed agriculture and industry and a population of more than 37 million [72]. Previous work found that the water quality in downstream regions is significantly lower than upstream regions [34,73]. This may result from the increase in untreated industrial wastewater and urban sewage from the upper reaches where Beijing and Tianjin are located, two highly urbanized and industrialized areas. Besides, a large number of sewage treatment plants and industrial parks (using mechanical, chemical and metallurgical processes) are also located in the downstream region. Therefore, Cluster 1 refers to the sites impacted by domestic sewage and industrial wastewater in this typical urban area.

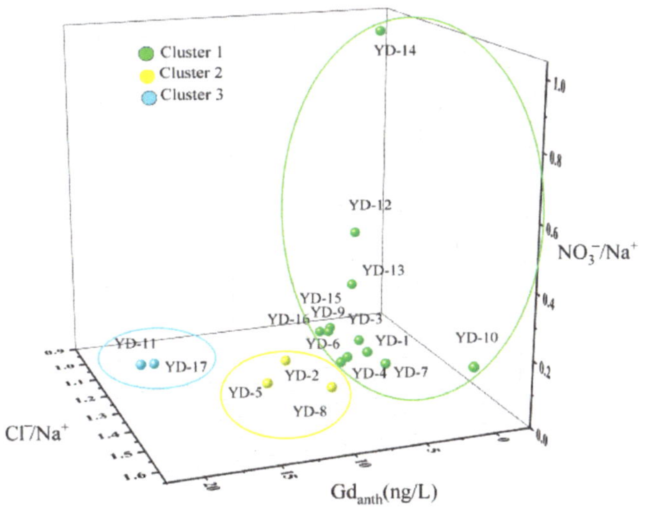

Figure 7. The cluster results of water samples using Gd_{anth}, NO_3^-/Na^+ and Cl^-/Na^+ as co-indicators in the Yongding River basin.

Cluster 2 includes sites in the middle to lower reaches of the mainstream which have high concentrations of Gd_{anth}, high Cl^-/Na^+ and low NO_3^-/Na^+. As discussed above, a significant positive relationship exists between the number of MRI instruments and Gd_{anth} concentrations. In this study, the hospital density near YD-5 and YD-8 is high, and Gd_{anth} may originate from medical wastewater discharged to nearby sewage treatment plants. However, no hospital was found near YD-2, so the high concentrations may be attributed to the leakage of urban underground pipes and the direct release of untreated sewage. High ratios of Cl^-/Na^+ may also originate from domestic sewage [74]. Therefore, Cluster 2 corresponds to the sites with surface water affected by urban sewage and medical wastewater.

Cluster 3 is located in the upper-middle reach and displayed high Gd_{anth}, low Cl^-/Na^+ and high NO_3^-/Na^+. Much higher Gd_{anth} levels were found near YD-11 and YD-17, which may result from strong point source pollution such as the discharge of medical wastewater. Low ratios of Cl^-/Na^+ indicate that this area is rarely affected by urban sewage. Besides, high ratios of NO_3^-/Na^+ may be associated with frequent agricultural activities within this area, since the proportion of cultivated land within this region has increased based on local land use statistics (Table 3). Research has reported that nitrogen fertilizer made up >50% of nitrate pollution in river water, and excessive application of nitrogen fertilizer in agricultural activities within the basin has resulted in a total of 96 kg hm^{-2} nitrogen loss, leading to the increases in the nitrate content of river water in this region [75]. Therefore, Cluster 3 was considered to be affected by agriculture rather than industrial and urban sewage.

3.6. REE Toxicological Risk for Humans and Aquatic Life

The toxicity and concentration of rare earth elements are positively correlated. In addition, the toxicity of rare earth elements is also closely related to the exposure time and leakage mode [20]. Studies have shown that when the concentration of rare earths reaches a constant value, it can harm aquatic organisms and humans. For example, sea urchin (embryos and sperms) were exposed to Ce (IV) and La (III) for 72 h and 1 h, respectively, at a concentrations from 10^{-8} to 10^{-5} M. The results showed full mortality following exposure to Ce (IV) at a concentration of 10^{-5} M and full developmental defects with no mortality when exposed to La (III) at 10^{-5} M. At the highest concentration (10^{-5} M), sperm exposed to Ce (IV) and La (III) had a decreased success rate in fertilization [76].

In this study, we mainly consider the impact of artificial Gd concentrations on aquatic organisms and humans. Concentrations of individual lanthanides measured in the Yongding River are much lower than the REEs reported in other eco-toxicological studies [77]. We also conclude that the total amount of all dissolved REEs with same patterns is also lower than the levels reported elsewhere [78]. Therefore, this study infers that the toxicological risk of dissolved REEs in the Yongding River to human and aquatic organisms is limited.

4. Conclusions

This work explored the spatial characteristics of REEs in a typical urban river (Yongding River) and the correlations between land use and human-derived REEs in river water. The ΣREE contents of river water were 6.27~86.26 ng L^{-1}, with an average value of 31.05 ng L^{-1}. The PAAS-normalized REE patterns exhibited Gd anomalies, varying from 1.73 to 14.97 and revealed pronounced Gd_{anth} in all river samples from 0.30 to 20.16, confirming the occurrence of anthropogenic Gd and its origin in the MRI diagnostic systems of hospitals. Correlation analysis revealed that the proportion of cultivated land was positively related to most REE concentrations, which are derived from applications of fertilizers containing artificial REEs in agricultural activities. By applying Gd_{anth}, NO_3^-/Na^+ and Cl^-/Na^+ as co-indicators, the relative impacts of sewage treatment plants, hospitals and agriculture on river water were accurately identified. Combined with hydro-chemical indicators and artificial REEs, this study revealed the influence of urban human activities on the river system and accurately tracked the source of man-made pollutants throughout the urban river basin.

Author Contributions: Conceptualization, X.G. and G.H.; Data curation, X.G.; Formal analysis, X.G.; Funding acquisition, G.H.; Methodology, X.G.; Resources, G.H.; Software, X.G.; Validation, X.G., G.H. and J.L.; Visualization, X.G. and J.L.; Writing—original draft, X.G.; Writing—review and editing, X.G., G.H. and S.Z. All authors have read and agreed to the published version of the manuscript.

Funding: This research was funded jointly by the National Natural Science Foundation of China (No. 41661144029; 41325010).

Institutional Review Board Statement: Not applicable.

Informed Consent Statement: Not applicable.

Data Availability Statement: The data presented in this study are available on request from the corresponding author.

Acknowledgments: The authors gratefully acknowledge Jie Zeng for assistance with field sampling and analysis.

Conflicts of Interest: The authors declare no conflict of interest.

References

1. Qian, Y.; Zheng, L.; Jiang, C.; Chen, X.; Chen, Y.; Xu, Y.; Chen, Y. Environmental geochemical characteristics of rare-earth elements in surface waters in the Huainan coal mining area, Anhui Province, China. *Environ. Geochem. Health* **2022**, *44*, 3527–3539. [CrossRef] [PubMed]
2. De Carlo, E.H.; Green, W.J. Rare earth elements in the water column of Lake Vanda, McMurdo Dry Valleys, Antarctica. *Geochim. Cosmochim. Acta* **2002**, *66*, 1323–1333. [CrossRef]
3. Han, G.; Liu, M.; Li, X.; Zhang, Q. Sources and geochemical behaviors of rare earth elements in suspended particulate matter in a wet-dry tropical river. *Environ. Res.* **2023**, *218*, 115044. [CrossRef] [PubMed]
4. Han, G.; Liu, C.-Q. Dissolved rare earth elements in river waters draining karst terrains in Guizhou Province, China. *Aquat. Geochem.* **2007**, *13*, 95–107. [CrossRef]
5. Polyakov, O. Chapter 18-Technology of Ferroalloys with Rare-Earth Metals. In *Handbook of Ferroalloys*; Gasik, M., Ed.; Butterworth-Heinemann: Oxford, UK, 2013; pp. 459–469.
6. Ebrahimi, P.; Barbieri, M. Gadolinium as an emerging microcontaminant in water resources: Threats and opportunities. *Geosciences* **2019**, *9*, 93. [CrossRef]
7. Novotny, E.; Sander, A.; Mohseni, O.; Stefan, H. Chloride ion transport and mass balance in a metropolitan area using road salt. *Water Resour. Res.* **2009**, *45*, W12410. [CrossRef]
8. Sholkovitz, E.R. The geochemistry of rare earth elements in the Amazon River estuary. *Geochim. Cosmochim. Acta* **1993**, *57*, 2181–2190. [CrossRef]
9. Zhou, J.; Wu, Q.; Gao, S.; Zhang, X.; Wang, Z.; Wu, P.; Zeng, J. Coupled controls of the infiltration of rivers, urban activities and carbonate on trace elements in a karst groundwater system from Guiyang, Southwest China. *Ecotoxicol. Environ. Saf.* **2023**, *249*, 114424. [CrossRef]
10. Kalender, L.; Aytimur, G. REE Geochemistry of Euphrates River, Turkey. *J. Chem.* **2016**, *2016*, 1012021. [CrossRef]
11. Kumar, M.; Goswami, R.; Awasthi, N.; Das, R. Provenance and fate of trace and rare earth elements in the sediment-aquifers systems of Majuli River Island, India. *Chemosphere* **2019**, *237*, 124477. [CrossRef]
12. Négrel, P. Water–granite interaction: Clues from strontium, neodymium and rare earth elements in soil and waters. *Appl. Geochem.* **2006**, *21*, 1432–1454. [CrossRef]
13. Hatje, V.; Bruland, K.W.; Flegal, A.R. Increases in Anthropogenic Gadolinium Anomalies and Rare Earth Element Concentrations in San Francisco Bay over a 20 Year Record. *Environ. Sci. Technol.* **2016**, *50*, 4159–4168. [CrossRef]
14. Jones, N. Materials science: The pull of stronger magnets. *Nature* **2011**, *472*, 22–23. [CrossRef]
15. Mondal, S.; Nguyen, V.T.; Park, S.; Choi, J.; Vo, T.M.T.; Shin, J.H.; Kang, Y.-H.; Oh, J. Rare earth element doped hydroxyapatite luminescent bioceramics contrast agent for enhanced biomedical imaging and therapeutic applications. *Ceram. Int.* **2020**, *46*, 29249–29260. [CrossRef]
16. Goodenough, K.M.; Wall, F.; Merriman, D. The Rare Earth Elements: Demand, Global Resources, and Challenges for Resourcing Future Generations. *Nat. Resour. Res.* **2018**, *27*, 201–216. [CrossRef]
17. Gao, S.; Wang, Z.; Wu, Q.; Wang, W.; Peng, C.; Zeng, J.; Yuqing, W. Urban geochemistry and human-impacted imprint of dissolved trace and rare earth elements in a high-tech industrial city, Suzhou. *Elem. Sci. Anth.* **2021**, *9*, 00151. [CrossRef]
18. Qu, R.; Han, G. Potassium isotopes of fertilizers as potential markers of anthropogenic input in ecosystems. *Environ. Chem. Lett.* **2022**. [CrossRef]
19. Wang, J.; Deng, P.; Wei, X.; Zhang, X.; Liu, J.; Huang, Y.; She, J.; Liu, Y.; Wan, Y.; Hu, H.; et al. Hidden risks from potentially toxic metal(loid)s in paddy soils-rice and source apportionment using lead isotopes: A case study from China. *Sci. Total Environ.* **2023**, *856*, 158883. [CrossRef]
20. Gwenzi, W.; Mangori, L.; Danha, C.; Chaukura, N.; Dunjana, N.; Sanganyado, E. Sources, behaviour, and environmental and human health risks of high-technology rare earth elements as emerging contaminants. *Sci. Total Environ.* **2018**, *636*, 299–313. [CrossRef]
21. Zhang, H.; Feng, J.; Zhu, W.; Liu, C.; Xu, S.; Shao, P.; Wu, D.; Yang, W.; Gu, J. Chronic toxicity of rare-earth elements on human beings. *Biol. Trace Elem. Res.* **2000**, *73*, 1–17. [CrossRef] [PubMed]

22. Prince, M.R.; Zhang, H.; Morris, M.; MacGregor, J.L.; Grossman, M.E.; Silberzweig, J.; DeLapaz, R.L.; Lee, H.J.; Magro, C.M.; Valeri, A.M. Incidence of nephrogenic systemic fibrosis at two large medical centers. *Radiology* **2008**, *248*, 807–816. [CrossRef] [PubMed]
23. Thomsen, H.S. Are the increasing amounts of gadolinium in surface and tap water dangerous? *Acta Radiol.* **2017**, *58*, 259–263. [CrossRef]
24. Song, H.; Shin, W.-J.; Ryu, J.-S.; Shin, H.S.; Chung, H.; Lee, K.-S. Anthropogenic rare earth elements and their spatial distributions in the Han River, South Korea. *Chemosphere* **2017**, *172*, 155–165. [CrossRef]
25. Jiang, C.; Li, Y.; Li, C.; Zheng, L.; Zheng, L. Distribution, source and behavior of rare earth elements in surface water and sediments in a subtropical freshwater lake influenced by human activities. *Environ. Pollut.* **2022**, *313*, 120153. [CrossRef] [PubMed]
26. Zeng, J.; Han, G.; Zhang, S.; Liang, B.; Qu, R.; Liu, M.; Liu, J. Potentially toxic elements in cascade dams-influenced river originated from Tibetan Plateau. *Environ. Res.* **2022**, *208*, 112716. [PubMed]
27. Perrat, E.; Parant, M.; Py, J.-S.; Rosin, C.; Cossu-Leguille, C. Bioaccumulation of gadolinium in freshwater bivalves. *Environ. Sci. Pollut. Res.* **2017**, *24*, 12405–12415. [CrossRef]
28. Parant, M.; Sohm, B.; Flayac, J.; Perrat, E.; Chuburu, F.; Cadiou, C.; Rosin, C.; Cossu-Leguille, C. Impact of gadolinium-based contrast agents on the growth of fish cells lines. *Ecotoxicol. Environ. Saf.* **2019**, *182*, 109385. [CrossRef]
29. Bispo, F.H.A.; de Menezes, M.D.; Fontana, A.; Sarkis, J.E.d.S.; Gonçalves, C.M.; de Carvalho, T.S.; Curi, N.; Guilherme, L.R.G. Rare earth elements (REEs): Geochemical patterns and contamination aspects in Brazilian benchmark soils. *Environ. Pollut.* **2021**, *289*, 117972. [CrossRef]
30. Carpenter, D.; Boutin, C.; Allison, J.E.; Parsons, J.L.; Ellis, D.M. Uptake and Effects of Six Rare Earth Elements (REEs) on Selected Native and Crop Species Growing in Contaminated Soils. *PLoS ONE* **2015**, *10*, e0129936. [CrossRef]
31. Silva, L.F.O.; Oliveira, M.L.S.; Crissien, T.J.; Santosh, M.; Bolivar, J.; Shao, L.; Dotto, G.L.; Gasparotto, J.; Schindler, M. A review on the environmental impact of phosphogypsum and potential health impacts through the release of nanoparticles. *Chemosphere* **2022**, *286*, 131513. [CrossRef]
32. Liu, Y.; Wu, Q.; Jia, H.; Wang, Z.; Gao, S.; Zeng, J. Anthropogenic rare earth elements in urban lakes: Their spatial distributions and tracing application. *Chemosphere* **2022**, *300*, 134534. [CrossRef] [PubMed]
33. Zhang, H.H.; Brown, D.F. Understanding urban residential water use in Beijing and Tianjin, China. *Habitat Int.* **2005**, *29*, 469–491. [CrossRef]
34. Jiang, J.; Zhao, G.; Xu, Y.; Zhao, J.; Liu, L.; Liu, C.; Wang, D.; Li, Y. Occurrence and distribution characteristics of heavy metals in the surface water of Yongding River Basin, China. *Environ. Sci. Pollut. Res.* **2022**, *29*, 17821–17831. [CrossRef]
35. Zhao, J.; Liu, Q.; Lin, L.; Lv, H.; Wang, Y. Assessing the comprehensive restoration of an urban river: An integrated application of contingent valuation in Shanghai, China. *Sci. Total Environ.* **2013**, *458-460*, 517–526. [CrossRef]
36. Wang, X.; Li, J.; Li, Y.; Shen, Z.; Wang, X.; Yang, Z.; Lou, I. Is urban development an urban river killer? A case study of Yongding Diversion Channel in Beijing, China. *J. Environ. Sci.* **2014**, *26*, 1232–1237. [CrossRef]
37. Wang, H.; Li, X.; Long, H.; Xu, X.; Bao, Y. Monitoring the effects of land use and cover type changes on soil moisture using remote-sensing data: A case study in China's Yongding River basin. *CATENA* **2010**, *82*, 135–145. [CrossRef]
38. Jiang, B.; Wong, C.P.; Lu, F.; Ouyang, Z.; Wang, Y. Drivers of drying on the Yongding River in Beijing. *J. Hydrol.* **2014**, *519*, 69–79. [CrossRef]
39. Sun, K.; Hu, L.; Guo, J.; Yang, Z.; Zhai, Y.; Zhang, S. Enhancing the understanding of hydrological responses induced by ecological water replenishment using improved machine learning models: A case study in Yongding River. *Sci. Total Environ.* **2021**, *768*, 145489. [CrossRef]
40. Li, X.; Han, G.; Liu, M.; Liu, J.; Zhang, Q.; Qu, R. Potassium and its isotope behaviour during chemical weathering in a tropical catchment affected by evaporite dissolution. *Geochim. Cosmochim. Acta* **2022**, *316*, 105–121. [CrossRef]
41. McLennan, S.M. Chapter 7. Rare Earth Elements in Sedimentary Rocks: Influence of Provenance and Sedimentary Processes. In *Geochemistry and Mineralogy of Rare Earth Elements*; Bruce, R.L., McKay, G.A., Eds.; De Gruyter: Berlin, Germany, 1989; pp. 169–200.
42. Bolhar, R.; Van Kranendonk, M.J. A non-marine depositional setting for the northern Fortescue Group, Pilbara Craton, inferred from trace element geochemistry of stromatolitic carbonates. *Precambrian Res.* **2007**, *155*, 229–250. [CrossRef]
43. Kulaksız, S.; Bau, M. Anthropogenic dissolved and colloid/nanoparticle-bound samarium, lanthanum and gadolinium in the Rhine River and the impending destruction of the natural rare earth element distribution in rivers. *Earth Planet. Sci. Lett.* **2013**, *362*, 43–50. [CrossRef]
44. Wang, L.; Han, X.; Ding, S.; Liang, T.; Zhang, Y.; Xiao, J.; Dong, L.; Zhang, H. Combining multiple methods for provenance discrimination based on rare earth element geochemistry in lake sediment. *Sci. Total Environ.* **2019**, *672*, 264–274. [CrossRef] [PubMed]
45. Smith, C.; Liu, X.-M. Spatial and temporal distribution of rare earth elements in the Neuse River, North Carolina. *Chem. Geol.* **2018**, *488*, 34–43. [CrossRef]
46. Kulaksız, S.; Bau, M. Rare earth elements in the Rhine River, Germany: First case of anthropogenic lanthanum as a dissolved microcontaminant in the hydrosphere. *Environ. Int.* **2011**, *37*, 973–979. [CrossRef] [PubMed]
47. Ma, L.; Wang, W.-X. Dissolved rare earth elements in the Pearl River Delta: Using Gd as a tracer of anthropogenic activity from river towards the sea. *Sci. Total Environ.* **2023**, *856*, 159241. [CrossRef]

48. Han, R.; Wang, Z.; Shen, Y.; Wu, Q.; Liu, X.; Cao, C.; Gao, S.; Zhang, J. Anthropogenic Gd in urban river water: A case study in Guiyang, SW China. *Elem. Sci. Anthr.* **2021**, *9*, 147. [CrossRef]
49. Zaharescu, D.G.; Burghelea, C.I.; Dontsova, K.; Presler, J.K.; Maier, R.M.; Huxman, T.; Domanik, K.J.; Hunt, E.A.; Amistadi, M.K.; Gaddis, E.E.; et al. Ecosystem Composition Controls the Fate of Rare Earth Elements during Incipient Soil Genesis. *Sci. Rep.* **2017**, *7*, 43208. [CrossRef]
50. Shiller, A.M. Dissolved rare earth elements in a seasonally snow-covered, alpine/subalpine watershed, Loch Vale, Colorado. *Geochim. Cosmochim. Acta* **2010**, *74*, 2040–2052. [CrossRef]
51. Lerat-Hardy, A.; Coynel, A.; Schafer, J.; Marache, A.; Pereto, C.; Bossy, C.; Capdeville, M.-J.; Granger, D. Impacts of Highway Runoff on Metal Contamination Including Rare Earth Elements in a Small Urban Watershed: Case Study of Bordeaux Metropole (SW France). *Arch. Environ. Contam. Toxicol.* **2022**, *82*, 206–226. [CrossRef]
52. Zeng, J.; Han, G.; Zhang, S.; Xiao, X.; Li, Y.; Gao, X.; Wang, D.; Qu, R. Rainwater chemical evolution driven by extreme rainfall in megacity: Implication for the urban air pollution source identification. *J. Clean. Prod.* **2022**, *372*, 133732. [CrossRef]
53. Zeng, J.; Han, G.; Zhang, S.; Xiao, X.; Li, Y.; Gao, X.; Wang, D.; Qu, R. Response of dissolved organic carbon in rainwater during extreme rainfall period in megacity: Status, potential source, and deposition flux. *Sustain. Cities Soc.* **2023**, *88*, 104299. [CrossRef]
54. Zeng, J.; Han, G.; Zhang, S.; Qu, R. Nitrate dynamics and source identification of rainwater in Beijing during rainy season: Insight from dual isotopes and Bayesian model. *Sci. Total Environ.* **2023**, *856*, 159234. [CrossRef] [PubMed]
55. Ji, Z.; Cui, Y.; Zhang, S.; Chao, W.; Shao, J. Evaluation of the Impact of Ecological Water Supplement on Groundwater Restoration Based on Numerical Simulation: A Case Study in the Section of Yongding River, Beijing Plain. *Water* **2021**, *13*, 3059. [CrossRef]
56. Liu, M.; Han, G.; Zeng, J.; Liang, B.; Zhang, Q. Effect of cascade reservoirs on geochemical characteristics of rare earth elements in suspended particle matter in Lancangjiang River, Southwest China. *Aquat. Sci.* **2022**, *84*, 19. [CrossRef]
57. Bayon, G.; Toucanne, S.; Skonieczny, C.; Andre, L.; Bermell, S.; Cheron, S.; Dennielou, B.; Etoubleau, J.; Freslon, N.; Gauchery, T.; et al. Rare earth elements and neodymium isotopes in world river sediments revisited. *Geochim. Cosmochim. Acta* **2015**, *170*, 17–38. [CrossRef]
58. Pereto, C.; Coynel, A.; Lerat-Hardy, A.; Gourves, P.-Y.; Schäfer, J.; Baudrimont, M. Corbicula fluminea: A sentinel species for urban Rare Earth Element origin. *Sci. Total Environ.* **2020**, *732*, 138552. [CrossRef]
59. Wei, J.; Zheng, X.; Liu, J.; Zhang, G.; Zhang, Y.; Wang, C.; Liu, Y. The Levels, Sources, and Spatial Distribution of Heavy Metals in Soils from the Drinking Water Sources of Beijing, China. *Sustainability* **2021**, *13*, 3719.
60. Han, G.; Xu, Z.; Tang, Y.; Zhang, G. Rare Earth Element Patterns in the Karst Terrains of Guizhou Province, China: Implication for Water/Particle Interaction. *Aquat. Geochem.* **2009**, *15*, 457. [CrossRef]
61. Han, G.; Yang, K.; Zeng, J. Spatio-Temporal Distribution and Environmental Behavior of Dissolved Rare Earth Elements (REE) in the Zhujiang River, Southwest China. *Bull. Environ. Contam. Toxicol.* **2022**, *108*, 555–562. [CrossRef]
62. Möller, P.; Morteani, G.; Dulski, P. Anomalous Gadolinium, Cerium, and Yttrium Contents in the Adige and Isarco River Waters and in the Water of Their Tributaries (Provinces Trento and Bolzano/Bozen, NE Italy). *Acta Hydrochim. Hydrobiol.* **2003**, *31*, 225–239. [CrossRef]
63. Kandarakis, I.; Cavouras, D.; Panayiotakis, G.S.; Triantis, D.; Nomicos, C.D. Europium-activated phosphors for use in X-ray detectors of medical imaging systems. *Eur. Radiol.* **1998**, *8*, 313–318. [CrossRef] [PubMed]
64. Bau, M.; Dulski, P. Distribution of yttrium and rare-earth elements in the Penge and Kuruman iron-formations, Transvaal Supergroup, South Africa. *Precambrian Res.* **1996**, *79*, 37–55. [CrossRef]
65. Guo, W.; Fu, Y.; Ruan, B.; Ge, H.; Zhao, N. Agricultural non-point source pollution in the Yongding River Basin. *Ecol. Indic.* **2014**, *36*, 254–261. [CrossRef]
66. Birka, M.; Wehe, C.A.; Telgmann, L.; Sperling, M.; Karst, U. Sensitive quantification of gadolinium-based magnetic resonance imaging contrast agents in surface waters using hydrophilic interaction liquid chromatography and inductively coupled plasma sector field mass spectrometry. *J. Chromatogr. A* **2013**, *1308*, 125–131. [CrossRef] [PubMed]
67. Wiche, O.; Zertani, V.; Hentschel, W.; Achtziger, R.; Midula, P. Germanium and rare earth elements in topsoil and soil-grown plants on different land use types in the mining area of Freiberg (Germany). *J. Geochem. Explor.* **2017**, *175*, 120–129. [CrossRef]
68. Han, G.; Li, F.; Tang, Y. Organic Matter Impact on Distribution of Rare Earth Elements in Soil Under Different Land Uses. *CLEAN Soil Air Water* **2017**, *45*, 1600235. [CrossRef]
69. Hoshyari, E.; Hassanzadeh, N.; Keshavarzi, B.; Jaafarzadeh, N.; Rezaei, M. Spatial distribution, source apportionment, and ecological risk assessment of elements (PTEs, REEs, and ENs) in the surface soil of shiraz city (Iran) under different land-use types. *Chemosphere* **2023**, *311*, 137045. [CrossRef]
70. Ullah, Z.; Khan, H.; Waseem, A.; Mahmood, Q.; Farooq, U. Water quality assessment of the River Kabul at Peshawar, Pakistan: Industrial and urban wastewater impacts. *J. Water Chem. Technol.* **2013**, *35*, 170–176. [CrossRef]
71. Keeney, D.R.; DeLuca, T.H. Des Moines River Nitrate in Relation to Watershed Agricultural Practices: 1945 Versus 1980s. *J. Environ. Qual.* **1993**, *22*, 267–272.
72. Yang, Y.; Liu, Y.; Li, Y.; Li, J. Measure of urban-rural transformation in Beijing-Tianjin-Hebei region in the new millennium: Population-land-industry perspective. *Land Use Policy* **2018**, *79*, 595–608. [CrossRef]
73. Liu, C.; Pan, C.; Chang, Y.; Luo, M. An integrated autoregressive model for predicting water quality dynamics and its application in Yongding River. *Ecol. Indic.* **2021**, *133*, 108354. [CrossRef]

74. Hao, Q.; Shao, J.; Cui, Y.; Xie, Z. Applicability of artificial recharge of groundwater in the Yongding River alluvial fan in Beijing through numerical simulation. *J. Earth Sci.* **2014**, *25*, 575–586. [CrossRef]
75. Fu, Y.; Gao, T.; Yan, L.; Zhang, A.; Ruan, B. Agro-ecological compensation standard based on emergy analysis in Yongding River basin. *Nongye Gongcheng Xuebao Trans. Chin. Soc. Agric. Eng.* **2013**, *29*, 209–217. [CrossRef]
76. Malhotra, N.; Hsu, H.-S.; Liang, S.-T.; Roldan, M.J.M.; Lee, J.-S.; Ger, T.-R.; Hsiao, C.-D. An Updated Review of Toxicity Effect of the Rare Earth Elements (REEs) on Aquatic Organisms. *Animals* **2020**, *10*, 1663. [CrossRef] [PubMed]
77. González, V.; Vignati, D.A.L.; Pons, M.-N.; Montarges-Pelletier, E.; Bojic, C.; Giamberini, L. Lanthanide ecotoxicity: First attempt to measure environmental risk for aquatic organisms. *Environ. Pollut.* **2015**, *199*, 139–147. [CrossRef]
78. Romero-Freire, A.; Minguez, L.; Pelletier, M.; Cayer, A.; Caillet, C.; Devin, S.; Gross, E.M.; Guérold, F.; Pain-Devin, S.; Vignati, D.A.L.; et al. Assessment of baseline ecotoxicity of sediments from a prospective mining area enriched in light rare earth elements. *Sci. Total Environ.* **2018**, *612*, 831–839. [CrossRef]

water

Article

Are UK Rivers Getting Saltier and More Alkaline?

Shan Jiang *, Xuan Wu, Sichan Du, Qin Wang and Dawei Han

Department of Civil Engineering, University of Bristol, Bristol BS8 1TR, UK
* Correspondence: og20481@alumni.bristol.ac.uk; Tel.: +86-191-5105-1570

Abstract: River salinisation and alkalinisation have become one of the major environmental problems threatening the safety of global freshwater resources. With the accelerated climate change and aggravating anthropogenic influences, it is important to identify the trends and causes of river salinisation and alkalinisation so that better mitigation measures could be taken. This study has focused on the UK rivers because there has been insufficient investigation on this topic. To understand the salinisation and alkalinisation trends and causes of rivers in the UK over the past 20 years from a vertical (analysis of each river) and horizontal (comparison of all rivers) perspective, this study uses the Theil-Sen regression and Mann-Kendall test to deal with the trends of conductivity (proxy on salinisation) and pH (proxy on alkalinisation), obtains outliers of conductivity and pH by boxplot, and calculates the Pearson's and the Kendall's Tau correlation coefficients (α = 0.05) between the water quality data and the potential factors (potential road salting, normalized difference vegetation index (NDVI), river discharge, agricultural and urban lands). The results show that the UK rivers are becoming more alkaline with a median pH increase of 0.05 to 0.40, but less salty with a median conductivity decrease of 0.06 to 0.11 mS/cm. And the changes in conductivity and pH have seasonality and regionality, which shows that there are usually greater changes in trends and medians of them in winter or through reaches with more anthropogenic disturbance. Furthermore, from a vertical perspective, the conductivity of more than 50% of rivers in this study is negatively correlated with NDVI and river discharge, and positively correlated with potential road salting, and the pH of that is positively correlated with agricultural lands. While from a horizontal perspective, NDVI and agricultural lands are positively correlated with pH, and potential road salting and urban lands are positively correlated with conductivity. Therefore, road salting, urbanisation, agricultural lands, river discharge and vegetation cover can be considered to affect river salinisation and alkalinisation in the UK.

Keywords: UK rivers; salinisation; alkalinisation; pH; conductivity

Citation: Jiang, S.; Wu, X.; Du, S.; Wang, Q.; Han, D. Are UK Rivers Getting Saltier and More Alkaline? *Water* **2022**, *14*, 2813. https://doi.org/10.3390/w14182813

Academic Editors: Guilin Han and Xiaolong Liu

Received: 8 August 2022
Accepted: 7 September 2022
Published: 9 September 2022

1. Introduction

River salinisation and alkalinisation are becoming one of the important factors affecting the global river ecosystem [1], which leads to the decline of biodiversity, the degradation of agricultural lands and freshwater areas, and the corrosion of infrastructure, thus threatening human living conditions [2–4]. Studying the changes in river salinisation and alkalinisation and analysing their causes will help to better protect freshwater resources and cope with the shortage of freshwater resources.

Generally, river salinisation refers to the accumulation of ions, leading to an increase in conductivity or total dissolved solids [5], which can be caused by natural (e.g., climate change, weathering and seawater intrusion) or anthropogenic (e.g., road salt, mining, agriculture and groundwater abstraction) processes [6–8]. The research on river salinisation can be divided into two periods. Before the 21st century, most researchers regarded that the decline of soil permeability caused by agricultural activities made rivers tend to be salinised in semi-arid and arid areas [9,10]. However, recent studies show that the salinity of rivers in many humid regions is also increasing [11,12], which may be attributed to the heavy salting of roads for winter de-icing (e.g., more extreme cold spells due to climate change), chemical weathering, mining, seawater inversion and agricultural activities [13]. Among them, the long-term use of road salt in North America leads to many inputs of Na^+ and Cl^-, which is one of the main reasons for salinisation aggravation in many rivers. For example, in the eastern USA, the concentration of Na^+ and Cl^- in rivers near New York State has increased approximately twice in 50 years as a result of winter de-icing [14,15], and a large amount of Ca^{2+} and Mg^{2+} has been observed to be displaced from the soil by Na^+, and finally discharged into rivers through surface runoff, which intensifies the salinisation of rivers in these areas [16,17]; Similar phenomena have been observed in southeastern Canada, both in Ontario and Quebec with the highest winter road salting, resulting in varying degrees of salinity increases in their freshwater areas [18]. On the other hand, chemical weathering due to acidic deposition is the main cause of river salinisation in warm and humid areas of North America. For instance, Davies [19] and Kaushal [20] have found that urban impervious surfaces release Ca^{2+} and K^+ under the influence of chemical weathering, thereby increasing the salinity of urban groundwater and surrounding watersheds. In addition, the salinity of the Rhine has been not only affected by potash mining, but threatened by seawater intrusion caused by climate warming [21,22]. In Australia, salinity in soils and rivers has also been increasing as saline groundwater levels rise with the increase in agricultural land use and fertilisers [23,24].

The alkalinisation of rivers refers to the increase of pH in the river, which is usually affected by temperature, aquatic biological activities, acidic deposition, chemical weathering, sewage discharge and land use, which is similar to the causes of river salinisation [1,13,25]. In the past, river alkalinisation was usually seen only as a recovery from acidification, which had not been studied much until the last 20 years. Currently, these phenomena are observed in many regions, such as the eastern USA and north-western France [26–28]. In the USA, there is a strong link between river salinisation and alkalinisation, which manifests as a simultaneous upward trend in conductivity, pH and base cations in the rivers. This phenomenon is called freshwater salinisation syndrome by Kaushal [13]. It may be explained as anthropogenic salt inputs, chemical weathering and agriculture, which promote base cation exchange in soil and rock, and the precipitated cations are channeled through surface runoff, thus altering the concentration of base cations as well as bicarbonates in the rivers and exacerbating salinisation and alkalinisation [22,29]. Its severity is also related to the hydrological and meteorological conditions of the study site [30,31].

However, despite the importance of the topic, there was relatively little research on river salinisation and alkalinisation in the UK, which was mainly carried out in the 1980s as a supplement to acid rain research [32]. As one of the countries with the highest urbanisation in Europe, it is important to study whether the UK rivers are threatened by salinisation and alkalinisation, which will not only help the UK better protect freshwater security, but also may be a significant reference for understanding the salinisation and alkalinisation of rivers in other European countries with a temperate climate and high urbanisation. This study is intended to test a hypothesis that the salinisation and alkalinisation of rivers in the UK has also shown an upward trend in recent years (e.g., over the last 20 years), as it has in other countries. To verify the hypothesis, this study selects a few representative river basins and collects the related water quality data on river salinisation and alkalinisation, then analyses the trends, similarities and differences of the relevant data. At the same time, this study also explores the possible causes and hazards of river salinisation and alkalinisation trends.

2. Materials and Methods

2.1. Study Sites

Considering the high level of urbanisation in the UK, to make the results more representative, better reflect the trends of river salinisation and alkalinisation in different parts of the UK and explore their causes, this study took into account factors such as the level of anthropogenic disturbance, types of landcover, climate, geological environment, and availability of water quality data, and selected the 50 Environmental Agency's Water Quality Data Archive (EAWQDA) freshwater monitoring sites of the River Tame, River Trent, River Mersey and River Avon (Figure 1) with differences in these factors as the research objects near the major cities in the UK. In addition, to gain a deeper understanding of the salinisation and alkalinisation of each river, 30 special sampling sites were selected from the 50 EAWQDA sites, which had full length data, closeness to roads as well as special areas and located in different river reaches. For instance, the special sampling sites in the midstream of the River Tame are at the entrance and outlet of Lake Lea Marston, which has a purifying effect on water quality [33]; The special sampling sites in the middle and lower River Mersey are near the entrance and exit of the river into the Manchester Canal, which is an important channel with serious pollution [34].

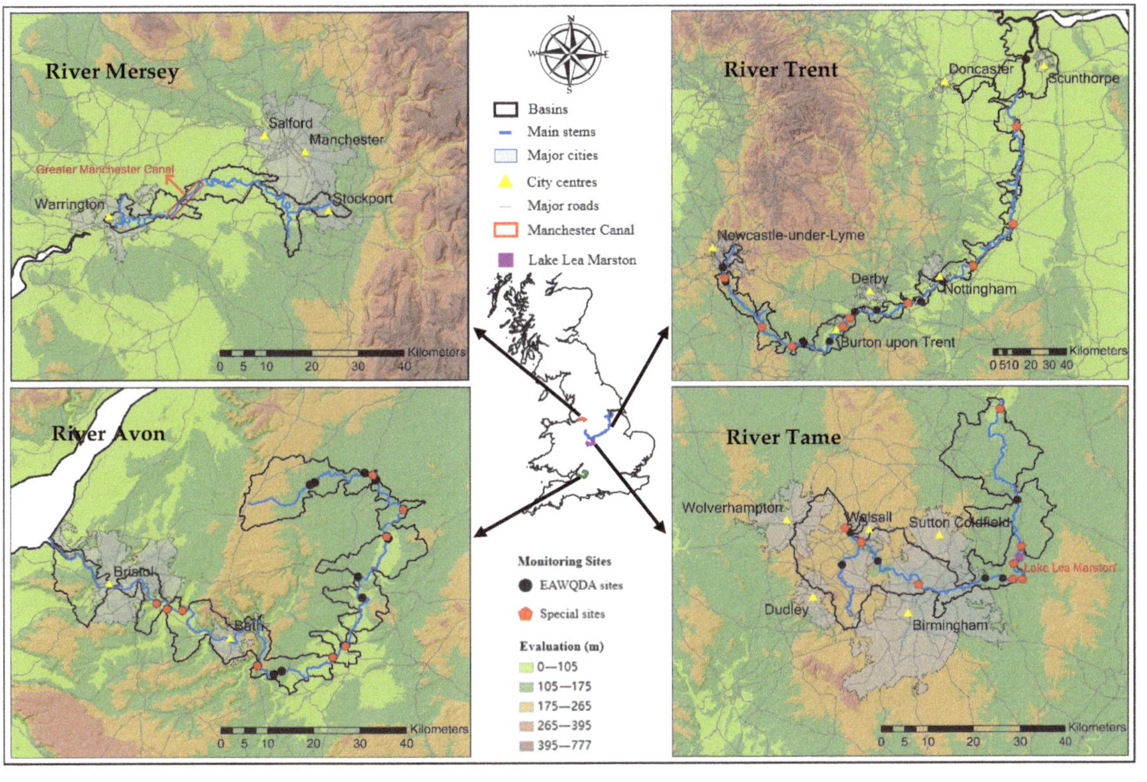

Figure 1. Main stems of the four rivers, with yellow triangles representing the main cities around the basins, grey lines for the major roads, black dots for the EAWQDA freshwater monitoring sites, and red pentagons for the special sampling sites.

2.2. Data Collection

The indicators used to represent river salinisation and alkalinisation mainly include pH, conductivity, base cations, dissolved oxygen (DO), dissolved inorganic carbon (DIC) and dissolved organic carbon (DOC), etc., [12,29]. However, only the monthly data of conductivity and pH were available from the 50 EAWQDA freshwater monitoring sites over a long period (at least 10 years). Thus, these two types of water quality data were selected for this study and used to represent river salinisation and alkalinisation, respectively.

As mentioned before, although there were many natural and anthropogenic factors affecting the river salinisation and alkalinisation, considering the availability of data, this study planned to use agricultural lands, urban lands, NDVI, river discharge and road salting data over the past 20 years to explore the causes of conductivity and pH trends in the four rivers. Among them, the annual data of agricultural and urban lands (km^2), the 16-day data of NDVI, and the daily data of river discharge (m^3/s) were obtained from the MCD12Q1, the MOD13A2 (https://search.asf.alaska.edu, accessed on 5 July 2021), and the National River Flow Archive (https://nrfa.ceh.ac.uk/, accessed on 16 May 2021), respectively. Unfortunately, there were no accurate records of road salting in the UK, so this study collected the daily air temperature, the annual total length of the classified roads (major and minor roads), the average width of the classified roads, usage of salting per square meter, and empirical difference of air-to-road temperature from CAMELS-GB [35], OS OpenData (https://osdatahub.os.uk/, accessed on 6 July 2021) and Met Office (https://wow.metoffice.gov.uk/, accessed on 13 June 2021) for each river, respectively, to simulate a new variable—the annual potential road salting (winter)—to replace the real road salting. The basis of this simulation includes: (1) The UK Highways Act 1980 stipulates that road salting is required in advance when there is a risk of freezing on the road; (2) The road freezing is related to road surface temperature (RST), which has about a 4 °C difference from the air temperature between October to April (https://www.metoffice.gov.uk/services/transport/road/opensite-resources, accessed on 15 June 2021); (3) If the RSTs are going to fall below 0.5 °C, the company responsible for salting will salt on the classified roads, including all major roads ('A' roads and motorways (M roads)) as well as some minor roads ('B' and 'C' roads), and the amount of salting is about 16 g/m^2 (https://www.buckinghamshire.gov.uk/parking-roads-and-transport/, accessed on 15 June 2021); (4) The usual width of the classified roads with one lane is 3.65 m (https://www.gov.uk/government/publications/uk-road-width-restrictions-foi, accessed on 15 June 2021). Its specific calculations are shown later.

2.3. Data Pre-Processing

Generally, water quality data have the characteristics of the irregular monitoring period, non-normality, measurement errors and outliers [36]. To avoid their confounding effects when simulating the trends of river conductivity and pH, this study averaged their original data by season for all sites firstly, defined as spring (March, April, May), summer (June, July, August), autumn (September, October, November) and winter (December, January, February) [13]. Moreover, for each river at the whole-basin scale, to calculate the correlation coefficients between the data of water quality and the factors more conveniently from vertical and horizontal perspectives later, this study calculated the annual average of pH, conductivity, agricultural and urban lands, the seasonal (winter and summer) average of pH, conductivity, NDVI, river discharge, the winter consumption of potential road salting, and the median level of these data over the past 20 years so that they were equal in quantity. The potential road salting is calculated as shown in Equations (1) to (3):

$$S_j = \frac{D_j R_j R_w S_u}{1000000},\ 2000 \leq j \leq 2015 \tag{1}$$

$$D_j = length(find(T_{ji})),\ i = 12\ Or\ 1 \leq i \leq 2 \tag{2}$$

$$T_{ji} = A_{ji} - \beta \leq 0.5 \tag{3}$$

where S_j is the potential road salting in the j year (t); D_j is the number of days to salt in the j year (d); R_j is the total length of roads in the basins in the j year (m); R_w is the average width of the major roads (m); S_u is the usage of salting per square meter (g/m^2); *length* is a counting function; *find* is a positioning function; T_{ji} is the daily RST in the i month of the j year (°C); A_{ji} is the daily air temperature in the i month of the j year (°C); β is the empirical difference of air-to-road temperature (set $\beta = 4$).

2.4. Water Quality Data Analysis

For the seasonal averages of conductivity and pH for all sites, this study used the Theil-Sen regression and the Mann-Kendall test to identify their trends and statistical significances for each river at the whole-river scale, and for different seasons and reaches. The Theil-Sen regression is a non-parametric trend analysis that is able to resist the influence of outliers and provide robust slope estimates by calculating the median slope of paired data, but the slope will be greatly affected if there are many non-detects in the dataset [37,38] (Equation (4)). However, this study solved this problem well by calculating the seasonal averages of conductivity and pH.

$$S_N = Median\left(\frac{N_j - N_i}{j - i}\right),\ 2000 \le i \le j \le 2021 \tag{4}$$

where S_N is the estimate of trend slope in the data N; N_j is the data N in the j year; N_i is the data N in the i year. The N has upward and downward trends when $S_N > 0$ and $S_N < 0$, respectively.

The Mann-Kendall test is a two-sided non-parametric trend test, which is normally used in conjunction with the Theil-Sen regression to obtain a full understanding of the trend. It is calculated as Equations (5) to (8):

$$S = \sum_{j=1}^{n-1} \sum_{i=j+1}^{n} sgn(N_j - N_i) \tag{5}$$

$$sgn(N_j - N_i) = \begin{cases} 1,\ N_j - N_i > 0 \\ 0, N_j - N_i = 0 \\ -1,\ N_j - N_i < 0 \end{cases} \tag{6}$$

$$s(S) = \frac{n(n-1)(2n+5)}{18} \tag{7}$$

$$Z = \begin{cases} \frac{S-1}{\sqrt{s(S)}},\ S > 0 \\ 0,\ S = 0 \\ \frac{S+1}{\sqrt{s(S)}},\ S < 0 \end{cases} \tag{8}$$

where S is the Mann-Kendall statistic; sgn is a step function; N_j and N_i are the data N in the j and i year; $s(S)$ is the variance of S; n is the length of time series; and Z is the standardised test statistics. This study set the null hypothesis (H_0) and alternative hypothesis (H_1), meaning that the trend of data was insignificant and significant, respectively ($\alpha = 0.05$). When $|Z| \ge Z_{1-\alpha/2}$ ($Z_{1-\alpha/2} = 1.96$), the H_1 is accepted and recorded as 1, and the trend of data is significant (the greater the $|Z|$, the higher the significance) [39,40].

Apart from the trends of conductivity and pH, their extreme values are also helpful to understanding the status and hazards of salinisation and alkalinisation across each river. Therefore, this study made boxplots for the conductivity and pH of all sampling sites in each river to observe their extreme values more visually [41]. The aforementioned data pre-processing and analysis were processed by MATLAB R2021a.

2.5. Correlation Coefficient Calculation

Before calculating the correlation between the data of the water quality and the factors, a Shapiro–Wilk test was carried out for all variables ($\alpha = 0.05$) [42], which proved to be good at testing the normal distribution for small samples so that it accorded with this study. When the two variables conformed to the normal distribution, the Pearson's correlation coefficient ($\alpha = 0.05$) was chosen for them, which was widely used in the measurement of the degree of linear dependence between two variables [43]. Otherwise, Kendall's Tau correlation coefficient ($\alpha = 0.05$) was chosen, which was less influenced by the outliers and small samples and only used for small datasets influenced by outliers strongly [44]. The ranges of these two correlation coefficients are $-1 < r < 1$, and the two variables have linear positive, negative and no correlation when $r > 0$, $r < 0$ and $r = 0$, respectively. All correlation coefficients and tests were calculated by SPSS 26.0.

3. Results

3.1. Trends of Conductivity and pH

From the vertical perspective, the results for water quality data at the whole-basin scale show that the conductivities of the three rivers (River Tame, Trent, and Mersey) have significant downward trends with a median decrease of 0.06 to 0.11 mS/cm (Table 1, Figure 2), and the pH values of three of the four rivers (River Tame, Trent, and Avon) have significant upward trends with a median increase of 0.05 to 0.40 (Table 2, Figure 3), Furthermore, the conductivity and pH in each river also demonstrate different median and significant upward or downward trends in the specific river reaches and seasons, which can be called the regionality and seasonality of conductivity and pH. On the other hand, it is worth noting that the three rivers (River Tame, Trent, and Mersey) experienced extreme conductivity in 2010 and 2013.

Table 1. The test result ($|Z| \geq 1.96$) of conductivity trends for each river, each river in winter and summer, different reaches, and different reaches in winter and summer.

	River Tame		River Trent		River Mersey		River Avon	
	h	Z	h	Z	h	Z	h	Z
River	1.00	−3.14	1.00	−2.85	1.00	−2.35	0.00	0.03
River (winter)	1.00	−3.01	1.00	−3.12	1.00	−2.10	0.00	−0.97
River (summer)	0.00	−0.54	0.00	0.20	0.00	−1.15	1.00	2.57
Upstream	1.00	−2.75	1.00	−2.31	0.00	−0.91	0.00	−1.01
Midstream	1.00	−3.93	0.00	−0.41	0.00	0.18	0.00	0.93
Downstream	1.00	−2.74	1.00	−2.01	1.00	−1.99	0.00	0.38
Upstream (summer)	0.00	−1.25	0.00	0.02	0.00	−0.34	0.00	0.27
Upstream (winter)	1.00	−2.21	1.00	−2.89	0.00	−0.75	0.00	0.73
Midstream (summer)	0.00	−1.15	0.00	1.02	0.00	0.31	0.00	1.52
Midstream (winter)	1.00	−2.30	0.00	−0.66	0.00	0.43	0.00	0.23
Downstream (summer)	0.00	−1.17	0.00	−1.10	0.00	−1.55	0.00	1.75
Downstream (winter)	1.00	−2.17	0.00	−0.38	1.00	−2.05	0.00	−1.46

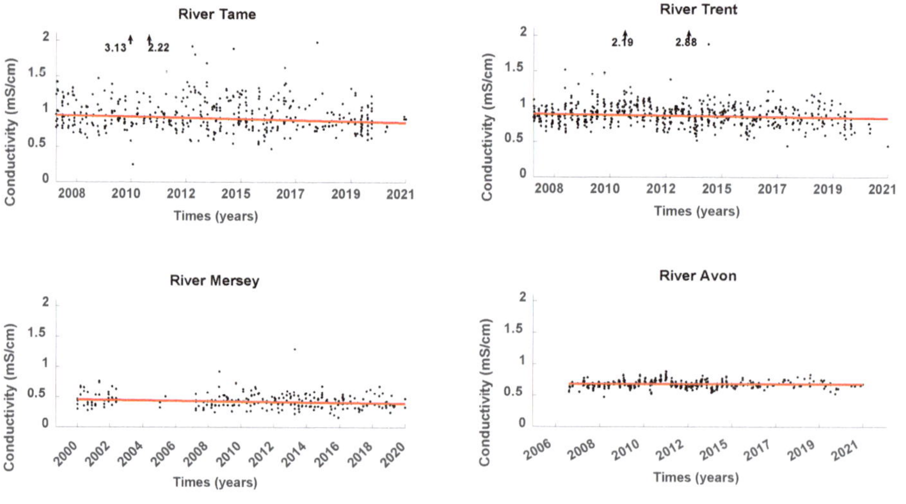

Figure 2. Conductivity of the whole rivers (The arrows represent outliers).

Table 2. The test result ($|Z| \geq 1.96$) of pH trends for each river, each river in winter and summer, different reaches, and different reaches in winter and summer.

	River Tame		River Trent		River Mersey		River Avon	
	h	Z	h	Z	h	Z	h	Z
River	1.00	13.56	1.00	9.29	0.00	−0.16	1.00	2.14
River (winter)	1.00	5.79	1.00	6.59	0.00	1.74	1.00	2.61
River (summer)	1.00	5.48	1.00	3.14	0.00	−0.52	0.00	−0.38
Upstream	1.00	7.62	1.00	6.14	0.00	1.50	0.00	1.15
Midstream	1.00	4.27	1.00	3.86	0.00	0.85	0.00	1.63
Downstream	1.00	7.16	1.00	2.32	1.00	2.27	0.00	1.24
Upstream (summer)	1.00	3.58	1.00	2.87	0.00	1.04	0.00	−0.25
Upstream (winter)	1.00	2.77	1.00	3.02	0.00	1.88	0.00	−0.53
Midstream (summer)	0.00	1.75	0.00	0.09	0.00	0.18	0.00	0.25
Midstream (winter)	0.00	1.71	1.00	2.69	0.00	0.22	1.00	2.33
Downstream (summer)	0.00	1.86	0.00	0.34	0.00	0.04	0.00	0.54
Downstream (winter)	1.00	3.84	1.00	2.24	1.00	2.38	0.00	1.50

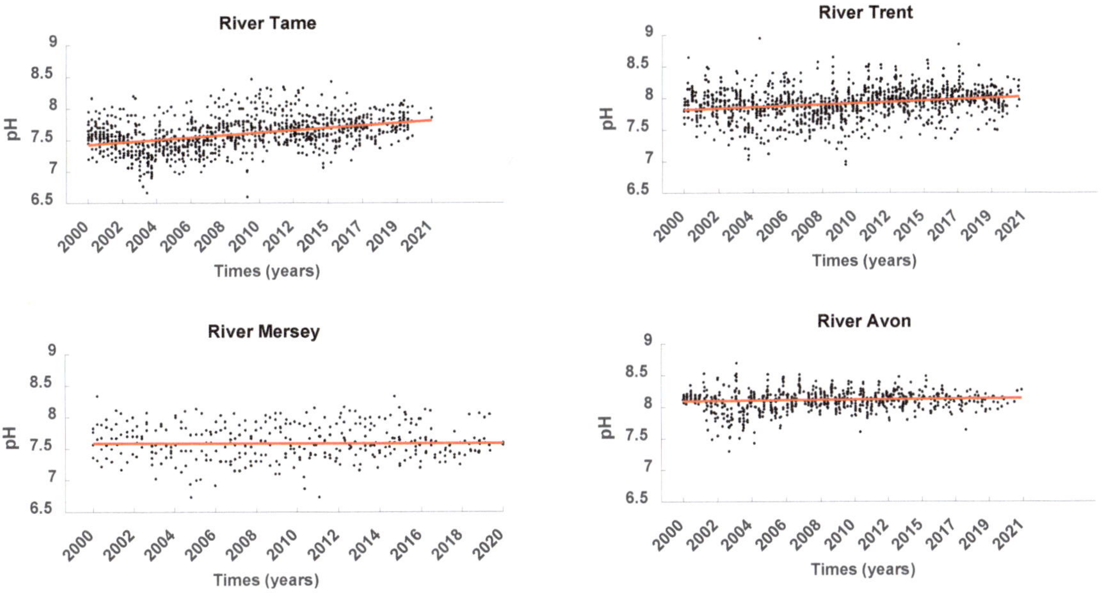

Figure 3. pH of the whole rivers.

3.1.1. The Regionality of Conductivity

When the rivers pass through the reaches with many anthropogenic disturbances, the conductivity usually changes greatly. The median of conductivity in the midstream of the River Tame flowing through the Lake Lea Marston is about 0.2 mS/cm lower than that in the upper reach near Birmingham (Figures 1 and 4), which is likely due to the purification of the lake [33] and the difference in urban lands between them. Furthermore, the conductivity of the River Mersey has the highest median level (0.45 to 0.55 mS/cm) and the most significant downward trend ($Z = -1.99$) in the lower reach (Figure 4, Table 1), which is explored later.

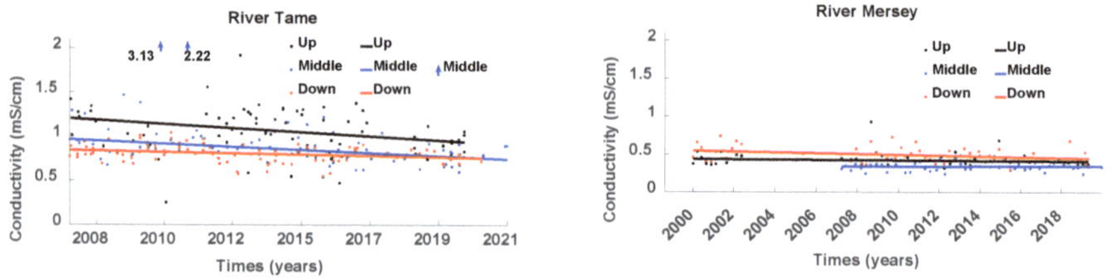

Figure 4. Examples of comparison of conductivity in the upper, middle, and lower reaches of the rivers.

3.1.2. The Seasonality of Conductivity

There are also some points worthy of attention about the seasonality of conductivity, especially in the River Tame and River Avon. Normally, the discharge of UK rivers in winter is much greater than that in summer [45], so the conductivity in summer is expected to be higher than that in winter. However, the conductivity of the River Tame in winter was higher than that in summer from 2008 to 2017 (Figure 5), which might indicate that there was more external input in winter. Different from the other three rivers, the conductivity of the River Avon in summer has a significant upward trend and its median in summer is higher than that in winter in all reaches (Figures 5 and 6), which is discussed later.

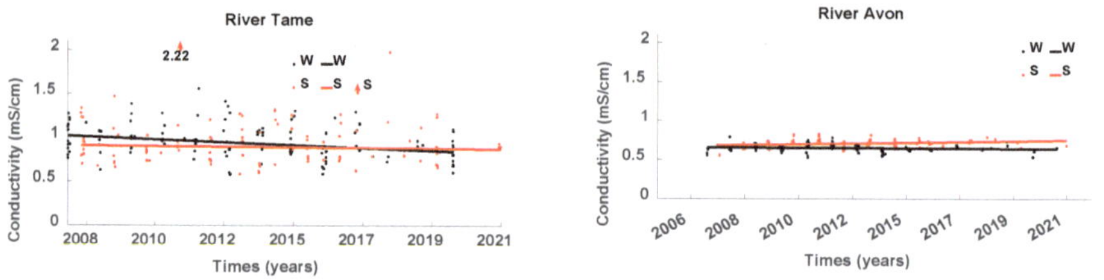

Figure 5. Examples of comparison of conductivity in summer and winter. (W: winter; S: summer).

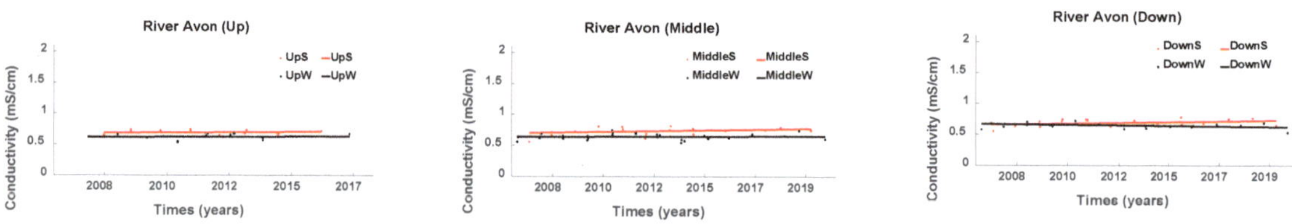

Figure 6. Comparison of conductivity in the upper, middle, and lower reaches of the river in winter and summer.

3.1.3. The Regionality of pH

Like conductivity, the pH of each river also has regionality. For instance, the median pH in the upstream River Tame is about 0.2 lower than that in the midstream (Figure 7); The median pH in the lower River Trent dominated by agricultural lands is about 0.1 higher than that of the middle reaches near Nottingham (Figures 1 and 7). Notably, the pH in the upstream River Trent has the most significant upward trend ($Z = 6.14$) (Table 2), and the median of it is only 0.02 different from that in the lower reaches in 2019 (Figure 7), which may reflect the impact of anthropogenic disturbance.

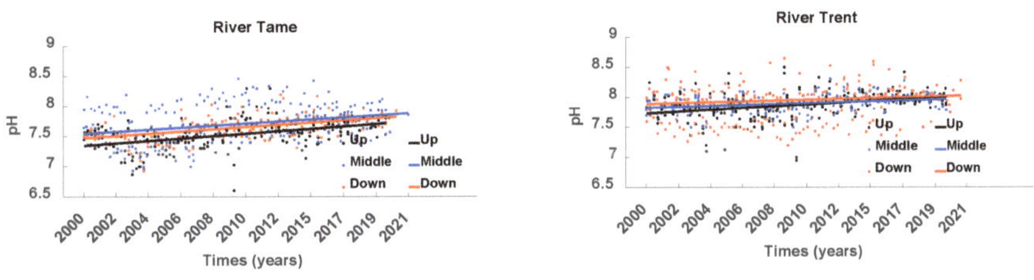

Figure 7. Examples of comparison of pH in the upper, middle, and lower river reaches.

3.1.4. The Seasonality of pH

Generally, in the UK summer, due to the higher water temperature, lower SO_2 in the air, and more vigorous photosynthesis of aquatic plants, the concentration of CO_2 and SO_2 in the rivers are lower than those in winter [46,47], resulting in the pH of the river in summer being higher than that in winter. Therefore, the information in Figure 8 is interesting—the pH of the River Trent conforms to this law, while that of the River Avon is contrary to it. However, not all reaches of the River Trent accord with the law, as the median pH in winter at the lower reach has been higher than that in summer since 2010 (Figure 9). In addition, the pH trends of both rivers in winter are more significant than in summer (Table 2). The causes for these phenomena are worthy of discussion later.

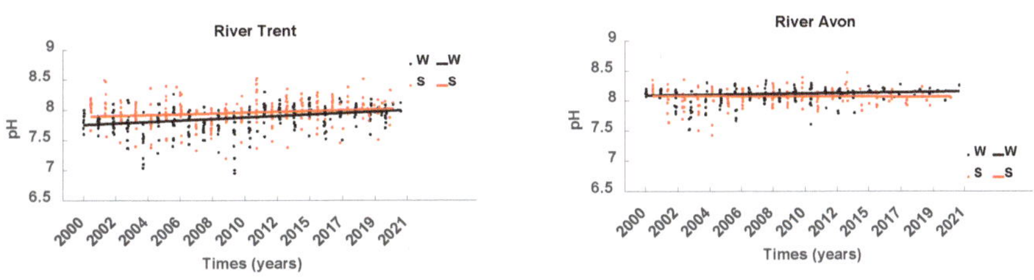

Figure 8. Examples of comparison of pH in summer and winter.

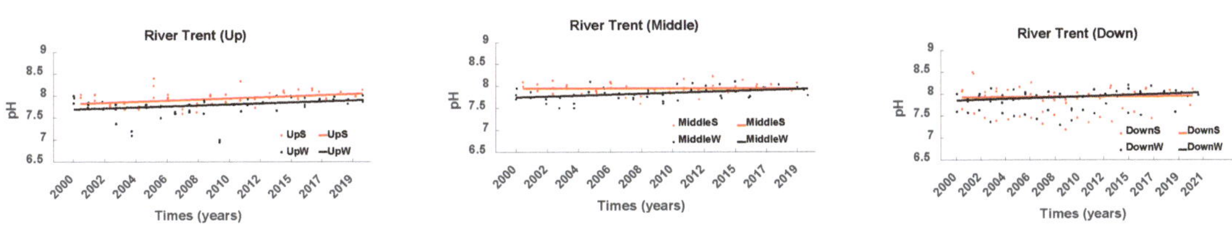

Figure 9. Examples of comparison of pH in the upper, middle, and lower reaches of the river in winter and summer.

3.2. Comparison of Conductivity and pH between All Rivers

From the horizontal point of view, there are some interesting points when comparing the conductivity and pH differences of the four rivers in this study. For the conductivity, the absolute value of conductivity trend significance in the three rivers is greater than that in the River Avon (Table 1). In addition, in Figure 10, the smallest fluctuation range (0.55–0.82 mS/cm) of conductivity is in the River Avon, while the largest fluctuation range is in the River Tame (0.48–1.47 mS/cm) likely due to different level of anthropogenic disturbances. Surprisingly, the River Mersey receiving external inputs from the Greater Manchester Canal [48] has the lowest conductivity level, the reasons for which are worth exploring later.

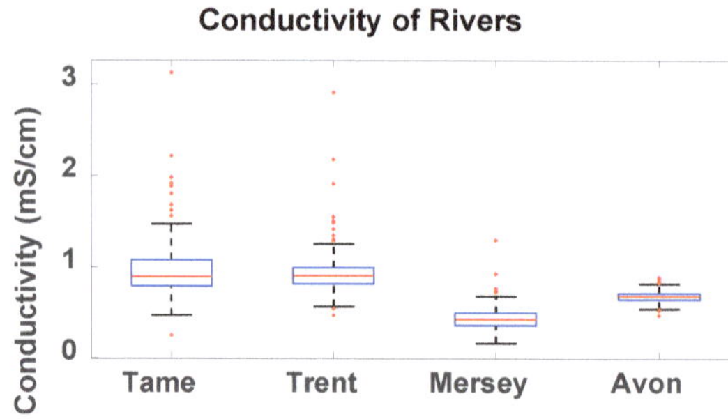

Figure 10. Differences in conductivity of the different rivers.

For the pH, the median pH in the three rivers (River Tame, Trent, and Mersey) are lower than that in the River Avon (Figure 11), which may indicate that the anthropogenic disturbance in the River Avon is the lowest. Moreover, although the medians of pH in the River Tame (7.60) and River Mersey (7.59) are extremely close, they have the largest and smallest trend significance (Table 2), respectively, which is worthy of discussion later.

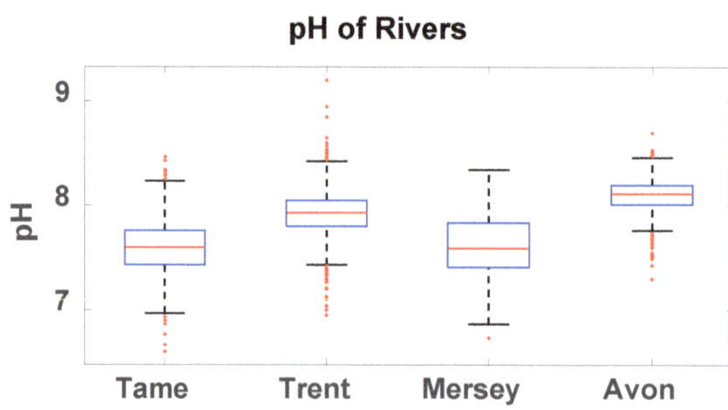

Figure 11. Differences in pH of the different rivers.

4. Discussion and Conclusions

4.1. Reasons for the Trends of River Conductivity

The difference between the results and the hypothesis comes from the salinisation trends. According to the positive correlation between conductivity and potential road salting (e.g., $r = 0.71$, $p = 0.03$) in Table 3, road salting is declining in the UK (but it is still one of the main external inputs of river salinity in winter, which may lead to higher medians and greater trend slopes of conductivity in winter than in summer). This is contrary to the trends shown in some other regions of the world, especially the USA (the main basis for the hypothesis). As mentioned before, road salting is related to climate and road mileage, so the reasons for this phenomenon may be due to: (1) The colder winter and more frequent snowstorms in the USA [49]; (2) The growth rate of road mileage in the USA (6.1%) [50] is much higher than that in the UK (2.1%) [51] over the past 20 years. Similarly, in the UK, compared with other rivers, the potential road salting of the River Mersey is the smallest due to the shortest major road mileage (107.9 km^2) in the basin, the small gap between the annual average number of days to salt (no more than 15 days), and the higher river discharge in winter (ranks second), which may be one of the reasons for the lowest median of conductivity in the River Mersey (Figure 10). In addition, the large amount of road salting caused by extreme cold weather in 2010 and 2013 [52] may be the main reason for the outliers of conductivity of some rivers in these two years (Figure 2), which is helpful to understand the effect of road salting on conductivity.

Table 3. The results of the correlation coefficients (the Kendall's Tau correlation coefficients are only calculated between the water quality data and landcover data) between the annual or seasonal average of the water quality and factors for each river in the past 20 years ('*' means significant).

Water quality	Factors	Tame		Trent		Mersey		Avon	
		r	*p*	r	*p*	r	*p*	r	*p*
pH (annual)	Agricultural lands	0.33 *	0.03 *	0.39 *	0.02 *	0.49	0.22	0.08	0.68
	Urban lands	−0.33	0.11	0.01	0.97	0.13	0.52	−0.08	0.71
pH (summer)	Discharge (summer)	−0.01	0.96	−0.47	0.07	0.46	0.07	0.38	0.15
	NDVI (summer)	−0.15	0.53	−0.08	0.74	0.13	0.60	−0.12	0.61
pH (winter)	Potential road salting (winter)	0.18	0.51	−0.18	0.51	−0.18	0.51	0.15	0.57
	Discharge (winter)	0.08	0.76	0.22	0.41	0.26	0.33	0.04	0.88
	NDVI (winter)	0.16	0.48	0.42	0.06	0.04	0.86	0.35	0.12
Conductivity (annual)	Agricultural lands	0.21	0.42	0.21	0.37	−0.25	0.25	0.07	0.79
	Urban lands	−0.21	0.42	0.33	0.14	−0.01	1.00	0.00	1.00
Conductivity (summer)	Discharge (summer)	−0.59	0.12	−0.56	0.11	−0.50	0.10	−0.86 *	0.00 *
	NDVI (summer)	−0.39	0.17	−0.37	0.24	−0.48	0.06	0.10	0.75
Conductivity (winter)	Potential road salting (winter)	0.71 *	0.03 *	0.54 *	0.04 *	0.83 *	0.03 *	0.25	0.52
	Discharge (winter)	0.00	1.00	−0.81 *	0.02 *	−0.71 *	0.01 *	−0.64	0.07
	NDVI (winter)	−0.71 *	0.01 *	−0.41	0.15	−0.58 *	0.02 *	−0.41	0.15

Furthermore, the results in Tables 3 and 4 also show that the river conductivity trends in the UK may be affected by river discharge, NDVI and urban lands. Specifically, the conductivity is negatively correlated with river discharge (e.g., $r = -0.81$, $p = 0.02$), and similar phenomena have been observed globally—less river discharge concentrates salt ions in the rivers [31], which may explain the significant rise of conductivity in the River Avon in the summer (Table 1, Figure 5).

Table 4. The results of the correlation coefficients between the median level (at a 20-year scale) of the water quality and factors in all rivers ('*' means significant).

Water Quality	Factors	r	*p*
pH (annual)	Agricultural lands	0.90 *	0.04 *
	Urban lands	−0.34	0.66
pH (summer)	Discharge (summer)	0.29	0.71
	NDVI (summer)	0.72	0.28
pH (winter)	Potential road salting (winter)	0.01	0.99
	Discharge (winter)	0.31	0.69
	NDVI (winter)	0.99 *	0.01 *
Conductivity (annual)	Agricultural lands	0.53	0.47
	Urban lands	0.84 *	0.04 *
Conductivity (summer)	Discharge (summer)	0.41	0.59
	NDVI (summer)	−0.33	0.67
Conductivity (winter)	Potential road salting (winter)	0.95 *	0.04 *
	Discharge (winter)	0.37	0.63
	NDVI (winter)	0.03	0.97

At the same time, the negative correlation between NDVI and conductivity has been observed in the Mississippi River [53] and many coastal areas [54], usually because higher soil and river conductivity may damage the health of plants (decrease of NDVI), which seems to be applicable to the results in Table 3 for the River Tame ($r = -0.71$, $p = 0.01$) and River Mersey ($r = -0.58$, $p = 0.02$).

Interestingly, while this study does not find a significant correlation between conductivity and urban lands on the vertical comparison, it is observed on the horizontal comparison ($r = 0.84$, $p = 0.04$), which is consistent with Carpenter's [55] finding that rivers receive more salt as they flow through cities. In addition, the results of the horizontal comparison may be used to account for the differences in conductivity between different reaches [12], for instance, the conductivity in the upper reaches of the River Tame flowing through Birmingham (more urban lands) is much higher than that in the middle and lower reaches dominated by agricultural lands (Figure 4). However, there is not a significant correlation between conductivity and agricultural lands, which is different from the phenomena observed in the central USA, whose causes include direct input of salt ions from fertilisers, and enhanced nitrification of fertilisers leading to the acceleration of the soil ion exchange rate [13].

Of course, the results in Tables 3 and 4 do not explain the causes for the trends of river conductivity in the UK comprehensively. In fact, the treatment measures may also play a key role in influencing the trends of conductivity. For example, since the 1980s, as a result of a series of legislative changes, the pollutants discharged into the upstream River Tame have decreased, and the closure of some plants and construction of the Lake lea Marston in the middle reach has led to the deeper ameliorating of the water quality [56], which is consistent with the most significant downward trend ($Z = -3.93$) of conductivity in the midstream of the River Tame (Table 1). Similarly, the River Mersey, one of the most polluted rivers in the UK's history [34], has had a great ameliorating of the river aquatic ecosystem since the Mersey Basin Campaign in 1985 [57], which is dedicated to river cleaning and may be one of the main reasons for the significant downward trend in the lower River Mersey (Table 1).

4.2. *Reasons for the Trends of River pH*

Current studies indicate that agricultural activities can accelerate rock weathering and base actions leaching, resulting in the alkalinisation of rivers [27,28], which seems to explain the positive correlation between pH and agricultural lands in the River Tame and the River Trent (Table 3). On the other hand, the results in Table 4 show a significant positive correlation between pH and NDVI in winter ($r = 0.99$, $p = 0.01$), which is consistent with the phenomenon that the growth of NDVI usually means the improvement of water quality, especially in the acidified rivers [53,58]. Unfortunately, this study does not observe the significant correlations between pH and urban lands, river discharge and road salting, which are usually considered to be the factors affecting river acidification [13,59–61]. The reasons for this may include: (1) The period of the observed data is probably insufficient; (2) The correlation may be nonlinear [12].

In addition to the aforementioned reasons, some factors are also considered to have an impact on the river pH, such as acidic deposition, geological conditions and treatment measures. For the UK, acidic deposition is one of the reasons for river acidification, particularly in the regions with serious industrial pollution, which may explain the phenomena that the River Mersey and River Tame, near heavy industrial cities (e.g., Manchester and Birmingham), have a lower median pH (Figures 1 and 11). Also, acidic deposition in the UK has seasonality, the reasons for which include: (1) The rate of heating supply in winter is much higher than the utilisation rate of air conditioners in summer, resulting in higher sulfur emissions in winter than in summer; (2) There are lower temperature and more precipitation in winter than in summer, causing acidic deposition in winter more easily [62]. Fortunately, since the last century, with the promulgation of the Clean Air Act and the promotion of clean energy [63,64], the acidic deposition in the UK has decreased significantly, especially in winter [65], which is consistent with the more significant upward trends of pH in winter for each river (Table 2). However, even though the decrease in acidic deposition may cause the increase in river pH, it seems unlikely that river pH is higher in winter than that in summer due to differences in the solubility of CO_2 and SO_2 in the river between winter and summer [46,47], unless there are additional alkaline inputs or extreme hydrology events. In fact, there is a large amount of limestone in the middle and lower reaches of the River Trent and the middle and upper reaches of the River Avon. This geological structure usually makes alkaline substances enter the river with surface runoff under the influence of chemical weathering [66], which may be one of the reasons for the higher pH in winter in these rivers (Figures 8 and 10). In addition, some local treatments may also play a key role in mitigating river acidification. For example, the joint committee composed of local government and industry in Manchester has issued a series of bills to ameliorate the water quality in the downstream river [57], which may be one of the main reasons for the rise of pH there over the past 20 years (Table 2). Similarly, the treatments of the upper and middle River Tame mentioned above [56] may also apply here. Considering that there are a lot of same anthropogenic disturbances in the upper and middle reaches of the River Tame and River Mersey (e.g., a high level of urbanisation and industrialisation, and the impact of the Manchester Canal [34]) (Figure 1), while the treatments of a single reach usually have a limited impact on the water quality at the whole-basin scale, which may explain the phenomenon that the pH of the River Tame has a significant upward trend at the whole basin scale, the River Mersey has no such trend (Table 2).

4.3. Potential Effects of Extreme River Conductivity and pH

Although the conductivity and pH for each river in this study are at 'Good' status ($6 < pH < 9$; annual average conductivity < 1 mS/cm) for most of the time according to the Water Framework Directive (Standards and Classification) Directions (England and Wales) 2015, some disturbing extremes are still observed. For example, the conductivity of the River Tame was as high as about 3.13 mS/cm in 2010 (Figures 2 and 10), and the extremes of pH in the River Trent in 2006 and the River Mersey in 2003 were 9.2 and 6.7, respectively (Figures 3 and 11). In general, it is necessary to pay attention to these extreme values, as they may imply damage to the river aquatic ecosystem. Some studies have shown that high river conductivity will compress the living space of aquatic organisms (e.g., invertebrates) and coastal plants [67,68], and too low a pH may also lead to the collapse of aquatic ecosystems [69]. In addition, the toxicity of the base cations is also different at different pH. For instance, the toxicity of NH_4^+ increases with pH exceeding 8 [70], while that of Zn^{2+} increases with pH drops below 8 [71]. Notably, in populated areas, high river conductivity may cause the leaching of metal ions from old pipes, and threaten the safety of drinking water [4], which may occur in the River Tame basin. Thus, for the UK, it is important to continue to strengthen the management of acid inputs and salt inputs, regulate the use of fertilisers and easily weathered materials, and control the area of agricultural land.

4.4. Limitations and Recommendations

The main limitations of this study come from the observation data, including (1) An insufficient period: Generally speaking, to reflect a long-term trend of data better, the length of data recording should be at least 25 years [12]. However, the water quality database of the Environment Agency and the UK Centre for Ecology and Hydrology cannot meet this requirement, which may cause some data to fail to show significant trends and correlations; (2) Fewer types: Limited by the dataset, this study only uses conductivity and pH to represent the river salinisation and alkalinisation, respectively. It is worth mentioning that there is not even an accurate record for road salting per year, which may cause uncertainties in this study; (3) Measurement uncertainties and errors: Although the seasonal averages are used to reduce the impact of uncertainties and errors on trends analysis in this study, it can not be avoided totally due to the monthly and irregular sampling period. (4) Policy changes: In the past 20 years, the UK's water quality monitoring standards have undergone three reforms, which caused a long-term gap in some monitor sites, which may cause uncertainties in the trends obtained in this study. Furthermore, this study only selected the data of four representative rivers for analysis, which are distributed in the central and western parts of England with anthropogenic disturbance factors. Therefore, the results of this study may not be fully representative of all UK rivers, especially the natural rivers with less anthropogenic disturbances.

Based on the aforementioned reasons, this study puts forward some suggestions: (1) To establish a comprehensive UK river water quality dataset as soon as possible; (2) Due to the limited data, this study does not discuss the degree to which the river is affected by acidic input and alkaline input. Future research may be able to make a more comprehensive quantitative analysis from the perspective of atmospheric emissions and changes in base cations; (3) This study does not explore the specific mechanism of the impact of river salinisation and alkalinisation on the environment, such as the chemical cycle of toxic ions in the process of salinisation. To better understand the hazards of salinisation and alkalinisation, this may be a useful direction; (4) Although most rivers in the UK are affected by anthropogenic disturbance (the level of urbanisation in the UK is very high), which means that the results of this study may be applied to more rivers in the UK, there are still some natural rivers that have not been considered. Future research can analyse the salinisation and alkalinisation trends of UK rivers on a broader scale; (5) As the salinisation and alkalinisation of rivers are mainly affected by anthropogenic disturbance and climate change, the results of this study may be applicable to the rivers of some European countries with temperate climate and high urbanisation. However, for countries with large areas and diverse climates such as the USA or China, the trends may be complex and different and deserve further exploration.

Author Contributions: Conceptualization, S.J. and D.H.; Methodology, S.J., D.H., Q.W., X.W. and S.D.; Writing—original draft, S.J. and X.W.; Writing—review and editing, D.H., Q.W. and S.D. All authors have read and agreed to the published version of the manuscript.

Funding: This publication has been funded by Dawei Han, a member of *Water* Editorial Board with one free publication.

Data Availability Statement: Not applicable.

Acknowledgments: The authors would like to thank the support provided by Jie Shi.

Conflicts of Interest: The authors declare that they have no conflict of interest.

References

1. Kaushal, S.S.; Likens, G.E.; Pace, M.L.; Reimer, J.E.; Maas, C.M.; Galella, J.G.; Utz, R.M.; Duan, S.; Kryger, J.R.; Yaculak, A.M.; et al. *Freshwater Salinization Syndrome: From Emerging Global Problem to Managing Risks*; Springer: Cham, Switzerland, 2021; Volume 0123456789, ISBN 0123456789.
2. Iglesias, M.C.A. A review of recent advances and future challenges in freshwater salinization. *Limnetica* **2020**, *39*, 185–211. [CrossRef]
3. Lopatina, T.; Anishchenko, O.; Oskina, N.; Zadereev, E. Threshold concentrations of the road salt for adverse effects on females and resting eggs of cladoceran Moina macrocopa. *Aquat. Ecol.* **2021**, *55*, 283–297. [CrossRef]
4. Kaushal, S.S. Increased Salinization Decreases Safe Drinking Water. *Environ. Sci. Technol.* **2016**, *50*, 2765–2766. [CrossRef] [PubMed]
5. Feistel, R.; Wielgosz, R.; Bell, S.A.; Camoes, M.F.; Cooper, J.R.; Dexter, P.; Dickson, A.G.; Fisicaro, P.; Harvey, A.H.; Heinonen, M.; et al. Metrological challenges for measurements of key climatological observables: Oceanic salinity and pH, and atmospheric humidity. Part 1: Overview. *Metrologia* **2015**, *53*, R1. [CrossRef]
6. Herbert, E.R.; Boon, P.; Burgin, A.J.; Neubauer, S.C.; Franklin, R.B.; Ardon, M.; Hopfensperger, K.N.; Lamers, L.P.M.; Gell, P. A global perspective on wetland salinization: Ecological consequences of a growing threat to freshwater wetlands. *Ecosphere* **2015**, *6*, 206. [CrossRef]
7. Oude Essink, G.H.P. Improving fresh groundwater supply—Problems and solutions. *Ocean Coast. Manag.* **2001**, *44*, 429–449. [CrossRef]
8. Keilholz, P.; Disse, M.; Halik, Ü. Effects of land use and climate change on groundwater and ecosystems at the middle reaches of the Tarim River using the MIKE SHE integrated hydrological model. *Water* **2015**, *7*, 3040–3056. [CrossRef]
9. Jacobsen, T.; Adams, R.M. Salt and silt in ancient mesopotamian agriculture. *Science* **1958**, *128*, 1251–1258. [CrossRef]
10. Cañedo-Argüelles, M.; Kefford, B.J.; Piscart, C.; Prat, N.; Schäfer, R.B.; Schulz, C.J. Salinisation of rivers: An urgent ecological issue. *Environ. Pollut.* **2013**, *173*, 157–167. [CrossRef]
11. Williams, W.D. Salinization: Unplumbed salt in a parched landscape. *Water Sci. Technol.* **2001**, *43*, 85–91. [CrossRef]

12. Kaushal, S.S.; Groffman, P.M.; Likens, G.E.; Belt, K.T.; Stack, W.P.; Kelly, V.R.; Band, L.E.; Fisher, G.T. Increased salinization of fresh water in the Northeastern United States. *Proc. Natl. Acad. Sci. USA* **2005**, *102*, 13517–13520. [CrossRef] [PubMed]
13. Kaushal, S.S.; Likens, G.E.; Pace, M.L.; Utz, R.M.; Haq, S.; Gorman, J.; Grese, M. Freshwater salinization syndrome on a continental scale. *Proc. Natl. Acad. Sci. USA* **2018**, *115*, E574–E583. [CrossRef] [PubMed]
14. Godwin, K.S.; Hafner, S.D.; Buff, M.F. Long-term trends in sodium and chloride in the Mohawk River, New York: The effect of fifty years of road-salt application. *Environ. Pollut.* **2003**, *124*, 273–281. [CrossRef]
15. Kelly, V.R.; Lovett, G.M.; Weathers, K.C.; Findlay, S.E.G.; Strayer, D.L.; Burns, D.J.; Likens, G.E. Long-Term Sodium Chloride Retention in a Rural Watershed: Legacy Effects of Road Salt on Streamwater Concentration. *Environ. Sci. Technol.* **2007**, *42*, 410–415. [CrossRef]
16. Raymond, P.A.; Oh, N.-H.; Turner, R.E.; Broussard, W. Anthropogenically enhanced fluxes of water and carbon from the Mississippi River. *Nature* **2008**, *451*, 449–452. [CrossRef]
17. Fortner, S.K.; Lyons, W.B.; Carey, A.E.; Shipitalo, M.J.; Welch, S.A.; Welch, K.A. Silicate weathering and CO_2; consumption within agricultural landscapes, the Ohio-Tennessee River Basin, USA. *Biogeosciences* **2012**, *9*, 941–955. [CrossRef]
18. Evans, M.; Frick, C. *The Effects of Road Salts on Aquatic Ecosystems*; Environment Canada: Gatineau, QC, Canada, 2001; 298p.
19. Davies, P.J.; Wright, I.A.; Jonasson, O.J.; Findlay, S.J. Impact of concrete and PVC pipes on urban water chemistry. *Urban Water J.* **2010**, *7*, 233–241. [CrossRef]
20. Kaushal, S.S.; McDowell, W.H.; Wollheim, W.M. Tracking evolution of urban biogeochemical cycles: Past, present, and future. *Biogeochemistry* **2014**, *121*, 1–21. [CrossRef]
21. Schulz, C.J.; Cañedo-Argüelles, M. Lost in translation: The German literature on freshwater salinization. *Philos. Trans. R. Soc. B Biol. Sci.* **2018**, *374*, 20180007. [CrossRef]
22. Vidic, R.D.; Brantley, S.L.; Vandenbossche, J.M.; Yoxtheimer, D.; Abad, J.D. Impact of shale gas development on regional water quality. *Science* **2013**, *340*, 1235009. [CrossRef]
23. Kefford, B.J.; chäfer, R.B.; Metzeling, L. Risk assessment of salinity and turbidity in Victoria (Australia) to stream insects' community structure does not always protect functional traits. *Sci. Total Environ.* **2012**, *415*, 61–68. [CrossRef] [PubMed]
24. Hart, B.T.; Lake, P.S.; Webb, J.A.; Grace, M.R. Ecological risk to aquatic systems from salinity increases. *Aust. J. Bot.* **2003**, *51*, 689–702. [CrossRef]
25. Raymond, P.A.; Cole, J.J. Increase in the export of alkalinity from North America's largest river. *Science* **2003**, *301*, 88–91. [CrossRef]
26. Rosfjord, C.H.; Webster, K.E.; Kahl, J.S.; Norton, S.A.; Fernandez, I.J.; Herlihy, A.T. Anthropogenically Driven Changes in Chloride Complicate Interpretation of Base Cation Trends in Lakes Recovering from Acidic Deposition. *Environ. Sci. Technol.* **2007**, *41*, 7688–7693. [CrossRef]
27. Whitmore, T.J.; Brenner, M.; Kolasa, K.V.; Kenney, W.F.; Riedinger-Whitmore, M.A.; Curtis, J.H.; Smoak, J.M. Inadvertent alkalization of a Florida lake caused by increased ionic and nutrient loading to its watershed. *J. Paleolimnol.* **2006**, *36*, 353–370. [CrossRef]
28. Aquilina, L.; Poszwa, A.; Walter, C.; Vergnaud, V.; Pierson-Wickmann, A.-C.; Ruiz, L. Long-Term Effects of High Nitrogen Loads on Cation and Carbon Riverine Export in Agricultural Catchments. *Environ. Sci. Technol.* **2012**, *46*, 9447–9455. [CrossRef] [PubMed]
29. Kaushal, S.S.; Likens, G.E.; Utz, R.M.; Pace, M.L.; Grese, M.; Yepsen, M. Increased river alkalinization in the eastern U.S. *Environ. Sci. Technol.* **2013**, *47*, 10302–10311. [CrossRef] [PubMed]
30. Mahmuduzzaman, M.; Ahmed, Z.U.; Nuruzzaman, A.K.M.; Ahmed, F.R.S. Causes of Salinity Intrusion in Coastal Belt of Bangladesh. *Int. J. Plant Res.* **2014**, *4*, 8–13. [CrossRef]
31. Jeppesen, E.; Brucet, S.; Naselli-Flores, L.; Papastergiadou, E.; Stefanidis, K.; Nõges, T.; Nõges, P.; Attayde, J.L.; Zohary, T.; Coppens, J.; et al. Ecological impacts of global warming and water abstraction on lakes and reservoirs due to changes in water level and related changes in salinity. *Hydrobiologia* **2015**, *750*, 201–227. [CrossRef]
32. Ling, K.A.; Ashmore, M.R.; MacRory, R.B. The use of word-based models to describe the development of UK acid rain policy in the 1980s. *Environ. Sci. Policy* **2000**, 3. [CrossRef]
33. Crabtree, B.; Hickman, M.; Martin, D. Integrated water quality and environmental cost-benefit modelling for the management of the River Tame. *Water Sci. Technol.* **1999**, *39*, 221–231. [CrossRef]
34. Jones, P.D. Water quality and fisheries in the Mersey estuary, England: A historical perspective. *Mar. Pollut. Bull.* **2006**, *53*, 144–154. [CrossRef]
35. Coxon, G.; Addor, N.; Bloomfield, J.P.; Freer, J.; Fry, M.; Hannaford, J.; Howden, N.J.K.; Lane, R.; Lewis, M.; Robinson, E.L.; et al. CAMELS-GB: Hydrometeorological time series and landscape attributes for 671 catchments in Great Britain. *Earth Syst. Sci. Data* **2020**, *12*, 2459–2483. [CrossRef]
36. Haq, S.; Kaushal, S.S.; Duan, S. Episodic salinization and freshwater salinization syndrome mobilize base cations, carbon, and nutrients to streams across urban regions. *Biogeochemistry* **2018**, *141*, 463–486. [CrossRef]
37. Theil, H. A rank-invariant method of linear and polynomial regression analysis, Part I. *Proc. R. Neth. Acad. Sci.* **1950**, *53*, 386–392.
38. Sen, P.K. Estimates of the Regression Coefficient Based on Kendall's Tau. *J. Am. Stat. Assoc.* **1968**, *63*, 1379–1389. [CrossRef]
39. Kendall, M.G. Rank Correlation Methods. *Biometrika* **1957**, *44*, 298. [CrossRef]
40. Mann, H.B. Nonparametric Tests Against Trend. *Econometrica* **1945**, *13*, 245–259. [CrossRef]
41. Dawson, R. How Significant is a Boxplot Outlier? *J. Stat. Educ.* **2017**, *19*, 143–162. [CrossRef]

42. Moore, R.H.; Eadie, W.T.; Drijard, D.; James, F.E.; Roos, M.; Sadoulet, B. Statistical Methods in Experimental Physics. *J. Am. Stat. Assoc.* **1973**, *68*, 494. [CrossRef]
43. Pearson, K., VII. Note on regression and inheritance in the case of two parents. *Proc. R. Soc. Lond.* **1895**, *58*, 240–242. [CrossRef]
44. Buck, O.; Niyogi, D.K.; Townsend, C.R. Scale-dependence of land use effects on water quality of streams in agricultural catchments. *Environ. Pollut.* **2004**, *130*, 287–299. [CrossRef]
45. Hannaford, J.; Buys, G. Trends in seasonal river flow regimes in the UK. *J. Hydrol.* **2012**, *475*, 158–174. [CrossRef]
46. Godoy, A.C.; Corrêia, A.F.; Boscolo, W.R.; Bittencourt, F.; Signor, A.; de Oliveira, J.D.; Feiden, A. Water Quality in a Reservoir used for Fish Farming in Cages in Winter and Summer Periods. *Water Air Soil Pollut.* **2018**, *229*, 63. [CrossRef]
47. Abdalmogith, S.S.; Harrison, R.M. An analysis of spatial and temporal properties of daily sulfate, nitrate and chloride concentrations at UK urban and rural sites. *J. Environ. Monit.* **2006**, *8*, 691–699. [CrossRef]
48. Vane, C.H.; Jones, D.G.; Lister, T.R. Mercury contamination in surface sediments and sediment cores of the Mersey Estuary, UK. *Mar. Pollut. Bull.* **2009**, *58*, 940–946. [CrossRef]
49. Shimkus, C.E.; Ting, M.; Booth, J.F.; Adamo, S.B.; Madajewicz, M.; Kushnir, Y.; Rieder, H. Winter storm intensity, hazards, and property losses in the New York tristate area. *Ann. N.Y. Acad. Sci.* **2017**, *1400*, 65–80. [CrossRef] [PubMed]
50. FHWA. *Highway Statistics Series*; FHWA: Washington, DC, USA, 2021; pp. 3–6.
51. Transport Department. Road Lengths in Great Britain: 2020. Available online: https://www.gov.uk/government/statistics/road-lengths-in-great-britain-2020 (accessed on 15 June 2021).
52. Rivett, M.O.; Cuthbert, M.O.; Gamble, R.; Connon, L.E.; Pearson, A.; Shepley, M.G.; Davis, J. Highway deicing salt dynamic runoff to surface water and subsequent infiltration to groundwater during severe UK winters. *Sci. Total Environ.* **2016**, *565*, 324–338. [CrossRef]
53. Griffith, J.A.; Martinko, E.A.; Whistler, J.L.; Price, K.P. Interrelationships among landscapes, NDVI, and stream water quality in the U.S. Central Plains. *Ecol. Appl.* **2002**, *12*, 1702–1718. [CrossRef]
54. Rasel, H.M.; Hasan, M.R.; Ahmed, B.; Miah, M.S.U. Investigation of soil and water salinity, its effect on crop production and adaptation strategy. *Int. J. Water Resour. Environ. Eng.* **2013**, *5*, 475–481. [CrossRef]
55. Carpenter, S.R.; Caraco, N.F.; Correll, D.L.; Howarth, R.W.; Sharpley, A.N.; Smith, V.H. Nonpoint pollution of surface waters with phosphorus and nitrogen. *Ecol. Appl.* **1998**, *8*, 559–568. [CrossRef]
56. Langford, T.E.L.; Shaw, P.J.; Howard, S.R.; Ferguson, A.J.D.; Ottewell, D.; Eley, R. Ecological recovery in a river polluted to its sources: The River Tame in the English Midlands. In *Ecology of Industrial Pollution*; Cambridge University Press: Cambridge, UK, 2012.
57. Kim, L.-H.; D'Arcy, B.J.; Ibanez, M.; Maniquiz-Redillas, M. Industrial estates as sources of water pollution. In *Wealth Creation without Pollution: Designing for Industry, Ecobusiness Parks and Industrial Estates*; IWA Publishing: London, UK, 2018.
58. Chu, H.-J.; Liu, C.-Y.; Wang, C.-K. Identifying the Relationships between Water Quality and Land Cover Changes in the Tseng-Wen Reservoir Watershed of Taiwan. *Int. J. Environ. Res. Public Health* **2013**, *10*, 478–489. [CrossRef] [PubMed]
59. Sarma, V.V.S.S.; Paul, Y.S.; Vani, D.G.; Murty, V.S.N. Impact of river discharge on the coastal water pH and pCO2 levels during the Indian Ocean Dipole (IOD) years in the western Bay of Bengal. *Cont. Shelf Res.* **2015**, *107*, 132–140. [CrossRef]
60. Ouyang, T.; Zhu, Z.; Kuang, Y. Assessing impact of urbanization on river water quality in the Pearl River Delta Economic Zone, China. *Environ. Monit. Assess.* **2006**, *120*, 313–325. [CrossRef] [PubMed]
61. Tang, W.; Jun Xu, Y.; Li, S. Rapid urbanization effects on partial pressure and emission of CO2 in three rivers with different urban intensities. *Ecol. Indic.* **2021**, *125*, 107515. [CrossRef]
62. Cichowicz, R.; Wielgosiński, G.; Fetter, W. Dispersion of atmospheric air pollution in summer and winter season. *Environ. Monit. Assess.* **2017**, *189*, 605. [CrossRef] [PubMed]
63. Greenstone, M. Did the Clean Air Act cause the remarkable decline in sulfur dioxide concentrations? *J. Environ. Econ. Manag.* **2004**, *47*, 585–611. [CrossRef]
64. Millar, M.A.; Burnside, N.M.; Yu, Z. District heating challenges for the UK. *Energies* **2019**, *12*, 310. [CrossRef]
65. Kirk, G.J.D.; Bellamy, P.H.; Lark, R.M. Changes in soil pH across England and Wales in response to decreased acid deposition. *Glob. Chang. Biol.* **2010**, *16*, 3111–3119. [CrossRef]
66. Passell, H.D.; Dahm, C.N.; Bedrick, E.J. Hydrological and geochemical trends and patterns in the upper rio grande, 1975 to 19991. *JAWRA J. Am. Water Resour. Assoc.* **2004**, *40*, 111–127. [CrossRef]
67. Hoagstrom, C.W. Causes and impacts of salinization in the lower pecos river. *Great Plains Res.* **2009**, *19*, 27–44.
68. Giosan, L.; Syvitski, J.; Constantinescu, S.; Day, J. Climate change: Protect the world's deltas. *Nature* **2014**, *516*, 31–33. [CrossRef] [PubMed]
69. Morrison, G.; Fatoki, O.; Persson, L.; Ekberg, A. Assessment of the impact of point source pollution from the Keiskammahoek Sewage Treatment Plant on the Keiskamma River—pH, electrical conductivity, oxygen-demanding substance (COD) and nutrients. *Water SA* **2001**, *27*, 475–480. [CrossRef]
70. Hickey, C.W.; Vickers, M.L. Toxicity of ammonia to nine native New Zealand freshwater invertebrate species. *Arch. Environ. Contam. Toxicol.* **1994**, *26*, 292–298. [CrossRef]
71. Li, X.F.; Wang, P.F.; Feng, C.L.; Liu, D.Q.; Chen, J.K.; Wu, F.C. Acute Toxicity and Hazardous Concentrations of Zinc to Native Freshwater Organisms Under Different pH Values in China. *Bull. Environ. Contam. Toxicol.* **1234**, *103*, 120–126. [CrossRef]

water

MDPI

Article

Distribution Characteristics and Genesis of Iron and Manganese Ions in Groundwater of Eastern Sanjiang Plain, China

Mingguo Wang [1], Min Wang [2], Li Yang [1,*], Tao Yang [1], Jingjie Li [1] and Yuanming Chen [3]

1 Center for Hydrogeology and Environmental Geology Survey, CGS, Baoding 071051, China; wangmingguo@mail.cgs.gov.cn (M.W.); yangtao@mail.cgs.gov.cn (T.Y.); lijingjie@mail.cgs.gov.cn (J.L.)
2 Zhoukou Vocational and Technical College, Zhengzhou 466000, China; wangmin2060@163.com
3 Hebei Branch of China National Geological Exploration Center of Building Materials Industry, Baoding 071051, China; Chenyuanming2023@163.com
* Correspondence: yl_b@mail.cgs.gov.cn

Abstract: Groundwater resources are an essential component of global water resources. Long-term consumption of groundwater exceeding the standard levels for iron and manganese can lead to chronic diseases, posing a threat to human health. The Sanjiang Plain is the largest swamp wetland area in China and has high levels of iron and manganese in the groundwater, but the cause still needs to be clarified. Based on the results of water quality tests of 41 groundwater samples in the Eastern Sanjiang Plain, this paper analyzes the distribution characteristics and causes of iron and manganese from the perspectives of the original strata environment, redox conditions, pH conditions, and hydrochemical indicator factors in the research area, using statistical methods and GIS technology. The results show that high iron and manganese content in groundwater is prevalent in the Eastern Sanjiang Plain, and the exceedance rate of manganese is higher than that of iron (87.80% and 82.90%, respectively). The primary sources of iron and manganese in groundwater are iron and manganese minerals in the original strata environment. Influenced by factors such as acidic conditions, reducing environment, and rich organic matter, insoluble high-valent iron and manganese oxides are reduced to low-valent and soluble divalent iron and manganese. At the same time, groundwater's high mineralization and evaporation concentration are conducive to increased iron and manganese content, while the influence of human activities is small.

Keywords: eastern Sanjiang plain; distribution characteristics; genesis; iron and manganese ions; groundwater

Citation: Wang, M.; Wang, M.; Yang, L.; Yang, T.; Li, J.; Chen, Y. Distribution Characteristics and Genesis of Iron and Manganese Ions in Groundwater of Eastern Sanjiang Plain, China. *Water* **2023**, *15*, 2068. https://doi.org/10.3390/w15112068

Academic Editor: Constantinos V. Chrysikopoulos

Received: 28 April 2023
Revised: 23 May 2023
Accepted: 25 May 2023
Published: 30 May 2023

1. Introduction

Iron and manganese are elements known for their variable valence and are richly present in the Earth's crust. In groundwater, the soluble divalent ions of these elements are widespread [1–5]. While iron and manganese are essential trace elements for human health, long-term exposure to high concentrations of these elements in groundwater has been linked to chronic poisoning, which poses a severe threat to human health. This can lead to several diseases affecting the digestive, nervous, and skeletal systems, as well as memory loss and inattentiveness [6–10]. At the same time, high iron and manganese content in water can affect crop yield, animal growth, and development [11,12]. To mitigate these risks, the Chinese Drinking Water Hygiene Standard (GB5749–2022) has set maximum permissible limits of 0.3 mg/L for iron and 0.1 mg/L for manganese [13]. The European Drinking Water Directive (2020/2184/UE) sets the upper limits of iron and manganese ions in drinking water at 0.2 mg/L and 0.05 mg/L, respectively [14]. The World Health Organization (WHO) recommends a critical value of manganese in water of 0.4 mg/L [15]. However, high-manganese groundwater remains a persistent problem worldwide, including in New

Zealand, Sweden, Bangladesh, Nigeria, India, and the United States [11,16–18]. Moreover, in several regions of China, such as northeastern Heilongjiang Province, southern Hebei, northern Ningxia, Shaanxi, Shanxi, Qinghai, Inner Mongolia, and the Pearl River Delta, the concentration of manganese in groundwater significantly exceeds the permissible level [19–24]. This is a cause for concern, as nearly 20% of China's groundwater resources are impacted by this issue, mainly in densely populated alluvial plains and inland basins. Notably, the primary reason for the high iron and manganese content in groundwater is primarily a natural outcome of the geographical and geological environment [25]. Given the potential adverse consequences of high iron and manganese levels in groundwater, the "water-quality type water shortage" phenomenon has become increasingly prevalent in some regions [26].

The Sanjiang Plain, located in northeastern Heilongjiang Province, China, has abundant water resources, with groundwater resources accounting for a significant portion, i.e., 27.4% of a total volume of 18.764 billion cubic meters. High iron and manganese levels in groundwater in the Sanjiang Plain have been reported for a long time. Jiamusi, the core city of the Sanjiang Plain, relies entirely on groundwater for its drinking water needs. Investigations by Wang Lan [27] in 2006 revealed that 100% of the iron and manganese content of the four groundwater drinking water sources in Jiamusi City exceeded the permissible limits. Similarly, Ma Hongwei [28] found that 77.78% of shallow groundwater samples collected from the eastern and northern regions of the Sanjiang Plain exceeded the permissible limit for iron, while 66.67% exceeded the permissible limit for manganese. However, there are few detailed studies on the influencing factors and genetic mechanism of Fe and Mn content in the groundwater of the Sanjiang Plain.

Trace elements in groundwater are mainly affected and restricted by factors such as aquifer media, groundwater flow field, REDOX conditions, pH, microorganisms, and anthropogenic activities [16,29–32]. High-Fe/Mn groundwater occurrence can primarily be attributed to natural factors, such as the hydrological environment, aquifer media, and groundwater recharge and discharge conditions. Owing to leaching, manganese minerals in soil or aquifer media enter groundwater [18,33,34]. Under low pH and Eh conditions, soluble forms of Fe^{2+} and Mn^{2+} prevail, leading to their presence in groundwater [35].

Statistical analysis, ion ratio, spatial analysis technology, correlation analysis, and principal component analysis are mainly used to analyze the distribution characteristics and causes of iron and manganese ions in groundwater. Wei et al. [21] employed statistical and spatial analysis techniques to investigate the formation of groundwater containing high levels of Fe and Mn in the Songnen Plain, northeast China. The study revealed that the properties of soils and aquifers were the contributing factors to the occurrence of Fe and Mn in the groundwater, with the use of fertilizer in Changchun resulting in an increased nitrate concentration and decreased Fe and Mn content. Rizwan et al. [36] evaluated groundwater quality in the Gadilam River basin using Geographical Information Technology (GIS), risk assessment, and statistical analysis, finding that higher Mn concentrations in tertiary formation were possibly caused by mining and related activities. Bacquart et al. [37] conducted spatial and correlation analyses and health effect evaluations of heavy metals, including Fe and Mn, in groundwater samples from Central Burma's Myingyan Township, which showed that there were multiple toxic inorganic contaminants present in well water, exceeding levels of concern for public health, which might have originated from natural geological deposits in the region. Solano S et al. [38] utilized hierarchical clustering analysis to study chemical concentrations and profile the groundwater quality of the Central Pacific, Costa Rica, which linked significant levels of Fe and Mn to the dissolution of hydrothermally altered volcanic rocks. In Luo et al. [39]'s research, pollution factors, comprehensive water quality index, PCA, and other methods were used to assess heavy metal elements in groundwater and pollutant sources.

The Sanjiang Plain is an essential ecological protection barrier and grain production area in northeast China. Still, the reason for high-iron and high-manganese groundwater formation in this area remains unclear. Therefore, this study took Shengli Farm and its surrounding areas in the east of the Sanjiang Plain as the research area, and 41 groups of groundwater samples were collected. Based on multivariate statistics and GIS spatial analysis techniques, the ion ratio, correlation analysis, and principal component analysis were adopted to achieve the following research objectives: (1) to determine the iron and manganese content level and the pollution status of groundwater in the study area; (2) to elucidate the primary factors and mechanisms affecting the migration and transformation of iron and manganese in groundwater in the study area; and (3) to reveal the source of Fe and Mn substances in groundwater with high Fe and Mn levels. The research results can provide theoretical support for protecting groundwater resources and preventing pollution.

2. Study Area

2.1. Overview

The Sanjiang Plain, situated in northeastern China, resulted from the accumulation of alluvial deposits from the Heilongjiang, Wusuli, and Songhua Rivers. As China's largest marsh wetland region, it covers an extensive area of 10.89×10^4 square kilometers. This area is well known for its rich soil, which produces 15 million tons of grain annually, making it a critical commodity grain base in China. The study area is primarily located in the eastern section of the Sanjiang Plain, with Shengli Farm positioned as the center. The research area was appropriately extended beyond the jurisdictional boundaries of Shengli Farm. Encompassing roughly 1000 square km and with a population of 20,000, the area is characterized as a classic agricultural site.

The study area is in a temperate humid and semi-humid continental monsoon climate. The area is characterized by short sunlight duration, severe cold temperatures, and dryness during the winter. Spring sees a rapid increase in temperature along with strong winds, rendering the area vulnerable to droughts and floods. The summer season is short and warm and tends to experience abundant rainfall, while high precipitation levels typify the autumn season. The mean annual precipitation in the study area falls between 500 and 550 mm, with 50–70% of total rainfall observed between May and September. Notably, precipitation between March and May constitutes merely 12.5% of the total annual rainfall. Water surface evaporation ranges from 580 to 650 mm, with the highest rates recorded between May and September and the lowest in December and January. The mean annual temperature in the area is 1.3 °C, with the coldest month seeing an average temperature of −16.2 °C, while the warmest month has an average temperature of 20.2 °C. The region experiences around 190 freezing days annually, with the frozen layer thickness ranging from 1.4 to 2.0 m [40].

The study area features an east-to-west sloping topography, with the western section of the region characterized by alluvial plains and terraces formed by rivers (Figure 1). The presence of extensive swamps and marshy meadows is also prevalent in the region. Conversely, the eastern part of the region is composed of residual hills. The Naoli River flows through the southeastern sector of the area, while the Bielahong River courses through its northwestern part. The central-eastern part of the region is typified by residual hills such as Shuangshan Mt., Xinan Mt., Fuyang Mt., Laochi Mt., Xiaogu Mt., Kaerka Mt., and other hills. The study area predominantly consists of outcrops of the Upper Triassic Dajiahe Formation siliceous rocks; Upper Triassic-Lower Jurassic Dalingqiao Formation sandstones, shales, and siliceous rocks; Upper Neogene basalt; and locally, Lower Cretaceous Dongshan Formation andesite and tuff. In the surrounding area, the sand layer of the unconsolidated Quaternary aquifer is distributed below the 0–18 m thick clay layer, which mainly comprises silty clay and viscous soil [41].

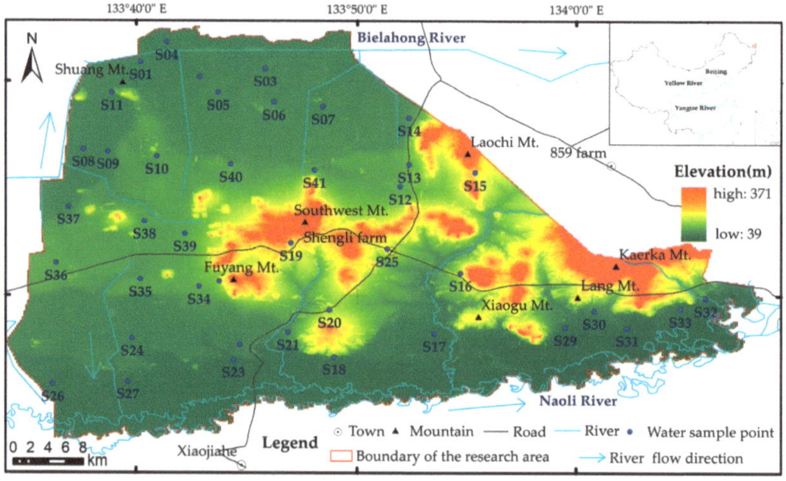

Figure 1. Topographic map of the research area and sample points' distribution.

2.2. *Groundwater*

The regional groundwater flows roughly from south to north, and the flow field in the research area is consistent with the trend of regional flow direction (Figure 2). The primary aquifer is the quaternary unconsolidated porous aquifer, with a typical groundwater depth ranging from 2.1 to 27.8 m. In the central and eastern residual mountainous areas, the aquifer is a fractured aquifer, which is typically low in water yield. From east to west in the study area, the water yield of the aquifer increases, and the groundwater depth becomes shallower. Quaternary groundwater is the primary source of drinking and irrigation water. Based on the single well water usage Q, the distribution of unconsolidated aquifer in the area can be divided into four types: extremely abundant (Q > 5000 m^3/d), abundant (5000 m^3/d > Q > 1000 m^3/d), moderate (1000 m^3/d > Q > 100 m^3/d), and scarce (Q < 100 m^3/d). The fractured rock porous aquifer in the study area has poor water yield (Figure 3).

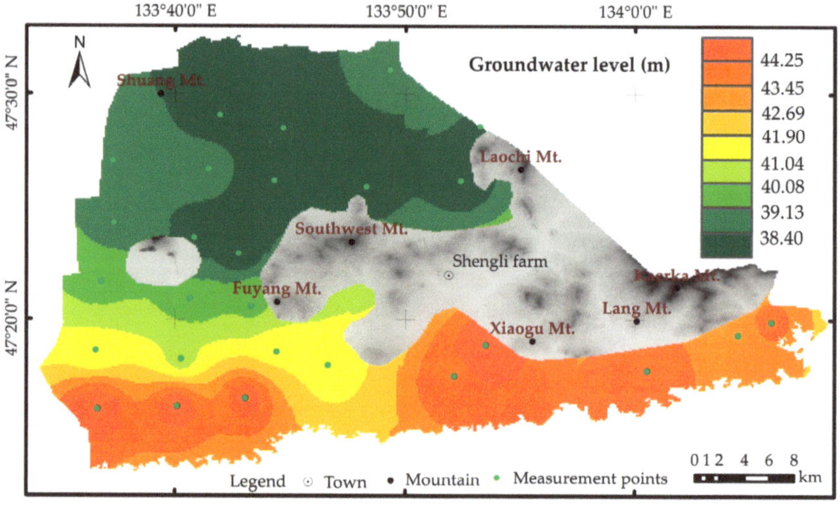

Figure 2. Diagram of the water table level in September in the study area.

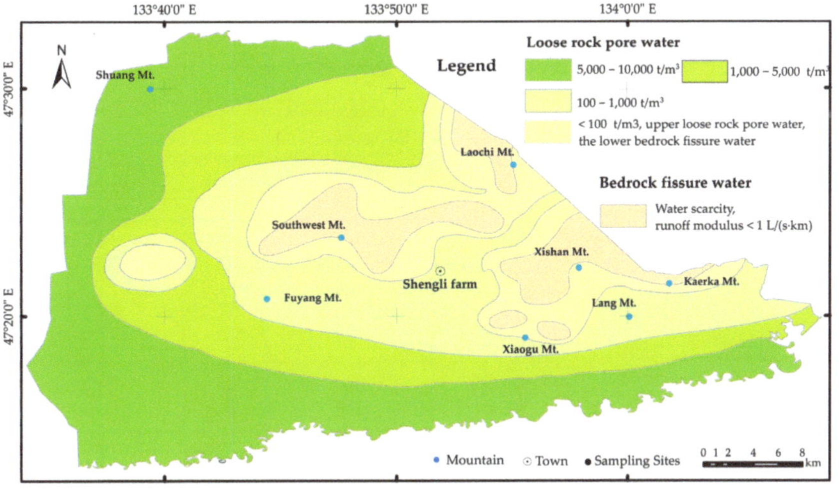

Figure 3. Hydrogeology setting of the study area.

Atmospheric precipitation infiltration is the most important source of recharge for unconsolidated rock porous water in the study area, and irrigation infiltration is also an essential component of groundwater recharge. Artificial extraction is the main means of groundwater discharge, while a small amount of evaporation discharge also exists. In addition, the northwest and south of the study area can receive surface water recharge from rivers during the flood season, while groundwater discharges into the rivers during the dry season.

3. Materials and Methods

3.1. Sampling and Analyzing

In this study, 41 groundwater samples were collected in September 2022, as illustrated in Figure 1. The sampling depths ranged from 2.1 to 27.8 m, and the samples were primarily obtained from the loose porous water in the Quaternary system. Before collection, the wells were purged for at least 10 min, and the pH and Eh values were measured on-site using a portable multi-parameter water quality analyzer (SmarTROLL MP, In-Situ, Fort Collins, CO, USA). The cation analysis samples were acidified with 1:1 nitric acid to a pH value of less than 2 and stored accordingly. The collection, storage, and transportation of groundwater samples were conducted in strict accordance with the Technical Specifications for Groundwater Environmental Monitoring (HJ164–2020) [42] and tested following the Hygienic Standard for Drinking Water (GB5750–2006) [43].

The samples were filtered using 0.45 μm acetate fiber membranes on site and sealed in 500 mL high-density Teflon plastic sample bottles. Samples for cation analysis were acidified with hydrochloric acid to pH less than 2, and all samples were stored under 4 °C sealed and protected from light. The concentrations of K^+, Na^+, Mg^{2+}, Ca^{2+}, and Fe^{2+} were measured using inductively coupled plasma optical emission spectrometry (ICP-OES) (Vista MPX, Varian Instruments, Walnut Creek, CA, USA). When conducting ICP-OES analysis, 1000 PPM Cs must be added to correct for ionization of Na and K in the plasma. The ion chromatography (IC) method was used to measure Cl^-, NO_3^-, and SO_4^{2-} (Dionex^{TM} ICS-1100, Thermo Fisher Scientific Inc., Waltham, MA, USA). The concentrations of HCO_3^- and CO_3^{2-} were determined by hydrochloric acid titration, and the total dissolved solids (TDS) were determined using the gravimetric method. All measurements were conducted at the Laboratory of the Center for Hydrogeology and Environmental Geology Survey.

Table 1 shows the precision and accuracy metrics for each chemical parameter. The precision was assessed using the relative standard deviation (RSD) and the relative deviation (RD), and all water samples had NICB values below 5%.

Table 1. Precision and accuracy metrics of each chemical parameter.

Parameter	Method	Wavelength (nm)	Detection Limit	RSD (%)	RD (%)
K^+	ICP-OES	766.490	0.04	6.72	0~0.76
Mg^{2+}	ICP-OES	279.077	0.006	4.52	0~1.20
Ca^{2+}	ICP-OES	315.887	0.02	3.12	0~0.27
Na^+	ICP-OES	589.592	0.15	2.69	0.6~3.42
Fe^{2+}	ICP-OES	259.939	0.013	3.42	0~1.36
Mn^{2+}	ICP-OES	257.610	0.008	2.25	0~2.25
Cl^-	IC		0.007	2.32	0~0.93
$SO_4{}^{2-}$	IC		0.018	2.10	0~0.91
$NO_3{}^-$	IC		0.016	1.81	0~3.12
$HCO_3{}^-$	VOL		5	2.80	0

3.2. Data Analysis

All statistical analyses and plots were made using SPSS v 17.0 and ArcGIS V 9.2. A statistical program, SPSS v 17.0, evaluated the relationships between various hydrochemical parameters and principal component analysis. ArcGIS V 9.2 was used to determine the spatial distributions of the pH, Eh, Fe, and Mn values. Ion ratio plots were finished using Microsoft® Excel.

4. Results and Discussion

4.1. Characteristics of Iron and Manganese Contents

The main hydrochemical parameters of the groundwater samples are presented in Table 2. The groundwater was primarily composed of $Ca{\cdot}Mg\text{-}HCO_3$, $Ca{\cdot}Mg\text{-}SO_4{\cdot}HCO_3$, $Ca{\cdot}Mg\text{-}HCO_3{\cdot}SO_4$, and $Ca\text{-}HCO_3$, with some samples containing $Mg{\cdot}Ca\text{-}HCO_3$, $Ca{\cdot}Mg\text{-}NO_3{\cdot}Cl$, $Ca{\cdot}Mg\text{-}SO_4$, $Ca{\cdot}Mg\text{-}HCO_3{\cdot}Cl$, and $Ca\text{-}HCO_3{\cdot}SO_4$. According to the Groundwater Quality Standard (GB/T14848-2017), the detection rate of Fe in the groundwater samples was 95.12%, with a concentration range of ND (not detected) to 31.55 mg/L, an average concentration of 3.78 mg/L, and a non-compliance rate of 87.80%. The detection rate of Mn was 92.68%, with a concentration range of ND–1.41 mg/L, an average concentration of 0.40 mg/L, and a non-compliance rate of 82.90%. The Drinking Water Standard for Residents (GB5749-2006) sets the limits for Fe and Mn concentrations in drinking water at 0.3 mg/L and 0.1 mg/L, respectively. Similar to the Groundwater Quality Standard (GB/T14848-2017), groundwater exceeding these limits is classified as high-Fe and high-Mn groundwater. In the study area, there were 36 and 34 groups of high-Fe and high-Mn groundwater, respectively, with non-compliance rates exceeding 80%, indicating a significant impact on residents' drinking water safety.

Table 2. Statistical table for hydrochemical parameters (in mg/L).

Parameters	PH	K^+	Na^+	Ca^{2+}	Mg^{2+}	$HCO_3{}^-$	$SO_4{}^{2-}$	$NO_3{}^-$	Cl^-	TDS	Fe^{2+}	Mn^{2+}	Eh
Mean	6.54	1.90	15.92	34.51	12.28	178.03	40.28	9.90	5.93	227.68	3.78	0.38	−15.76
Min	5.94	0.56	5.74	14.56	6.32	64.07	0.10	0.01	0.42	110.00	0.01	0.01	−78.30
Max	7.24	7.89	30.81	92.71	32.96	353.92	217.85	114.60	53.66	521.00	31.55	1.41	64.31
S.D. [a]	0.26	1.23	5.13	17.82	5.58	63.75	53.45	22.88	11.64	96.07	6.39	0.33	31.71
C.V. [b]	0.04	0.65	0.32	0.52	0.45	0.36	1.33	2.31	1.96	0.42	1.69	0.86	−2.01

Notes: [a] S.D.: standard deviation; [b] C.V.: coefficient of variation.

To investigate the spatial distribution characteristics of Fe and Mn concentrations in the groundwater, isopleth maps were generated using ArcGIS V9.2 software (Figure 4). The results showed that the overall background concentrations of Fe and Mn were high, with high-background areas distributed in Daohuzhuang, Shuang Mt., Laochi Mt., and west of Fuyang Mt. High-background areas of Mn were also observed in the southeast of Kaerka Mt. in the east. Low-value points were sporadically distributed in the central and western parts of the study area. Combined with Figure 2, along the direction of groundwater flow, both Fe and Mn concentrations showed a significant increasing trend, indicating a similar distribution pattern. Fe and Mn concentrations, with a broader distribution range, exceeded the standard limits.

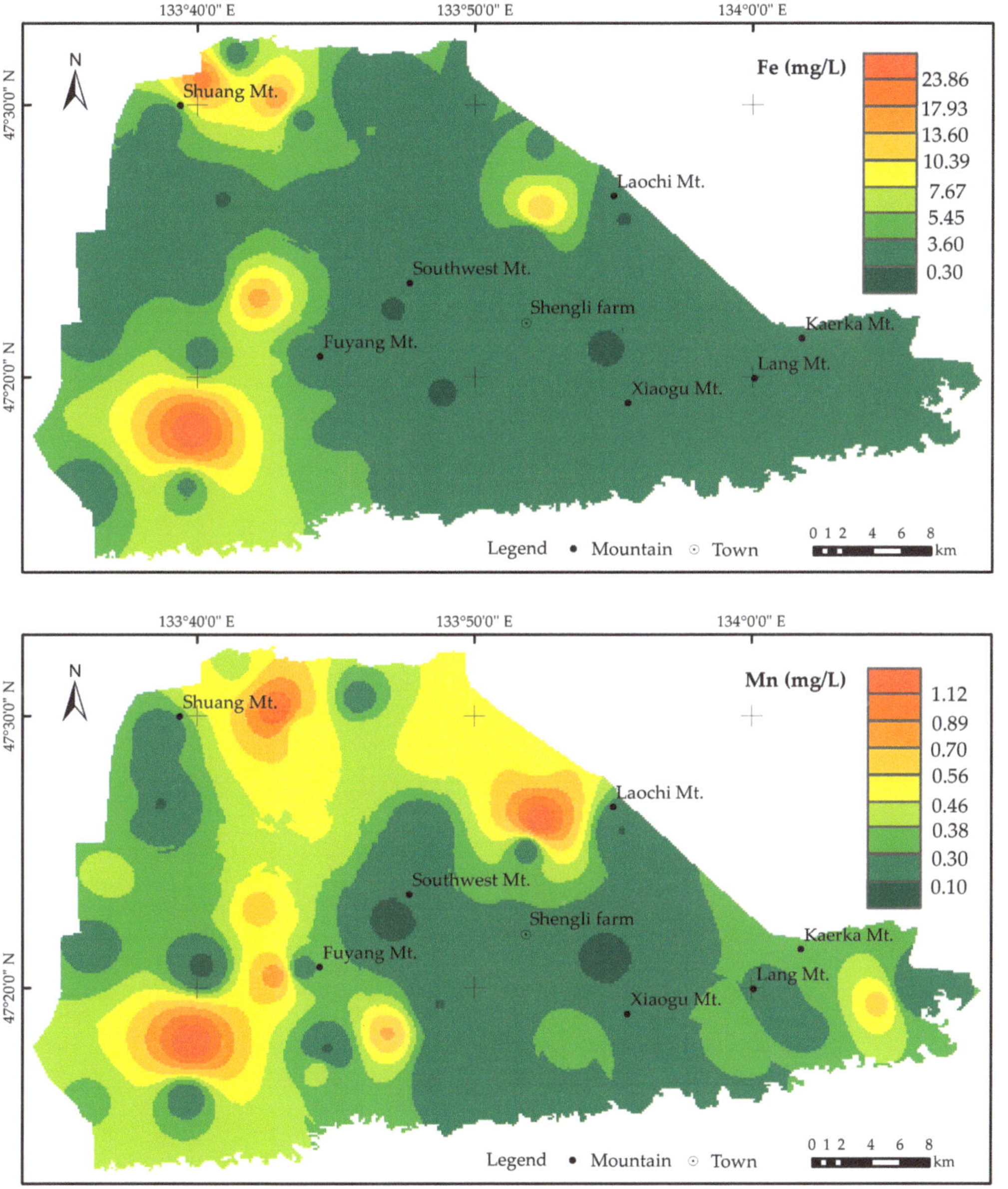

Figure 4. Spatial variation in iron and manganese content.

A particular relationship exists between the changes in the concentration of each ion in the same groundwater system. Correlation analysis can be used to investigate the source of ions in hydrochemistry research. The correlation between components from the same source is strong. Otherwise, the correlation could be poor [33,34].

The results of the correlation analysis are presented in Figure 5. There is a positive correlation between Fe^{2+}, Mn^{2+}, and $SO_4{}^{2-}$, with the correlations between Fe^{2+} vs. Mn^{2+}, Fe^{2+} vs. $SO_4{}^{2-}$, and Mn^{2+} vs. $SO_4{}^{2-}$ being 0.771, 0.600, and 0.642, respectively, indicating a common source. There is a significant negative correlation between Fe^{2+}, Mn^{2+}, and pH, meaning that a decrease in pH favors an increase in the content of Fe^{2+} and Mn^{2+} ions in groundwater. Usually, Cl^- and $NO_3{}^-$ can reflect the impact of human activities [33], but Fe^{2+} and Mn^{2+} have no significant correlation with them, indirectly indicating that the content of Fe^{2+} and Mn^{2+} is less affected by human activities.

	Na^+	K^+	Ca^{2+}	Mg^{2+}	NO_3^-	SO_4^{2-}	HCO_3^-	Cl^-	TDS	pH	Eh	Fe^{2+}
K^+	0.371 *											
Ca^{2+}	0.277	0.239										
Mg^{2+}	0.636 **	0.253	0.637 **									
NO_3^-	0.456 **	0.299	0.360*	0.261								
SO_4^{2-}	0.315	0.133	0.568 **	0.731 **	0.015							
HCO_3^-	0.076	−0.046	0.395 *	0.360 *	−0.142	−0.266						
Cl^-	0.523 **	0.596 **	0.314 *	0.231	0.878 **	−0.045	−0.098					
TDS	0.387 *	0.286	0.785 **	0.675 **	0.387 *	0.542 **	0.315 *	0.361 *				
pH	0.139	0.026	0.076	0.074	−0.005	−0.415 *	0.559 **	0.151	0.019			
Eh	0.400 **	0.078	0.103	0.109	0.298	−0.125	0.138	0.313 *	−0.058	0.271		
Fe^{2+}	0.059	0.088	0.395 *	0.335 *	−0.190	0.600 **	−0.128	−0.224	0.254	−0.496 **	−0.252	
Mn^{2+}	−0.092	−0.063	0.235	0.249	−0.331 *	0.642 **	−0.147	−0.371 *	0.216	−0.493 **	−0.293	0.771 **

Figure 5. Person correlation coefficients of groundwater hydrochemical parameters in the Eastern Sanjiang Plain. (** Significant at the 0.01 level. * Significant at the 0.05 level).

4.2. Cause Analysis of High-Iron and High-Manganese Groundwater in Northeastern Sanjiang Plain

4.2.1. Primary Stratigraphic Environment

High-Fe and high-Mn groundwater, as types of geogenic poor-quality water, result from long-term interactions between the groundwater and surrounding geological formations, which are influenced by various factors, such as aquifer media and soil chemical composition [19,44]. The Gibbs diagram can qualitatively determine the hydrochemical evolution mechanism controlled by atmospheric precipitation, rock filtration, and evaporation concentration [45]. The Gibbs diagram of groundwater in the study area (Figure 6) showed that the groundwater samples were mainly within the TDS concentration range of 10–1000 mg/L, with $Na^+/(Na^++Ca^{2+})$ ratios mainly distributed between 0.1 and 0.7, and $Cl^-/(Cl^- + HCO_3{}^-)$ ratios mainly distributed between 0.0 and 0.5, falling between "rock filtration control" and "evaporation concentration control", indicating that the hydrochemical evolution of groundwater was largely controlled by rock filtration and evaporation concentration, with some groundwater samples falling outside the Gibbs model, possibly due to human activities or cation exchange [46].

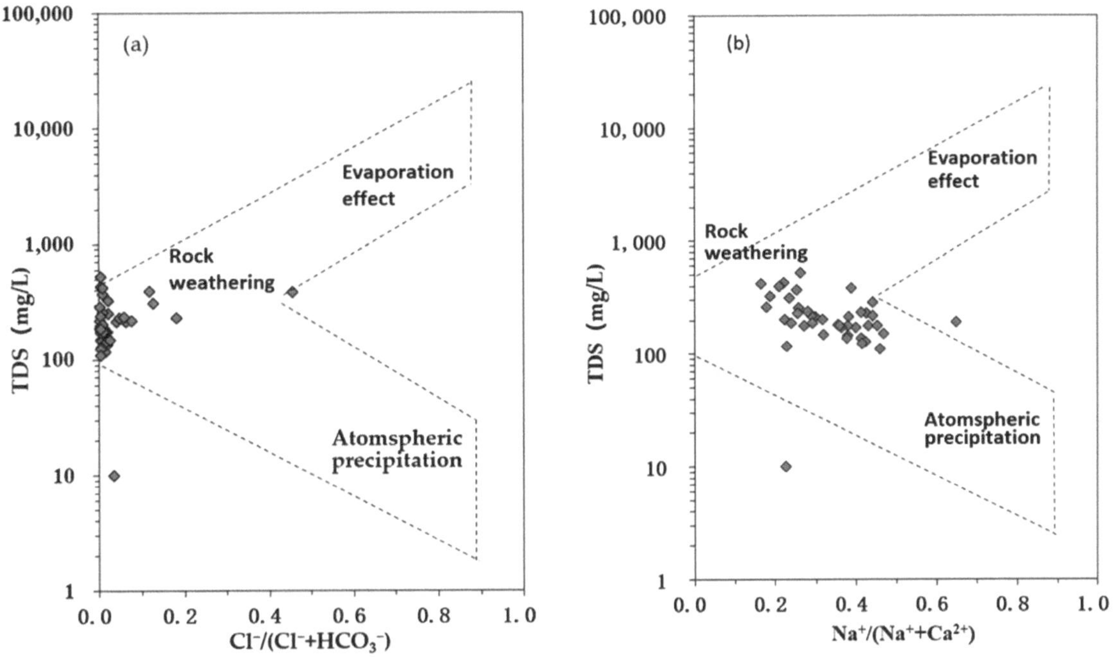

Figure 6. Gibbs plots of groundwater samples in the Eastern Sanjiang plain.

The northern and western peripheries of the study area are rich in iron and manganese mineral resources, with the ore deposit type being shallow sea sedimentary metamorphic. The proven geological reserves of iron ore are 54 million tons, with an average iron content of about 30%, mainly composed of pyrite and magnetite. At the same time, manganese minerals mainly include manganese oxide and manganese carbonate [38]. These minerals are weathered and undergo water–rock interactions and are ultimately transported by rivers to the plains, where some of their components dissolve into the groundwater, becoming the main source of Fe and Mn in the groundwater.

At the same time, the Sanjiang Plain has a strong Late Pleistocene pedogenesis, with the development of iron-rich aluminous soils with high magnetic susceptibility and leached soils rich in iron and manganese nodules since the Quaternary, which could also provide abundant sources of Fe and Mn for the groundwater [47].

4.2.2. REDOX Condition

In a relatively closed groundwater environment, a reducing environment (low Eh value) is conducive to dissolving and releasing Fe and Mn oxides in sediments. The insoluble high-valence Fe and Mn oxides in the formation will be reduced to lower-valence and more soluble Fe^{2+} and Mn^{2+}. In an oxidizing environment (high Eh value), Fe^{2+} and Mn^{2+} in groundwater will be oxidized and precipitated, decreasing Fe and Mn content in groundwater [48]. To analyze the relationship between Eh and the Fe and Mn content in groundwater, an Eh contour map of groundwater in the *Kaerka Mt.* area was drawn based on the sample's Eh value (Figure 7). The Eh values in the study area are mostly negative, ranging from −78.30 to 64.31 mV, with an average value of −15.76 mV. The runoff of the Bielahong River and the Naoli River flows from west to east in the study area, which dilutes the upstream groundwater, lowers the Eh value, favors the formation of Fe^{2+} and Mn^{2+}, and tilts towards a reducing environment. The points where iron and manganese exceed the standard in the groundwater of the plain area are mainly located in areas with developed water systems, such as alluvial plains and valleys, where the groundwater flow rate is slow (the groundwater environment is still in the transitional stage or reducing environment). In low-lying alluvial and alluvial areas, iron and manganese content in surface sediments is relatively high.

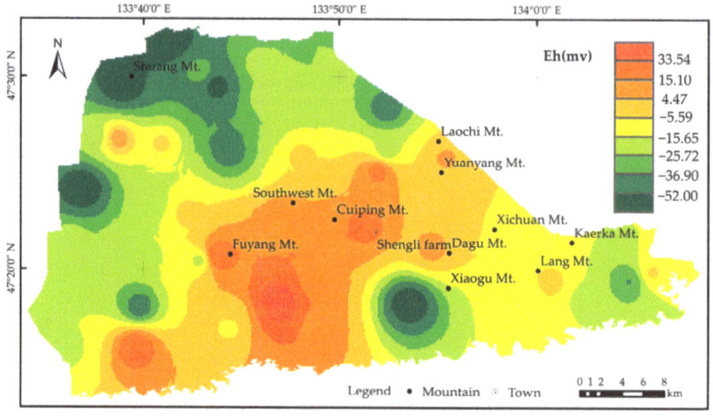

Figure 7. Isogram of Eh in groundwater in the Eastern Sanjiang Plain.

4.2.3. Acid–Base Conditions of Groundwater

Under normal circumstances, Fe^{2+} and Mn^{2+} can relatively easily enter groundwater from formations, while it is relatively difficult for Fe^{3+} and Mn^{4+} (iron and manganese oxides) to enter groundwater from formations. However, under acidic conditions, Fe^{2+}, Mn^{2+}, Fe^{3+}, and Mn^{4+} are more likely to enter groundwater [49]. The pH contour map of groundwater in the Eastern Sanjiang Plain (Figure 8) shows that the pH values of groundwater in the study area are between 5.94 and 7.24, and weakly acidic groundwater is mainly concentrated in the northeast and southwest of Shengli Farm. This is because the area is primarily marsh soil, and the soil (0–1.5 m) is also acidic. Under acidic conditions, Fe and Mn will undergo dissolution reactions, which are conducive to the dissolution and enrichment of Fe^{2+} and Mn^{2+}. The groundwater in the southern part of the study area is close to neutral, and the lithology is mainly limestone. The Eh–pH diagram of groundwater at Shengli Farm (Figure 9) shows that Mn and Fe mainly exist as Fe^{2+} and Mn^{2+} in groundwater, which causes severe and widespread exceedance of Fe and Mn content. As groundwater is closely related to human activities, it is generally believed that the excessive content of Fe and Mn in the aquifer is closely associated with human activities.

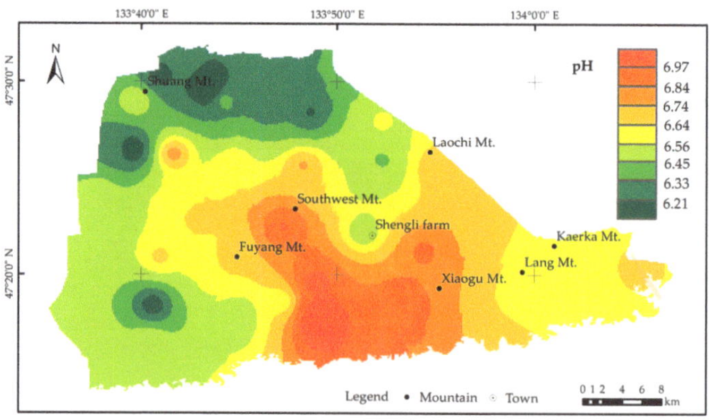

Figure 8. Isogram of pH in groundwater in the Eastern Sanjiang Plain.

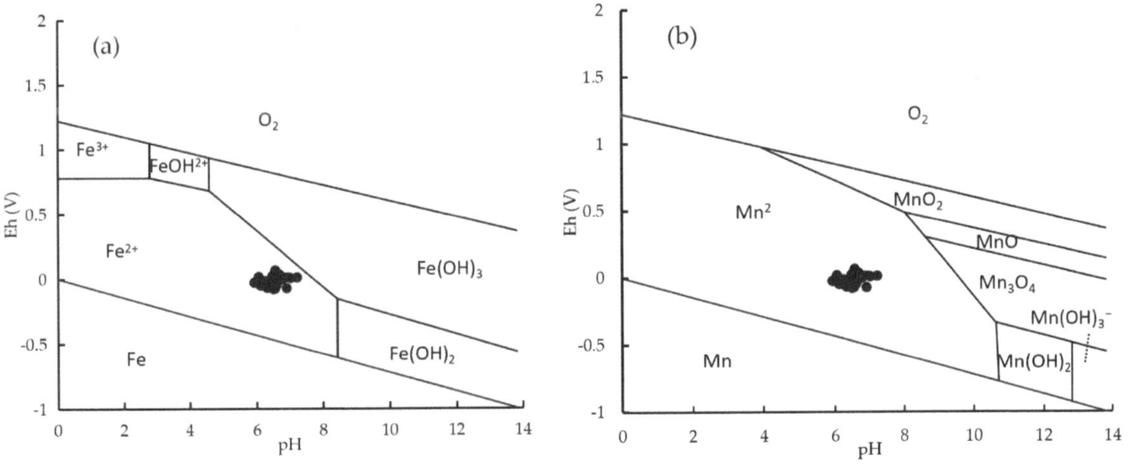

Figure 9. Eh–pH diagram of Fe (**a**) and Mn (**b**) of groundwater in the Northeastern Sanjiang Plain.

The overlying layers in the plain area of the study area are mainly white loess, meadow soil, and marsh soil, which contain abundant organic matter [50]. This is conducive to enriching Fe and Mn in groundwater through chemical reactions. Li Xiao et al.'s research shows that the average NH_4^+ concentration in shallow groundwater in the northeast of the Sanjiang Plain is 0.73 mg/L, which is higher than the groundwater quality standard of 0.5 mg/L [13]. The high NH_4^+ concentration indicates that groundwater tends to be in a reducing environment, which is conducive to forming Fe^{2+} and Mn^{2+}, resulting in iron and manganese exceedance.

4.3. Main Ion Source Analysis

The suitability of the hydrochemical parameters of groundwater in the Northeastern Sanjiang Plain for principal component analysis was tested using Bartlett's test of sphericity and the KMO test [37,51]. The results of the Bartlett sphericity test (significance 0.00 < 0.05) and KMO measurement test (0.516 > 0.5) indicated that principal component analysis was suitable for each element. After the K_{kaise} normalized factors were subjected to orthogonal rotation by Varimax, four principal components with eigenvalues greater than 1 were obtained, and their contribution rates reached 74.03%, which could reflect the main information of all heavy metals.

Based on the principal component (PC) loadings, 13 parameters were grouped into three PCs (F1–F3, Table 3). The main loadings of F1 were Mg^{2+}, Ca^{2+}, and TDS, which were strongly correlated with SO_4^{2-}, K^+, and Fe^{2+} (with loadings of 0.784, 0.601, and 0.543, respectively). This indicated that high mineralization and evaporation concentrations were conducive to increased Fe^{2+} content in groundwater. The main loading of F2 was Cl^-, which was strongly correlated with NO^{3-}, Fe^{2+}, Mn^{2+}, pH, and Eh (with loadings of 0.692, −0.666, −0.775, 0.572, and 0.560, respectively). The positive correlation between F2 and Fe^{2+} and Mn^{2+}, combined with Cl^- (Figure 8), suggested that human activities had a relatively small impact on the excessive iron and manganese content in groundwater. The main loading of F3 was HCO_3^-, which was well correlated with pH (with a loading of 0.618).

Table 3. Principal component analysis matrix.

Element	F1	F2	F3
Na^+	0.461	0.385	−0.162
K^+	0.601	0.448	−0.311
Mg^{2+}	0.940	−0.041	0.087
Ca^{2+}	0.837	0.043	0.427
NO_3^-	0.453	0.692	−0.269
SO_4^{2-}	0.784	−0.453	−0.118
HCO_3^-	0.074	0.171	0.926
Cl^-	0.439	0.809	−0.227
Fe^{2+}	0.543	−0.666	−0.043
Mn^{2+}	0.434	−0.775	−0.001
pH	-0.262	0.572	0.618
Eh	0.027	0.560	−0.111
TDS	0.827	0.075	0.310
Initial eigenvalue	3.393	2.624	1.385
variance contribution rate/%	33.928	26.243	13.854

5. Conclusions

This study analyzed 41 groundwater samples from the Eastern Sanjiang Plain to investigate their distribution characteristics and main controlling factors. The main conclusions are as follows:

(1) The range of iron content in groundwater in the study area is ND–31.55 mg/L, with a mean value of 3.78 mg/L and an excessive rate of 87.80%; the range of manganese content is ND–1.41 mg/L, with a mean value of 0.38 mg/L and an excessive rate of 82.90%; the excessive rate of iron in groundwater is higher than that of manganese, and the distribution range is wider. To ensure the safety of drinking water for residents, it is recommended that the groundwater within the area be appropriately treated before consumption.

(2) The Fe and Mn in groundwater in the study area exist in the form of Fe^{2+} and Mn^{2+}, respectively. The primary sources of Fe and Mn in groundwater are the Fe and Mn minerals in the original geological environment of the study area. Groundwater tends to be reduced due to surface runoff and soil organic matter, leading to excessive Fe and Mn content.

(3) High mineralization and evaporation concentrations are conducive to increased Fe content, as indicated by the factor analysis results of hydrochemical indicators. There is no significant correlation between Fe^{2+}, Mn^{2+} and Cl^-, NO_3^-. The excessive content of Fe^{2+} and Mn^{2+} in groundwater is relatively less affected by human activities.

This study systematically investigated and discussed the distribution, influencing factors, and sources of groundwater iron and manganese at Shengli Farm of the Eastern Sanjiang Plain, providing hydrochemical evidence for water safety, food security, and ecological security and offering a reference for the analysis of iron and manganese sources in similar regions. At the same time, further research and discussions are needed on the treatment methods and strategies for high-iron and high-manganese groundwater, and dynamic monitoring of iron and manganese content in groundwater should be strengthened. Research on the relationship between agricultural activities and iron and manganese content should be conducted to ensure sustainable agricultural development.

Author Contributions: Conceptualization, L.Y. and M.W. (Mingguo Wang); Data curation, M.W. (Min Wang); Formal analysis, M.W. (Mingguo Wang) and M.W. (Min Wang); Investigation, T.Y. and J.L.; Methodology, M.W. (Mingguo Wang) and L.Y.; Project administration, M.W. (Mingguo Wang); Resources, T.Y. and Y.C.; Software, T.Y. and J.L.; Supervision, L.Y.; Validation, M.W. (Mingguo Wang); Writing—original draft, L.Y.; Writing—review and editing, L.Y. All authors have read and agreed to the published version of the manuscript.

Funding: This research was funded by the Geological Survey project of China Geological Survey, No. DD20230456 and DD20221754. The APC was funded by China Geological Survey.

Data Availability Statement: Not applicable.

Conflicts of Interest: The authors declare no conflict of interest.

References

1. Rushdi, M.I.; Basak, R.; Das, P.; Ahamed, T.; Bhattacharjee, S. Assessing the health risks associated with elevated manganese and iron in groundwater in Sreemangal and Moulvibazar Sadar, Bangladesh. *J. Hazard. Mater. Adv.* **2023**, *10*, 100287. [CrossRef]
2. Zhai, Y.; Han, Y.; Xia, X.; Li, X.; Lu, H.; Teng, Y.; Wang, J. Anthropogenic Organic Pollutants in Groundwater Increase Releases of Fe and Mn from Aquifer Sediments: Impacts of Pollution Degree, Mineral Content, and pH. *Water* **2021**, *13*, 1920. [CrossRef]
3. Podgorski, J.; Araya, D.; Berg, M. Geogenic manganese and iron in groundwater of Southeast Asia and Bangladesh—Machine learning spatial prediction modeling and comparison with arsenic. *Sci. Total Environ.* **2022**, *833*, 155131. [CrossRef]
4. Guo, Y.M.; Huang, T.L.; Wen, G.; Cao, X. The simultaneous removal of ammonium and manganese from groundwater by iron-manganese co-oxide filter film: The role of chemical catalytic oxidation for ammonium removal. *Chem. Eng. J.* **2017**, *308*, 322–329. [CrossRef]
5. Shvartsev, S.L. Geochemistry of fresh groundwater in the main landscape zones of the Earth. *Geochem. Int.* **2008**, *46*, 1285–1398. [CrossRef]
6. Wei, M.; Huang, Q.; Dai, Y.; Zhou, H.; Cui, Y.; Song, W.; Di, D.; Zhang, R.; Li, C.; Wang, Q.; et al. Manganese, iron, copper, and selenium co-exposure and osteoporosis risk in Chinese adults. *J. Trace Elem. Med. Biol.* **2022**, *72*, 126989. [CrossRef]
7. Gupta, A. Iron overload in human disease. *N. Engl. J. Med.* **2012**, *366*, 1549.
8. Bhutiani, R.; Kulkarni, D.B.; Khanna, D.R.; Gautam, A. Water quality, pollution source apportionment and health risk assessment of heavy metals in groundwater of an industrial area in north India. *Expo. Health* **2016**, *8*, 3–18. [CrossRef]
9. Yadav, K.K.; Kumar, S.; Pham, Q.B.; Gupta, N.; Rezania, S.; Kamyab, H.; Yadav, S.; Vymazal, J.; Kumar, V.; Tri, D.Q.; et al. Fluoride contamination, health problems and remediation methods in Asian groundwater: A comprehensive review. *Ecotoxicol. Environ. Saf.* **2019**, *182*, 109362. [CrossRef] [PubMed]
10. Yunker, L.M.; Parboosingh, J.S.; Conradson, H.E.; Faris, P.; Bridge, P.J.; Buithieud, J.; Title, L.M.; Charbonneaub, F.; Verma, S.; Lonn, E.M.; et al. The effect of iron status on vascular health. *Vasc. Med.* **2006**, *8*, 3–18. [CrossRef]
11. Jaafar, M.; Shrivastava, A.; Bose, S.R.; Felipe-Sotelo, M.; Ward, N.I. Transfer of arsenic, manganese and iron from water to soil and rice plants: An evaluation of changes in dietary intake caused by washing and cooking rice with groundwater from the Bengal Delta, India. *J. Food Compos. Anal.* **2021**, *96*, 103748. [CrossRef]
12. Sharma, S.; Kaur, I.; Nagpal, A.K. Estimation of arsenic, manganese and iron in mustard seeds, maize grains, groundwater and associated human health risks in Ropar wetland, Punjab, India, and its adjoining areas. *Environ. Monit. Assess.* **2018**, *190*, 385. [CrossRef] [PubMed]
13. *GB5749-202022*; Ministry of Health of the People's Republic of China; China National Standardization Administration. Standards for Drinking Water Quality. Standard Press of China: Beijing, China, 2003. (In Chinese)
14. *Official J EU23.12.2020.435*; Directive 2020/2184 of the European Parliament and of the Council of 16 December 2020 on the Quality of Water Intended for Human Consumption (Recast). Official Journal of the European Union: Luxembourg, 2020.
15. World Health Organization. *Guidelines for Drinking-Water Quality (Edition, F)*; World Health Organization: Geneva, Switzerland, 2011.
16. Vega, M.A.; Kulkarni, H.V.; Johannesson, K.H.; Taylor, R.J.; Datta, S. Mobilization of co-occurring trace elements (CTEs) in arsenic contaminated aquifers in the Bengal Basin. *Appl. Geochem.* **2020**, *122*, 104709. [CrossRef]
17. Johnson, C.; Nandi, A.; Joyner, T.; Luffman, I. Iron and Manganese in Groundwater: Using Kriging and GIS to Locate High Concentrations in Buncombe County, North Carolina. *Groundwater* **2017**, *56*, 87–95. [CrossRef]
18. Sharma, G.K.L.; Jena, R.K.; Ray, P.; Yadav, K.K.; Moharana, P.C.; Cabral-Pinto, M.M.; Bordoloi, B. Evaluating the geochemistry of groundwater contamination with iron and manganese and probabilistic human health risk assessment in endemic areas of the world's largest River Island, India. *Environ. Toxicol. Pharmacol.* **2021**, *87*, 103690. [CrossRef]
19. Hu, M.; Zhou, P.; Chen, C. Spatial and temporal distribution and affecting factors of iron and manganese in the groundwater in the middle area of the Yangtze River Basin, China. *Environ. Sci. Pollut. Res. Int.* **2022**, *29*, 61204–61221. [CrossRef] [PubMed]
20. Li, T.; Zhang, F.; Meng, L. Occurrence, distribution, and prediction of iron and manganese in groundwater of opencast mines: An example from Inner Mongolia, China. *Environ. Monit. Assess.* **2021**, *193*, 544.
21. Zhang, Z.; Xiao, C.; Yang, W.; Adeyeye, O.A.; Liang, X. Effects of the natural environment and human activities on iron and manganese content in groundwater: A case study of Changchun city, Northeast China. *Environ. Sci. Pollut. Res. Int.* **2021**, *28*, 41109–41119. [CrossRef]
22. Huang, B.; Li, Z.; Chen, Z.; Chen, G.; Zhang, C.; Huang, J.; Nie, X.; Xiong, W.; Zeng, G. Study and health risk assessment of the occurrence of iron and manganese in groundwater at the terminal of the Xiangjiang River. *Environ. Sci. Pollut. Res. Int.* **2015**, *22*, 19912–19921. [CrossRef]

23. Yin, X.; Jiang, B.; Feng, Z.; Yao, B.; Shi, X.; Sun, Y.; Wu, J. Comprehensive evaluation of shallow groundwater quality in central and southern Jiangsu Province, China. *Environ. Earth Sci.* **2017**, *76*, 400. [CrossRef]
24. Huang, S.; Chen, L.; Li, J.; Xu, J.; Xie, W.; Zhang, C. The effects of colloidal Fe and Mn on P distribution in groundwater system of Jianghan Plain, China. *Sci. Total Environ.* **2023**, *854*, 158739. [CrossRef]
25. Zhai, Y.; Cao, X.; Xia, X.; Wang, B.; Teng, Y.; Li, X. Elevated Fe and Mn Concentrations in Groundwater in the Songnen Plain, Northeast China, and the Factors and Mechanisms Involved. *Agronomy* **2021**, *11*, 2392. [CrossRef]
26. Wu, A.; Hao, A.; Guo, H.; Liu, J.; Zhang, E.; Wang, H.; Wang, X.; Wen, X.; Zhang, C. Main progress and prospect for China's hydrogeological survey. *J. Groundw. Sci. Eng.* **2020**, *8*, 195–209.
27. Wang, L.; Meng, X.; Xu, H. Analysis of Causes of Superstandard Fe and Mn Content in source water of Catchment Area in Jiamusi. *Environ. Sci. Manag.* **2006**, *31*, 152–153. (In Chinese)
28. Ma, H.; Li, X.; Cui, J.; Yang, Z. A Preliminary Study on The Organic and Inorganic Pollution of Groundwater of Jiansanjiang Farm of Sanjiang Plain. *Geol. Resour.* **2014**, *23*, 38–41. (In Chinese)
29. Ilyushin, Y.V.; Asadulagi, M.-A.M. Development of a Distributed Control System for the Hydrodynamic Processes of Aquifers, Taking into Account Stochastic Disturbing Factors. *Water* **2023**, *15*, 770. [CrossRef]
30. Martirosyan, A.V.; Martirosyan, K.V.; Mir-Amal, A.M.; Chernyshev, A.B. Assessment of a Hydrogeological Object's Distributed Control System Stability. In Proceedings of the 2022 Conference of Russian Young Researchers in Electrical and Electronic Engineering (ElConRus), Saint Petersburg, Russia, 25–28 January 2022; pp. 768–771.
31. Liu, J.; Peng, Y.; Li, C.; Gao, Z.; Chen, S. Characterization of the hydrochemistry of water resources of the Weibei Plain, Northern China, as well as an assessment of the risk of high groundwater nitrate levels to human health. *Environ. Pollut.* **2021**, *268*, 115947. [CrossRef]
32. Xing, W.; Wei, L.; Ma, W.; Li, J.; Liu, X.; Hu, J.; Wang, X. Geochemistry and Sources Apportionment of Major Ions and Dissolved Heavy Metals in a Small Watershed on the Tibetan Plateau. *Water* **2022**, *14*, 3856. [CrossRef]
33. Abesser, C.; Robinson, R.; Soulsby, C. Iron and manganese cycling in the storm runoff of a Scottish upland catchment. *J. Hydrol.* **2006**, *326*, 59–78. [CrossRef]
34. Palmucci, W.; Rusi, S.; Di Curzio, D. Mobilisation processes responsible for iron and manganese contamination of groundwater in Central Adriatic Italy. *Environ. Sci. Pollut. Res. Int.* **2016**, *23*, 11790–11805. [CrossRef]
35. Ewusi, A.; Sunkari, E.D.; Seidu, J.; Coffie-Anum, J. Hydrogeochemical characteristics, sources and human health risk assessment of heavy metal dispersion in the mine pit water–surface water-groundwater system in the largest manganese mine in Ghana. *Environ. Technol. Innov.* **2022**, *26*, 102312. [CrossRef]
36. Rizwan, K.M.; Thirukumaran, V.; Suresh, M. Assessment and source identification of heavy metal contamination of groundwater using geospatial technology in Gadilam River basin, Tamil Nadu, India. *Appl. Water Sci.* **2021**, *11*, 102. [CrossRef]
37. Bacquart, T.; Frisbie, S.; Mitchell, E.; Grigg, L.; Cole, C.; Small, C.; Sarkar, B. Multiple inorganic toxic substances contaminating the groundwater of Myingyan Township, Myanmar: Arsenic, manganese, fluoride, iron, and uranium. *Sci. Total Environ.* **2015**, *517*, 232–245. [CrossRef] [PubMed]
38. Solano, S.C.; Ingrid, V.-A.; Rolando, C.-M.; Huapaya, R.P.S. Multivariate data analysis applied to groundwater geochemical characterization, Central Pacific, Costa Rica. *Appl. Geochem.* **2023**, *151*, 105599. [CrossRef]
39. Luo, M.; Zhang, Y.; Li, H.; Hu, W.; Xiao, K.; Yu, S.; Zheng, C.; Wang, X. Pollution assessment and sources of dissolved heavy metals in coastal water of a highly urbanized coastal area: The role of groundwater discharge. *Sci. Total Environ.* **2022**, *807*, 151070. [CrossRef]
40. Zheng, Y.; Wu, Y.; Pan, X.; Qi, P.; Dai, C.; Zhang, G. Evolution characteristics of meteorological and hydrological drought in Sanjiang Plain. *Res. Soil Water Conserv.* **2019**, *26*, 177–184+189. (In Chinese)
41. Shu, L.; Yin, X.; Yuan, Y.; Lu, Y.; Lu, C.; Liu, B. Temporal and spatial variation of water quantity exchange between surface water and ground water in typical district of Sanjiang Plain. *Shuilixuebao* **2021**, *52*, 1151–1162. (In Chinese)
42. *HJ164-2020*; Ministry of Ecology and Environment of the People's Republic of China. Technical Specifications for Environmental Monitoring of Groundwater. China Environment Publishing Group: Beijing, China, 2020. (In Chinese)
43. *GB5750-2006*; Ministry of Health of the People's Republic of China; China National Standardization Administration. Sanitary Standard Test Method for Drinking Water. Standard Press of China: Beijing, China, 2007. (In Chinese)
44. Chakraborty, M.; Mishra, A.K.; Mukherjee, A. Influence of hydrogeochemical reactions along flow paths on contrasting groundwater arsenic and manganese distribution and dynamics across the Ganges River. *Chemosphere* **2022**, *287*, 132144. [CrossRef]
45. Li, C.; Gao, X.; Wang, Y. Hydrogeochemistry of high-fluoride groundwater at Yuncheng Basin, northern China. *Sci. Total Environ.* **2015**, *508*, 155–165. [CrossRef]
46. Peng, C.; Liu, Y.; Chen, H.; Yuan, Q.; Chen, Q.; Mei, S.; Wu, Z. Analysis of Hydrogeochemical Characteristics of Tunnel Groundwater Based on Multivariate Statistical Technology. *Geofluids* **2021**, *2021*, 4867942. [CrossRef]
47. Sun, W.; Zhang, E.; Chen, R.; Shen, J. Lacustrine carbon cycling since the last interglaciation in northeast China: Evidence from n-alkanes in the sediments of Lake Xingkai. *Quat. Int.* **2019**, *523*, 101–108. [CrossRef]
48. Khozyem, H.; Hamdan, A.; Tantawy, A.A.; Emam, A.; Elbadry, E. Distribution and origin of iron and manganese in groundwater: Case study, Balat-Teneida area, El-Dakhla Basin, Egypt. *Arab. J. Geosci.* **2019**, *12*, 523. [CrossRef]

49. Carretero, S.; Kruse, E. Iron and manganese content in groundwater on the northeastern coast of the Buenos Aires Province, Argentina. *Environ. Earth Sci.* **2015**, *73*, 1983–1995. [CrossRef]
50. Mao, D.H.; Wang, Z.M.; Li, L.; Miao, Z.H.; Ma, W.H.; Song, C.C.; Ren, C.Y.; Jia, M.M. Soil organic carbon in the Sanjiang Plain of China: Storage, distribution and controlling factors. *Biogeosciences* **2015**, *12*, 1635–1645. [CrossRef]
51. Wang, M.; Yang, L.; Li, J.; Liang, Q. The Evaluation and Sources of Heavy Metal Anomalies in the Surface Soil of Eastern Tibet. *Minerals* **2023**, *13*, 86. [CrossRef]

water

MDPI

Article

Geochemistry and Sources Apportionment of Major Ions and Dissolved Heavy Metals in a Small Watershed on the Tibetan Plateau

Wencong Xing [1,2], Lai Wei [1,2], Wenmin Ma [1,2], Jun Li [1,*], Xiaolong Liu [1,*], Jian Hu [3] and Xiaoxia Wang [1,2]

1 Tianjin Key Laboratory of Water Resources and Environment, Tianjin Normal University, Tianjin 300387, China
2 School of Geography and Environmental Science, Tianjin Normal University, Tianjin 300387, China
3 Skate Key Laboratory of Urban and Regional Ecology, Research Center for Eco-Environmental Sciences, Chinese Academy of Sciences, Beijing 100085, China
* Correspondence: lijun5931@163.com (J.L.); xiaolong.liu@tjnu.edu.cn (X.L.)

check for updates

Citation: Xing, W.; Wei, L.; Ma, W.; Li, J.; Liu, X.; Hu, J.; Wang, X. Geochemistry and Sources Apportionment of Major Ions and Dissolved Heavy Metals in a Small Watershed on the Tibetan Plateau. *Water* **2022**, *14*, 3856. https://doi.org/10.3390/w14233856

Academic Editor: Ondra Sracek

Received: 25 October 2022
Accepted: 24 November 2022
Published: 26 November 2022

Abstract: Due to environmental sensibility and fragility, the water chemistry revolution and heavy metals accumulation influenced by natural and anthropogenic processes in the rivers on the Tibetan Plateau have recently become a global concern. However, targeted studies in small watersheds on the Tibetan Plateau are relatively limited. A study of surface waters in Duilong Qu (DLQ), a small watershed located on the Tibetan Plateau, have been conducted to assess the impact of natural and anthropogenic activities on the water environment of the DLQ by analyzing the major ions and heavy metals (Cd, Cr, Mn, Fe, Ni, Cu, Zn, Pb, and As) in the river waters. The results of the analysis of major ions showed that SO_4^{2-} and HCO_3^- were the dominant anions and Ca^{2+} was the dominant cation, indicating that the water chemistry of the river waters was mainly of the HCO_3-Ca type. The results of Piper diagram and Gibbs diagram analysis indicate that the water solute is mainly controlled by the weathering of carbonate rocks, followed by the influence of geothermal water confluence. Compared to the world river average, the concentrations of Cr, Pb, and As in the studied rivers were relatively high. The heavy metal concentrations satisfy the standards of WHO and GB (Chinese national standard) guidelines. The PCA-APCS-MLR model has been employed and evidenced as a reliable tool to identify the sources of the heavy metals in this study. The results revealed that the heavy metals in the DLQ are caused by natural sources, geothermal water, and mining operations. The primary sources of As (93.63%), Cr (93.07%), Mn (73.53%), Fe (59.54%), and Pb (58.28%) in the DLQ were geothermal water, while Zn (91.41%), Mn (20.67%), Fe (40.46%), and Pb (26.15%) originated mainly from natural sources. Additionally, Cu (91.41%) was primarily influenced by mining operations, and Ni originated from mining (53.61%) and geothermal water sources (46.39%), while Cd (97.88%) originated from unknown sources. In the high-flow season periods from 1992 to 2017, the As concentrations in the DLQ decreased significantly, which might result from increasing precipitation and runoff. Overall, the results of this study suggest that both natural and anthropogenic activities have jointly affected the solutes in small rivers on the Tibetan Plateau, and heavy metal pollution should be emphasized in the future.

Keywords: Duilong Qu; major ions; heavy metals; source appointment; PCA-APCS-MLR

1. Introduction

In the 21st century, many of the major problems facing humanity are related to water quality issues [1]. These problems will be exacerbated in the future by climate change due to increased water temperatures, melting glaciers, etc, and will further generate feedbacks to the deterioration of water environment globally. For example, according to a report developed by the Watershed Initiative, the water quality status of the Mississippi River basin was worsening because of agricultural activities, human sewage; and natural works [2]; in China, industrialization, urban development, and intensive agriculture are the major drivers of water quality worsening [3]; in India, 70% of surface water and groundwater is contaminated with pollutants [4]. However, most of the current research has focused on rivers that are highly influenced by anthropogenic influences, while little study has been conducted to clarify the water quality evolution of rivers that are influenced by natural processes [2], especially on remote small river watersheds. Relevant studies are urgently needed for assessing global water quality evolution, and, further, to provide more information for water protection policy makers.

The state of river water around the world has been gradually changing, caused by changes in climate and human activities [5–7]. According to data provided by the World Health Organization (WHO) in 2019, one-third of the global population lacks access to safe drinking water [8]. The retreat of the Tibetan Plateau (TP) glaciers due to global warming have shown a remarkable influence on the chemical composition of water and the level of heavy metal pollution in TP rivers [5]. The TP is the source of 10 major rivers in Asia and is known as the "Water Tower of Asia" [9]. Since it is the supply of water for about 40% of the world's population, understanding the water chemistry and quality of TP rivers is essential for society and ecosystems [5]. The primary controlling mechanism of river water chemistry on the TP is natural processes, including rock weathering, soil erosions, and solute inputs from groundwater, which are also characterized by its geological background and regional water cycling [9]. Specifically, soluble substances originating from the soil have been demonstrated as a critical factor dominating the river water chemistry, especially for small watersheds [10,11]. Besides, detailed information about hydrochemical processes would be helpful in studying and understand the formation, migration, transformation, and enrichment of heavy metals in river waters. Studies have reported that heavy metals in TP rivers are influenced not only by rock weathering and water–rock interaction [12,13], but also by artificial activities, such as input of geothermal resources [14], discharge of domestic sewage [9], traffic discharge [15], and mining activities [16,17]. In fact, more and more studies show that the source of some water bodies pollution can only be attributed to natural geological processes [18–20]. Rock minerals (e.g., serpentine minerals, spinel, pyroxene, and olivine) contain a large amount of heavy metal elements [18]. Heavy metals in minerals are released during water–rock action, which, in turn, causes pollution of the water environment [19,20]. Therefore, it is crucial to identify the different sources of heavy metal contaminations and quantify the contribution of the sources to heavy metal contaminations in rivers on the TP.

Since the TP plays an essential role in water resources in Asia, previous studies have been devoted to the influence of climate change on the hydrology and water resources of the TP [21]. In contrast, there has been little discussion on water chemistry and the water environment in TP rivers related to climate change. The study found that more than 80% of the water ion balance in the rivers of the TP consists of Ca^{2+}, Mg^{2+}, HCO_3^-, and SO_4^{2-} [9]. The primary controlling mechanisms of water chemistry are natural processes and exhibit apparent space heterogeneity [22]. For example, the water quality of rivers in the southern TP is dominated by the control of rock weathering [22], whereas the water quality of rivers in the north-central TP is also mainly influenced by evaporation-crystallization [9]. Moreover, the TP rivers are generally considered to be unpolluted due to sparse population and minimal industrial activity [5]. However, recent studies showed that concentrations of As, Pb, Cu, and Cd in southern TP rivers exceed the WHO drinking water guidelines due to anthropogenic processes (e.g., mining activities and sewage discharge) and geothermal water [9,23]. Thus far, however, the majority of previous research has concentrated on the major lakes and rivers of the TP, which often exhibit large runoffs [24,25]. Currently, there are relatively few studies on small watersheds, and the analysis of pollution sources is insufficient, although they are likely to play an essential role in the livelihood of residents. In fact, clarifying the characteristics and analyzing the heavy metal pollution sources in small watersheds not only enriches the water chemistry data of TP, but is also an important basis for accurate heavy metals contamination control in regional rivers. Therefore, the research on heavy metals pollution source analysis in small watersheds should receive more attention.

Geothermal water is formed in the subsurface rock and has a heat source [26]. It has been found that the temperature of geothermal water can be as high as 289 °C [14]. Therefore, the discharge of geothermal water will cause the temperature of the surrounding rivers to rise and reduce the dissolved oxygen content (DOC) of the rivers, resulting in the deterioration of water quality. The high temperature and water–rock interaction make the geothermal water dissolve abundant anions and cations, with a high degree of mineralization [27]. Previous studies have confirmed that the geothermal water inflow affects the water quality of surrounding rivers [28]. For example, the concentrations of K^+, Na^+, Cl^-, and SiO_2 of the river water downstream in the Zaotang River (China) increased due to the inflow of hot spring water from the Rehai geothermal field [28]. Thorsten et al. [29] showed that concentrations of Mn, Zn, and Cu in Río Vacas Heladas (Chile) exceeded WHO guideline values for drinking water. Timperley et al. [30] estimated that 78% of the Cl^- in the Waikato River (New Zealand) comes from geothermal fluids. More and more studies are proving the impact of geothermal water on rivers related to river runoff. The Büyük Menderes River (Turkey) runoff exceeded 26.651 m^3/s, and the emission of geothermal waters into the streams did not result in any environmental problems [31]. Blaine et al. [32] found that the fraction of the Yellowstone River (USA) from geothermal sources was about 0.2% during the high-flow season and 6.5% during the low-flow season. In addition, geothermal fluids contain very high concentrations of harmful chemical components, especially arsenic (As) [14]. Arsenic concentrations in water at El Tatio (Northern Chile) [33], Waikato River (New Zealand) [34], and Yellowstone National Park (USA) [35] were reported to be as high as 27,000 mg/L, 3800 mg/L, and 7800 mg/L, respectively. Thus, it was inferred that the discharge of geothermal water, which contains high As concentrations, might be a reason for the high As concentration and accumulation in the surrounding rivers. It is known that the long-term intake of high concentrations of As may induce endemic As poisoning [23]. Therefore, it is worth paying attention to the impact of geothermal water on the surrounding water quality. The TP is a region of intense geothermal activity in China [14]. The Lhasa River (LR) basin is situated in the mid-southern TP and is the main principal tributary of the middle Yarlung Tsangpo River [13]. As the core area of politics, economics, culture, transportation, and religion in the Tibet Autonomous Region, the LR basin is essential to Tibet's socioeconomic development [5,13]. The Duilong Qu (DLQ, Qu means river in the Tibetan language), a major tributary of the LR, is located downstream from the town of Yangbajing (YBJ) [12,23]. The YBJ geothermal field is the largest geothermal field on the TP, with the highest heat storage temperature (average temperature of 248 °C) in China [23]. Previous studies on YBJ geothermal water and surrounding areas have mainly focused on regional and geothermal geology [36]. However, it is still unclear how geothermal water discharge impacts the water quality of the DLQ. Thus, when the geothermal water and domestic sewage of surrounding residents flow into the DLQ [23], the water quality of the DLQ may be at risk of contamination [14,23]. As the water source for the production and living of residents [23], the water quality safety of the DLQ is a severe issue for the residents.

Most previous studies have focused on the evaluation of water quality by using methods such as the factor water quality identification index method and the water quality index (WQI) [37]. However, the investigation of pollution sources so far has been inadequate. Factor analysis (FA) and principal component analysis (PCA) have been used to identify the primary pollution sources in the environment [38]. However, the above methods have difficulty quantitatively describing the impact degree of major pollution sources [24]. The absolute principal component scores multiple linear regression model (APCS-MLR) based on PCA/FA has been proposed to quantify the contribution (in %) of identified pollutant sources [39]. The model has been effectively applied to identify pollution in air, soil, and surface water [38]. Previous studies have used the APCS-MLR model to quantify the contribution of pollution sources in rivers, showing that the analysis results are objective and accurate [38]. In this study, the major ions and heavy metals of river water in the DLQ were investigated. The aims of the study are to: (1) analyze the major ionic composition and water quality of the DLQ; (2) identify the natural and/or anthropogenic sources of heavy metals; (3) study the effect of geothermal hot springs and climate change on As concentrations in river water in the DLQ. This study would reveal the effects of natural and anthropogenic activities on the solutes of small rivers on the TP and provide a scientific basis for protecting the water environment of river water in DLQ under the background of climate change.

2. Materials and Methods

2.1. Study Area

The DLQ (29°38′ N–30°09′ N, 90°31′ E–90°59′ E) is a primary tributary in the middle reaches of the LR, originating from the southern foot of the Gangdese Mountain with a length of 137 km, a drainage area of 5093 km^2, and an elevation range of 3610–4130 m. The watershed is generally characterized by complex terrain with large portions of the region being deep valley. The DLQ valley plain stretches from northwest to southeast, with an average longitudinal gradient of about 5.63 ‰ and a riverbed width of 100~170 m. The climate of the DLQ valley plain is plateau temperate semi-arid monsoon, with an average annual temperature of 8 °C. The low-flow season is from November to April. About 89.1% of the yearly precipitation is primarily concentrated in the high-flow season (from June to September) [13]. The river runoff is mainly concentrated in summer, and the runoff from July to August accounts for 50% of the annual runoff. The annual mean runoff of the DLQ is 24 m^3/s, and the maximum peak runoff is 75 m^3/s [40]. The bedrock of the Lhasa valley area is dominated by Mesozoic limestone, metamorphic sandy slate, and Himalayan granite, in addition to Yanshanian granite, Quaternary clastic rocks, and eruptive rock.

2.2. Sampling and Analysis

The sampling protocol was based on the Technical Guidance on Water Quality Sampling (HJ 494-2009) and the geomorphological features of the watershed. In addition, we considered that the geothermal activity in Yangbajing (YBJ), the anthropogenic activity in the township, and the mining activity may cause changes in the water environment; therefore, we set up sampling points near these areas. A total of 23 surface water (depth < 0.5 m) samples were collected during the high-flow season (July–August) in 2015 (Figure 1), including 12 samples in the mainstream of the DLQ, 8 samples in tributaries of the DLQ, 2 samples in LR (two sampling sites are close to the confluence of two rivers and located at the upper stream and downstream of the confluence, respectively), and one geothermal water sample from the YBJ geothermal field.

The sampling protocol in this study follows a national standard method: the "Water Quality-Guidance on sampling techniques" (HJ 494-2009). Water samples used portable water samplers (Dongxiyi Co., Beijing, China) to collect surface water from where the water flow was most rapid to avoid the riparian effects, to ensure the water samples were representative. Pre-cleaned bottles were rinsed three times before being filled with water samples. Replicates and blank samples were collected and used to ensure the accuracy of the analysis and complete the determination. The pH, electrical conductivity (EC), water temperature (T), chlorophyll-a (Chl-a), and total dissolved solids (TDS) were measured in situ with a Multi-Parameter Water Quality Sonde (V2-4 6600, YSI, Yellow Springs, OH, USA) in the field. The collected water samples were filtered through disposable cellulose acetate membrane filters (0.45 μm) within 12 h of collection and then packed separately. The total alkalinity (Alk) was determined by the Gran titration method with 0.024 mol·L^{-1} HCl in situ. Based on Alk, pH, T, and other anion and cation concentrations, the concentrations of HCO_3^- and CO_3^{2-} were calculated by PHREEQC, which was explored by USGS [41]. The concentrations of Ca^{2+}, K^+, Mg^{2+}, Na^+, and Si were measured by plasma atomic emitted spectrometer (ICP-OES, Optima 5300DV, PerkinElmer, Waltham, MA, USA) according to the current valid standard Water Quality-Determination of 32 Elements-Inductively

Coupled Plasma Emission Spectrometry (HJ776-2015). The standard curves were plotted with a standard solution of multiple elements (Agilent Technologies Inc., Santa Clara, CA, USA, Part number 6610030700) prepared on the same day. The concentrations of Cl^-, NO_3^- and SO_4^{2-} were measured using an ion chromatograph (Dionex ICS-2100, Dionex, Sunnyvale, CA, USA) according to the current valid standard Water Quality-Determination of Inorganic Anions (F^-, Cl^-, NO_2^-, Br^-, NO_3^-, PO_4^{3-}, SO_3^{2-}, and SO_4^{2-}) -Ion Chromatography (HJ84-2016). The working curves were plotted using the standard solution (National Center for Reference Materials, Product number GNM-M042193-2013). The analysis errors were within ±5%. The concentrations of heavy metals, i.e., cadmium (Cd), chromium (Cr), manganese (Mn), iron (Fe), nickel (Ni), copper (Cu), zinc (Zn), and lead (Pb), were quantified by inductively-coupled plasma mass spectrometry (ICP-MS, X-7series, Thermo Fisher Scientific, Cleveland, OH, USA) according to the recommended method by Environmental Quality Standards for Surface Water (GB3838-2002). The standard curves were plotted with a standard solution of multiple elements (Agilent Technologies Inc, Part number 8500-6940). The relative standard deviations of the results were better than 3%. The concentrations of arsenic (As) were quantified by an atomic fluorescence spectrophotometer (LC-AFS, AFS-830, Jitian Instrument Co., Beijing, China), of which the relative standard deviation was less than 1%. Standard solutions were purchased from the National Center for Reference Materials. Determination of high arsenic samples such as geothermal water by step dilution was done to the concentration range of the standard curves. The correlation coefficients of the standard curves of each item were obtained by linear regression (r) ≥ 0.9999, which is in accordance with the requirements of Water Environment Monitoring Specification (SL219-2013). Replicates were also measured with water samples to check the stability. In addition, the standard solution corresponding to the measured sample content was added for quality control. All the precisions of measurements of the replicates and samples were better than ±3%. The acid used for the analysis was of high purity and the water was of high purity. Each batch of samples was blanked throughout the experiment to eliminate any contamination during sample processing and determination.

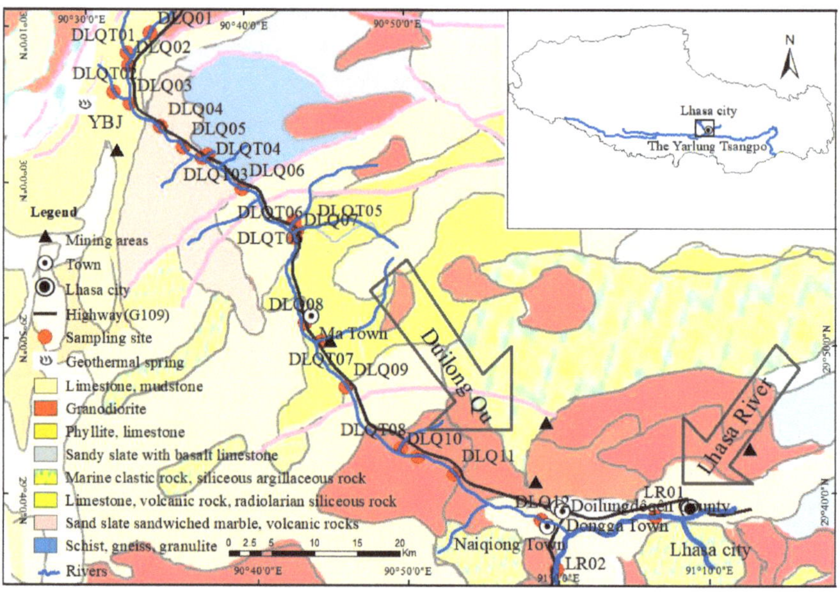

Figure 1. Distribution of Sampling sites in DLQ.

2.3. Analytical Methods

Sources of heavy metals in surface waters of the LR and DLQ were analyzed using correlation analysis, principal component analysis (PCA), and the principal component analysis-absolute principal component score-multiple linear regression receptor model (PCA-APCS-MLR). The APCS-MLR was performed for quantitative source assignment using the normalized factor scores and eigenvectors acquired by PCA. The main steps of the PCA-APCS-MLR are as follows [39]:

Step 1: All heavy metal concentration data are normalized using Equation (1):

$$Z_{ij}= (C_{ij} - \overline{C_i})/\sigma_i \tag{1}$$

where C_{ij} is the accumulation value of heavy metal i at location j, and C_i and σ_i are the mean concentration and the standard deviation for heavy metal i, respectively.

Step 2: The true zero for calculating the scores of individual factors is obtained by introducing an artificial sample with zero concentration of all variables by the formula:

$$(Z_0)_i= (0 - \overline{C_i})/\sigma_i \tag{2}$$

where $(Z_0)_i$ is established as the absolute zero concentration.

Step 3: The PCA approach can be used to obtain factor scores for variables from normalized heavy metals concentration data. The factor score of $(Z_0)_i$ was subtracted from the factor score of Z_{ij} to estimate the APCS for each heavy metal.

Step 4: The C_i of the source contribution was obtained with a multiple linear regression procedure:

$$C_i= b_{0i} + \sum_{p=1}^{n} (b_{pi} \times \mathrm{APCS}_p) \tag{3}$$

where C_i is the estimated value of heavy metal concentration, b_{0i} is the constant term of multiple regressions, b_{pi} is the regression coefficient of the multivariate linearity, and the mean of $b_{pi} \times \mathrm{APCS}_p$ is the contribution value of the source p.

3. Results and Discussion

3.1. Physicochemical Properties

Table 1 shows the physicochemical properties of the rivers. The pH value is a common indicator of natural water acidity, which can affect the form and activity of heavy metals [9,13]. The pH value of YBJ was 9.5, which was higher than that of natural water (6–9) [13], indicating that the effects of the deep thermal water were dominated by sodium chlorite fluid [42]. The pH values of river water in the LR and DLQ ranged from 7.0–8.4, indicating a strong buffering ability of the river water in the context of the inputs of alkaline geothermal water from the YBJ geothermal field [12,13]. In addition, heavy metals are more easily deposited in sediments in alkaline water environments. Thus, the actual concentrations of heavy metals in rivers may be underestimated [9]. The high temperature of YBJ geothermal water was up to 80 °C. However, the temperatures in the DLQ (20.0–21.1 °C) were not increased significantly. The highest concentration of Chl-a (23.9 μg·L^{-1}) was observed at the sampling site near Naiqiong town, which may be related to the bloom of algae in summer [43]. Naiqiong Town is well-developed in agriculture, and agricultural activities such as fertilization and planting can increase the accumulation of nutrients in rivers [10,43,44]. In addition, high precipitation and high runoff during the high-flow season may transport more nutrients to a river [45,46], contributing to the high Chl-a concentration in the river near Naiqiong town. The DO concentrations of river waters in the LR and DLQ ranged from 6.3 to 6.6 mg·L^{-1} (DO saturation ranged from 94.7% to 100.1%). The DO concentration of river water in the DLQ was higher when compared with rivers at the same altitude on the TP, such as the Requ River (6.0 mg·L^{-1}) and the Chumaer River (5.6 mg·L^{-1}), which may be caused by the accelerated mixing of water and oxygen in ambient air due to the fast flow rate of the river water in the DLQ [47]. The EC value of the water in YBJ was 2.9 mS·cm^{-1}, and the TDS was 1422 mg·L^{-1}, which belonged to brackish water (1–3 g·L^{-1}) [48]. The EC values of the LR and DLQ rivers ranged between 0.04 and 0.2 mS·cm^{-1}, and the TDS values of those rivers ranged from 33 to 150 mg·L^{-1}. The flow rate in the DLQ during the high-flow season can reach 75.1 m^3/s, while the geothermal water discharge is 0.389 m^3/s [49]. Therefore, YBJ geothermal water discharge has little effect on the solute level of the mainstream during the high-flow season, and the DLQ still exhibits typical freshwater characteristics (<1 g·L^{-1}) [13]. Compared with several large rivers on the TP, such as the Mekong River and the Yarlung Tsangpo River [5], the lower TDS and EC values in the DLQ rivers indicate that the influence of alkaline thermal springs on the hydrochemical parameters of the river water is limited.

Table 1. Physical-chemical parameters of surface water in the DLQ, LR, and YBJ geothermal water.

Sites		pH	T (°C)	DO (mg·L^{-1})	Chl-a (μg·L^{-1})	EC (mS·cm^{-1})	TDS (mg·L^{-1})	Reference
YBJ		9.5	80			2.9	1422	This study
LR 01		7.5	20.4	6.4	2.2	0.1	83	
LR 02		7.1	20.7	6.3	2.6	0.2	106	
DLQ (n = 20)	Min	7.0	20.0	6.3	0.5	0.04	33	
	Mean	7.6	20.5	6.4	10.1	0.1	84.5	
	Max	8.4	21.1	6.6	23.9	0.2	150	
Requ River (Source region of the Yellow River)	Mean	/	/	6.0	/	0.2	417	[50]
Chumaer River (Source region of the Yangtze River)	Mean	8.5	19.46	5.6	/	0.2	1884	[51]
Mekong River	Mean	/	/	/	/	0.3	302	[5]
Yarlung Tsangpo River	Mean	/	/	/	/	0.1	112	

Note(s): "/" means no detection.

3.2. Main Ionic Components and Hydrochemical Types

Table 2 shows the major ion concentrations in the study area. The major ions in YBJ geothermal water and rivers are markedly different, which is related to the difference in geological conditions. The concentrations of ions in YBJ geothermal water are higher than those in the rivers (LR and DLQ).

Table 2. Statistics of major ion concentrations in the DLQ, LR, and YBJ geothermal water (mg/L).

Sites		K^+	Na^+	Ca^{2+}	Mg^{2+}	Cl^-	SO_4^{2-}	HCO_3^-	NO_3^-	Si
YBJ		137.2	376.0	4.3	0.4	440.7	71.8	357.2	0.4	237.5
LR 01		1.4	6.8	31.25	5.5	14.4	21.6	86.5	0.6	3.2
LR 02		1.4	4.9	19.39	1.6	7.3	12.1	54.6	1.8	3.2
DLQ (n = 20)	Min	0.7	1.3	6.2	0.3	1.2	1.5	20.4	0.2	1.8
	Mean	1.6	4.2	18.6	1.66	4.3	16.1	41.5	1.2	3.0
	Max	4.5	8.2	33.0	2.8	13.4	52.1	75.8	3.2	4.7

The dominant anion and cation in YBJ geothermal water were Na^+ and Cl^-, respectively (Table 2), and the water chemical type was Cl-Na type (Figure 2). This might be because the alkaline solutions are prone to dissolving silicates, and the interaction of alkaline water with rock minerals (mainly granite in the deep layer of the YBJ geothermal field) lead to an increase in the concentrations of SiO_2, Na^+, and K^+ [26]. Ca^{2+} accounted for the highest total cationic equivalent charge of 63~87% in the river water of the DLQ, followed by Na^++ K^+ (4–26%) and Mg^{2+} (3–9%), while HCO_3^- (22–84%) was the dominant anion, followed by SO_4^{2-} (6–75%) and Cl^- (2–19%). Therefore, 78% of the water samples in the DLQ belong to the HCO_3^--Ca types, and 22% of the water samples are SO_4·Cl-Ca types. The water samples in the DLQ near the YBJ geothermal water belong to the SO_4·Cl-Ca type, indicating that the contribution of solutes in the geothermal field cannot be ignored. Previous studies have shown that geothermal water contains high concentrations of Cl^- [27,32]. In this study, the Cl^- concentration in YBJ was as high as 440.7 mg/L (Table 2). In addition, the highest concentrations of SO_4^{2-} (52.1 mg/L) were observed near the YBJ geothermal water, which is mainly due to the fact that the H_2S and SO_2 contained in the geothermal water are easily oxidized to form SO_4^{2-} during the upward migration process [27,52]. The water chemistry of the DLQ gradually changed from the SO_4·Cl-Ca type to the HCO_3-Ca type with the direction of water flow, mainly due to more dissolved carbonate flowing into the river and the dilution of SO_4^{2-} and Cl^- in the river. Ca^{2+} was the dominant cation of river water in LR, accounting for 69–71% of the cations, while HCO_3^- was the dominant anion, accounting for 71–74% of the anions. Thus, the water chemistry type of LR is HCO_3-Ca, which is consistent with previous studies [5,22], indicating the important contribution of carbonate weathering to the ionic composition of river water in LR [13].

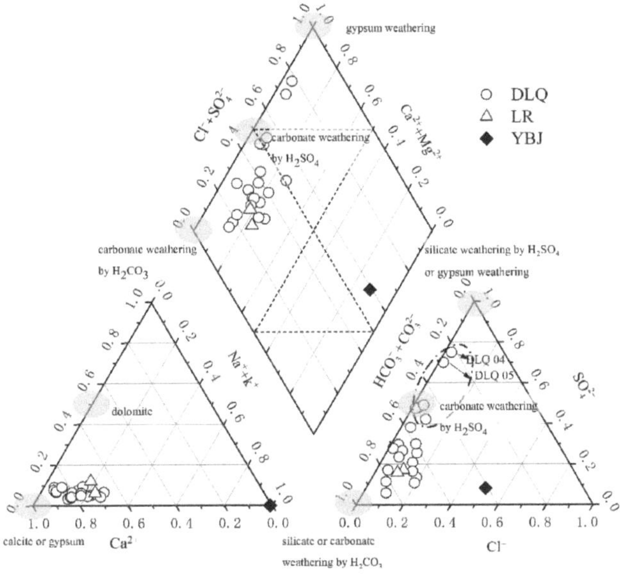

Figure 2. Piper diagram showing major ion compositions of the river waters. The end-members were taken from Spence and Telmer [53] and Li et al. [48].

On the Gibbs diagram (Figure 3), the scatter of major ions in the water of YBJ was outside the Gibbs diagram, which was mainly due to the high $Na^+/(Na^+ + Ca^{2+})$ weight ratio because of the high Na^+ concentrations (376.0 mg/L) and low concentrations of Mg^{2+} (0.4 mg/L) and Ca^{2+} (4.3 mg/L) in the geothermal water [27]. The $Na^+/(Na^+ + Ca^{2+})$ ratio in the LR and DLQ was less than 0.5, and the TDS concentration was moderate, being mainly controlled by rock weathering [54]. Similar findings were also found by Qin et al. [13] in the LR watershed. In the mixing diagram (Figure 2), most of the water samples in the LR and DLQ were located near the carbonate endmember, indicating that carbonate weathering driven by carbonic acid and sulfuric acid has a significant impact on the major ion chemistry of the LR and DLQ. Generally, the dissolution and weathering of carbonate rocks (calcite, dolomite, etc.) in natural water is the main source of Ca^{2+} and HCO_3^- [5,9,55], and the weathering of magnesium-bearing carbonate rocks also provide Mg^{2+} in the river [55]. In this study, Ca^{2+}, Mg^{2+}, and HCO_3^- showed a highly significant correlation ($p < 0.01$), which also proved that these ions mainly originated from the weathering of carbonate rocks [13,55]. The geology of the LR basin contains a higher proportion of Mesozoic limestone, a major weathering source of Ca, Mg, and HCO_3^- [13]. Furthermore, HCO_3^- showed positive correlation with Ca^{2+} ($R^2 = 0.82$, $p < 0.01$) and Mg^{2+} ($R^2 = 0.75$, $p < 0.01$), suggesting that dissolved calcite was the most significant contribution to river ions, followed by dolomite [55]. Basically, Cl^- is primarily generated from the dissolution of evaporites [5,9]. K^+ and Na^+ are mainly derived from the dissolution of evaporites and the weathering of silicate minerals [9,56]. Cl^-, K^+, and Na^+ were highly significantly correlated ($p < 0.01$), indicating that these elements had similar sources. In addition, Na^+ was found to be related to Si ($R^2 = 0.51$, $p < 0.05$). Therefore, Cl^-, K^+, Na^+, and Si were generated primarily by the dissolution of evaporites and/or the weathering of silicate minerals [9,55]. The source of ions in the river water in the DLQ is primarily influenced by rock weathering and dissolution.

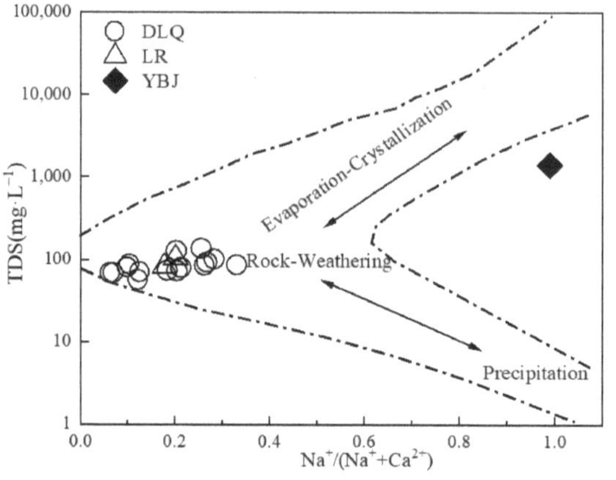

Figure 3. The Gibbs graph of major ion compositions in DLQ, LR, and YBJ geothermal water.

Long-term interactions between geothermal fluids and host rocks lead to As accumulation in hydrothermal systems [57]. As mentioned before, high As concentrations have been found in rivers around the world [33]. Thus, arsenic is often regarded as a characteristic element of geothermal water [23,52]. In this study, a significantly positive association was found between SO_4^{2-} and As ($R^2 = 0.67$, $p < 0.01$), indicating that SO_4^{2-} might have similar origins to As, and which can be attributed to geothermal water. In addition, oxidative dissolution of As-containing sulphides may also result in SO_4^{2-} and As enrichment [58]. Moreover, SO_4^{2-} showed a significantly positive correlation with Ca^{2+} ($R^2 = 0.70$, $p < 0.01$). This could be due to: (1) the dissolution of gypsum [5]; (2) the erosion of the carbonate rocks driven by the sulfuric acid, which was produced from the oxidation of H_2S in geothermal water [27]. However, gypsum is rarely distributed in the DLQ watershed; therefore, the contribution of gypsum dissolution is negligible [59]. Generally, NO_3^- is an important water chemistry parameter, which mainly originates from agricultural fertilizers, soil organic nitrogen, and domestic sewage discharge [9,55,60]. Previous studies reported that Lhasa city has 182,000 inhabitants [5], and the ecology of river waters in the LR is damaged by domestic, industrial, and agricultural wastewater discharged into the river [9,13]. Hence, slightly higher NO_3^- concentrations at sampling sites around Lhasa city indicate the influence of urban sewage and agricultural activities on the LR (Table 2) [13].

3.3. Exposure Level of Heavy Metals and Assessment of Water Quality

Since these rivers are the major drinking water sources for residents, the water quality of the study area was evaluated. The criteria used for evaluation include the Environmental Quality Standard for Surface Water (GB3838-2002), Sanitation Standard for Drinking Water (GB5749-2006), World Health Organization (WHO), National Primary Drinking Water Regulations (NPDWR), and National Secondary Drinking Water Regulations (NSDWR) (Table 3).

Table 3. Descriptive statistical results of heavy metals in surface water (mean ± standard deviation, μg/L).

	Cd	Cr	Mn	Fe	Ni	Cu	Zn	Pb	As	Reference
YBJ	0.10	1.53	38.85	358.79	2.15	8.13	17.94	4.88	3532.21	This study
LR 01	0.04	1.45	5.19	39.27	0.75	2.89	4.20	0.32	3.06	
LR 02	0.05	1.99	13.43	47.19	1.10	0.91	2.44	0.50	5.05	
DLQ (n = 20)	0.05 ± 0.01	1.65 ± 0.42	7.30 ± 3.65	78.36 ± 40.93	0.84 ± 0.4	1.03 ± 0.62	5.26 ± 3.58	0.53 ± 0.27	3.33 ± 2.98	
Source region of the Yangtze River	0.02	1.83	/	/	1.73	5.87	3.52	1.06	10.02	[16]
Natural waters in Tibet		4.13	0.77	95.57	1.01	0.38	4.91	0.03	6.11	[61]
World average	0.08	0.7	34	66	0.8	1.48	10	0.03	0.62	[62]
WHO	3	50 (P)	400 (C)	3000	70	2000	3000	10	10	[63]
GB I/V	1	10	/	/	20	10	50	10	50/100	[64]
GB5749-2006	5	50	100	300	/	1000	1000	10	50	[65]
NPDWR	5	100		/	/	/	/	15	10	[66]
NSDWR	-	50	300	/	1000	5000	/	100	/	

Note(s): "/" means no detection. P: provisional guideline value because of uncertainties in the health database; C: concentrations of the substance at or below the health-based guideline value may affect the appearance, taste, or odor of the water, leading to consumer complaints.

The results show that heavy metal concentrations in YBJ geothermal water meet the water quality standards (or guidelines), except for As and Fe. The concentration of Fe in YBJ geothermal water was 358.79 mg/L, which exceeded the GB standard for drinking water (300 mg/L). The As concentration in YBJ geothermal water (3532.21 mg/L) was much higher than the limit value of class V of the GB standard level for Surface Water (100 g/L). Since high concentrations of As were found in YBJ geothermal water, it can be said that the influx of geothermal water impacts river water quality [14]. A previous study found that the discharge of YBJ geothermal water resulted in As contamination in the DLQ during the low-flow season [12]. However, the concentrations of heavy metals of river waters in the LR and DLQ meet the water quality standards (or guidelines) in this study, indicating a weak impact of geothermal water on the water quality of river waters in the LR and DLQ, due to the strong dilution effect of the river during the high-flow season [12,25]. This is similar to the results of heavy metals during the high-flow season of DLQ by Li et al. [23]. Moreover, it is noted that the concentrations of Cr, Pb, and As in the rivers of the LR and DLQ are significantly above average compared to other rivers [62]. The concentrations of Mn, Pb, and Cu in river waters in the LR and DLQ were much higher than other natural waters in Tibet (0.77 mg/L for Mn, 0.03 mg/L for Pb, and 0.38 mg/L for Cu) [61]. The concentrations of heavy metals in river waters in the LR and DLQ were comparable with those in river water of the source area in the Yangtze River [16], except for Cd. In addition to rock weathering, geothermal water inflow, and discharge of urban sewage and agricultural wastewater [12,13,24], mining activities increase heavy metal concentrations in the river waters in the DLQ and LR due to the abundant mineral deposits (e.g., Cu and Pb-Zn mines) in the LR watershed [17,41]. The results showed that the confluence of geothermal water has not led to heavy metal contamination in the studied river waters. However, it is noteworthy that the occurrence of sudden geological disasters may cause abruptly high concentrations of heavy metals in rivers in a short period of time, especially for As. Therefore, it is important to develop geothermal water resources rationally while minimizing or avoiding the hazards caused by unexpected events. This will better ensure the sustainable use of geothermal water resources.

3.4. Spatio-Temporal Distribution of Heavy Metals

3.4.1. Spatio Distribution of Heavy Metals

The spatial distribution of different heavy metals of river water in DLQ demonstrated that (Figure 4), except for the YBJ geothermal water, the concentrations of heavy metals in other Sampling sites are generally low (Table 3). Overall, the concentrations of As, Fe, Mn, Zn, Pb, and Cu of river water in DLQ increase after the inflow of YBJ geothermal water, indicating the contribution of geothermal water, while the decrease, afterward, was mainly due to the water self-purification and the inflowing of the tributaries. In addition, the concentration of heavy metals (except As) of river water in DLQ increased again after passing near the towns of Dongga and Naiqiong (DLQ12), which reflects the influence of sewage input from the towns.

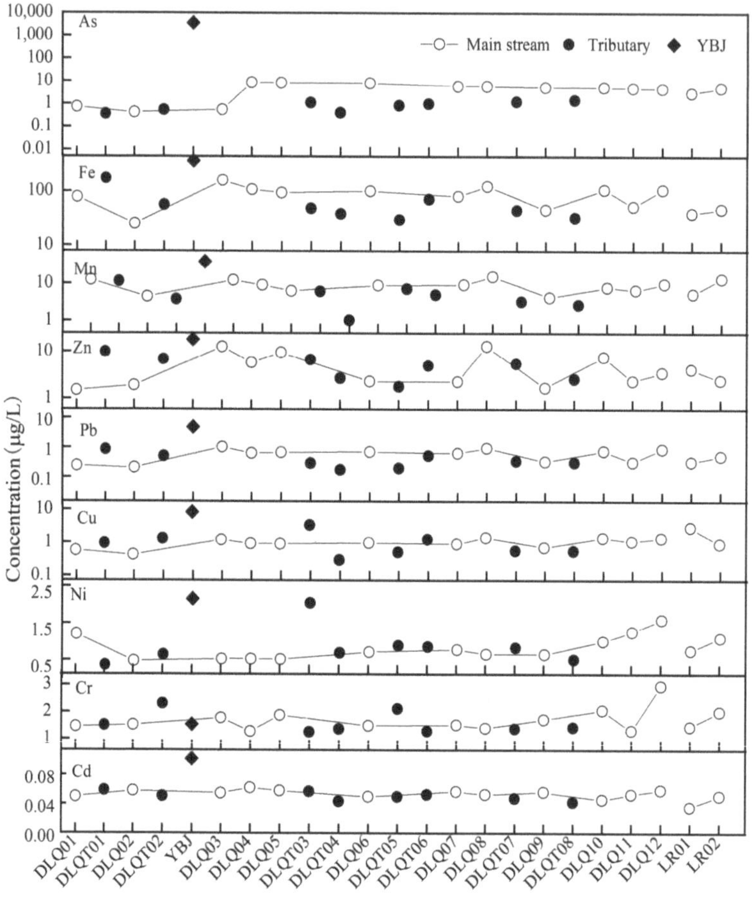

Figure 4. Spatial variations in heavy metals along the flow direction in DLQ.

The concentration of As gradually decreased from DLQ04 to DLQ12, indicating that it may be influenced by river water dilution and sediment adsorption in the riverbed [12]. The flow rate of the mainstream in DLQ can reach 75.1 m^3/s during the high-flow season [40], and concentrations of As in the tributaries stream of DLQ are generally lower than those in the mainstream (Figure 4). Hence, the decrease in As concentration was mainly attributed to the dilution effect of runoff [12]. In addition, adsorption from streambed sediments, especially by iron oxyhydroxides [58], may play an important role in reducing arsenic concentrations in the river [14]. A study in May 2007 showed that most of the As in rivers could be removed from the river by adsorption by river sediments [57]. However, the runoff rate in July and August (73.6–75.1 m^3/s) of DLQ is much larger than the runoff rate in May (10.1 m^3/s). Therefore, sediment sorption is limited due to increased runoff rates, resulting in the desorption of As from the sediment [57]. The curves of Fe, Mn, Zn, and Pb were very similar (Figure 4), indicating that familiar sources exist between them [12]. High concentrations of Fe (123.61 mg/L), Mn (12.28 mg/L), Zn (12.96 mg/L), and Pb (0.95 mg/L) were observed in DLQ08, which is related to the proximity of the sampling site to agricultural and grazing areas (Ma Town). Livestock grazing in agricultural and pastoral regions can lead to the release of metal elements in the surrounding shallow water sediments [12,13]. In addition, Wang et al. [15] found that transportation activities caused elevated Zn and Pb concentrations in soils on the highway G109 on the Qinghai-Tibet Plateau and affected the area beyond 100 m. Nevertheless, Highway G109 is less than 100 m away from some DLQ sampling sites, indicating that Zn and Pb may be influenced by traffic activities. The highest concentrations of Cu (3.31 mg/L) and Ni (2.04 mg/L) existed in DLQT03, indicating the influence of mining activities in the surrounding metal deposits [13,17]. However, this was not observed in the mainstream, indicating a strong dilution effect of surface runoff [13]. This study has shown that the heavy metals in fertilizers are mainly Cr and Co, and agricultural activities can lead to elevated Cr concentrations in the river [25]. In this study, a high Cr concentration (2.96 mg/L) was observed in the site near the town of Naiqiong, where agriculture has been well-developed, and this was associated with the heavy use of pesticides and fertilizers. Li et al. [25], likewise, observed highly concentrated Cr in the Nianchu River, a grain producing region in Tibet. Except for Zn and Cu, the heavy metal concentrations of LR02 were higher than those of LR01, indicating that the influx of DLQ had an impact on the water quality of LR. It is worth noting that the concentrations of As (5.05 mg/L) and Mn (13.43 mg/L) of LR02 were higher than those of DLQ12 (4.75 mg/L for As and 9.65 mg/L for Mn), indicating that LR is also affected by mining activities, domestic waste, and municipal sewage [13,23,25]. In summary, the spatial distribution of heavy metals in the LR and DLQ was closely related to geothermal resources and human activities.

3.4.2. Impact of Climate Change on As Concentration in DLQ

Since the 1980s, the LR basin has been affected by global warming, and the annual average temperature and precipitation in the basin have been increasing [25,67]. Since the precipitation and flow data of DLQ are lacking, the meteorological and hydrological data of Lhasa city were selected as a reference in this study. The annual precipitation in Lhasa increased at a rate of 28.81 mm/10 a, and the summer precipitation increased at 32.18 mm/10 a from 1980 to 2019 [68]. Although the hydrological data of Lhasa city cannot directly represent the precipitation conditions of DLQ, they still have some reference values for the city's precipitation variation trend. The trend of the average river runoff in DLQ is consistent with precipitation since precipitation is the primary source of recharge in the DLQ river's precipitation [23,68]. In addition, small-scale glaciers have developed at the headwaters of the DLQ tributaries, and meltwater from the ice and snow due to climate warming will also cause an increase in river runoff [69].

The seasonal and spatial variation in different trace metals differed because of their complex governing factors [12]. Unlike other elements, As concentrations in river water affected by geothermal activity exhibit significant seasonal variation [12]. Li et al. [23] observed that the As level of the LR during the low-flow season was about three times higher than during the high-flow season. This is due to the more significant contribution of geothermal water during the low-flow season and the stronger dilution process during the high-flow season [25,70]. Thus, the As level is highly dependent on the river's hydrological state [23]. Some studies have shown that, although As levels in rivers are also affected by pH and adsorption/desorption processes [52], changes in As concentrations are mainly influenced by river runoff [23]. Therefore, we collected the As data in river water of the DLQ from 1992 to 1994 [40], 1997 to 1999 [49], 2010 [23], and 2017 [71], respectively, to analyze the effects of precipitation and runoff changes on As (Figure 5).

From 1992 to 2017, the As concentrations in the river showed a decreasing trend (35.5 g/L~1.2 g/L) during the high-flow season, which may be related to a significant increase in precipitation and runoff due to climate change. In addition, from 1992 to 2010, the most significant reduction was found in the high-flow season (35.5 g/L~12.9 g/L) compared to the normal-flow season (59.9 g/L~40 g/L) and the low-flow season (292 g/L~205.6 g/L), which was primarily caused by the stronger dilution of the river in the high-flow season [25]. Therefore, based on the variations in As concentrations in this study, it can be inferred to some extent that the recent increase in precipitation and runoff in LR caused by climate change may lead to a decrease in the concentration of heavy metals in river water. Similarly, metal concentrations in the Rhine River were diluted by the river due to climate change as observed by Zwolsman et al. [72].

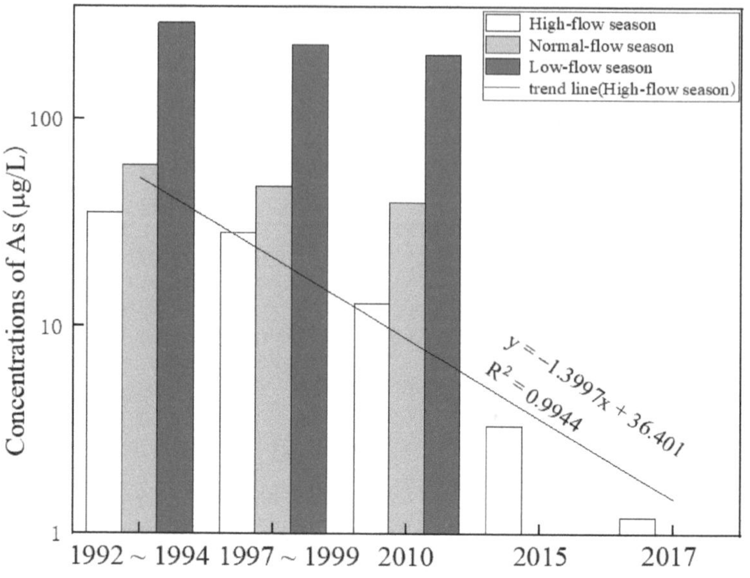

Figure 5. The seasonal variations in As in DLQ during the period from 1992 to 2017. Data were referred to [23,40,49,71].

3.5. Pollution Source Identification Based on the Pearson Correction, PCA and PCA-MLR

3.5.1. Correlation Analysis

The results of Pearson correlation coefficients between the measured heavy metals are listed in Table 4. A positive correlation was found among Mn-Fe ($p < 0.01$, $R^2 = 0.67$), Mn-Pb ($p < 0.01$, $R^2 = 0.64$), Fe-Pb ($p < 0.01$, $R^2 = 0.90$), Pb-Zn ($p < 0.01$, $R^2 = 0.72$), and Fe-Zn ($p < 0.01$, $R^2 = 0.71$), suggesting a common source of these heavy metals. Fe and Mn are mainly derived from the crust [12,25]. Therefore, Mn, Fe, Pb, and Zn are mainly derived from a natural geological background. Moreover, a significant positive correlation was observed between Cd and Fe ($p < 0.01$, $R^2 = 0.43$), which may be related to the migration and transformation between them, because they were not similar in terms of spatial distribution pattern (Figure 4). The abundance of Cu in the LR basin indicates that Cu is mainly derived from mining activities [12,25]. Ni showed positive correlations with Cu ($p < 0.05$, $R^2 = 0.52$), K^+ ($p < 0.05$, $R^2 = 0.48$), and Na^+ ($p < 0.05$, $R^2 = 0.47$) in the DLQ, indicating that Ni is influenced by mining activities [25] and natural processes [24]. Cr and As had no significant correlation with other heavy metals, indicating that they are not controlled by a single factor in the study area [12]. Arsenic showed a significant association with NO_3^- ($p < 0.01$, $R^2 = 0.58$), suggesting that As might be related to geothermal activity, agricultural operations, and grazing [12,25].

3.5.2. Principal Component Analysis

PCA was performed for nine heavy metals (Table 5), with a KMO value of 0.561 (> 0.5) and Bartlet's sphericity test result of 0.000 (0.000 < 0.05), showing that the heavy metals were appropriate for PCA analysis. Three principal components (PCs) with eigenvalues greater than one were extracted, the cumulative variance contribution was 57.85%, and control factors for each PC were explained as follows.

Table 4. The Pearson correlation analysis of heavy metals in DLQ.

	Cd	Cr	Mn	Fe	Ni	Cu	Zn	Pb	As	Ca^{2+}	K^+	Mg^{2+}	Na^+	Si	Cl^-	NO_3^-	SO_4^{2-}
Cr	0.12																
Mn	0.34	0.14															
Fe	0.43 *	0.11	0.67 **														
Ni	0.06	0.18	0.09	−0.21													
Cu	−0.12	−0.06	0.07	0.04	0.52 *												
Zn	0.22	0.00	0.39	0.71 **	−0.22	0.29											
Pb	0.39	0.31	0.64 **	0.90 **	−0.18	0.11	0.72 **										
As	0.26	0.07	0.29	0.24	−0.05	0.02	0.04	0.39									
Ca^{2+}	−0.21	−0.20	0.18	−0.09	0.00	0.17	−0.02	0.03	0.52 *								
K^+	−0.04	0.52 *	0.17	−0.06	0.48 *	−0.08	−0.31	−0.07	−0.21	0.03							
Mg^{2+}	−0.47 *	−0.23	0.03	−0.22	−0.06	0.33	−0.14	−0.16	0.35	0.84 **	−0.01						
Na^+	−0.46 *	0.29	−0.08	−0.37	0.47 *	0.10	−0.38	−0.30	−0.32	0.10	0.71 **	0.21					
Si	−0.33	−0.13	0.31	−0.08	0.35	0.06	−0.13	−0.02	0.09	0.59 **	0.33	0.40	0.51 *				
Cl^-	−0.53 *	−0.01	−0.02	−0.23	0.16	0.10	−0.31	−0.34	−0.28	0.19	0.54 **	0.42	0.61 **	0.26			
NO_3^-	0.24	−0.17	0.15	0.00	−0.08	−0.18	−0.23	0.12	0.58 **	0.57 **	−0.25	0.30	−0.26	0.28	−0.30		
SO_4^{2-}	0.17	−0.02	0.04	0.04	−0.13	0.04	0.14	0.12	0.67 **	0.70 **	−0.06	0.54 **	−0.18	0.06	−0.10	0.39	
HCO_3^-	−0.51 *	−0.17	0.18	−0.22	0.19	0.26	−0.19	−0.12	0.14	0.82 **	0.20	0.75 **	0.43 *	0.78 **	0.439 *	0.37	0.19

Note(s): * significant at $p < 0.05$ levels; ** significant at $p < 0.01$ levels.

Table 5. Principal component analysis matrix of heavy metals in DLQ.

	PC1	PC2	PC3
Cd	0.297	**0.606**	−0.100
Cr	−0.049	**0.643**	0.143
Mn	**0.624**	0.457	0.108
Fe	**0.902**	0.273	−0.104
Ni	−0.256	0.251	**0.866**
Cu	0.261	−0.235	**0.864**
Zn	**0.906**	−0.148	0.052
Pb	**0.866**	0.397	−0.038
As	0.185	**0.558**	−0.074
Eigenvalue	3.46	1.56	1.25
% of variance	33.71	18.61	17.34
Cumulative %	33.71	52.32	69.66

Note(s): Bolded values are > 0.5.

PC1 accounted for 33.71% of the total variance and was mainly contributed by Fe, Zn, Pb, and Mn. These heavy metals were highly significantly correlated (Table 4) and had similar spatial distribution patterns (Figure 4); therefore, Fe, Mn, Pb, and Zn may have similar sources or chemical characteristics [25]. Generally, Fe and Mn are abundant in the earth's crust [12], and the background values of Pb and Zn in the watershed are relatively high [9,73]. Therefore, PC1 was mainly influenced by natural sources. In addition, Wang et al. [15] showed that the Pb and Zn concentrations in the surface soils were high in the major transportation routes of the Qinghai-Tibet Plateau, while the DLQ is close to the national highway G109 (Figure 1), suggesting that the concentrations of Pb and Zn in river water in the DLQ may be influenced by transportation. In general, the Pb contained in car exhaust enters the water through dry and wet deposition, which increases the concentration of Pb in the river water [15]. At the same time, Zn, Pb, and Mn particles are released by the abrasion of vehicle components and incomplete combustion of fuel [73]. Therefore, PC1 was mainly influenced by natural sources and, secondly, may be influenced by traffic activities.

PC2 accounted for 11.91% of the total variance and was mainly contributed by Cd, Cr, and As. Based on the spatial distribution of As in this study area (Figure 4), it could be observed that the high concentration of As mainly occurred after the inflow of geothermal water. The As concentration of river water in DLQ is primarily influenced by geothermal wastewater discharge [12], which is consistent with previous studies [23]. Therefore, PC2 was primarily related to the input of geothermal springs. In addition, the towns of Dongga and Naiqiong are important agricultural bases in the DLQ watershed, and pesticides and chemical fertilizers used in agricultural production are also important sources of As, Cd, and Cr [25]. Therefore, PC2 was mainly affected by geothermal activities and may be affected by agricultural activities secondarily.

PC3 accounted for 11.91% of the total variance and was mainly contributed by Ni and Cu. The LR watershed was rich in mineral resources (mainly Cu, Zn, and Pb ores) [9,41]. Studies have shown that Ni and Cu are influenced by the rich mineral resources and previously managed mining activities in the basin [12,41]. Therefore, PC3 was primarily affected by mining activities.

3.5.3. Source Apportionment Using APCS-MLR

Following identified potential sources, the contributions of heavy metals were quantified with the APCS-MLR. As shown in Table 6, the ratio of the mean observed and estimated values of heavy metals observed indicated a high credibility of the APCS-MLR analysis.

Table 6. Contribution of heavy metals pollution sources in DLQ.

	PC1	PC2	PC3	Unidentified Sources (%)	Estimated Mean Concentration (mg/L)	Observed Mean Concentration (mg/L)	Ratio (E/O)
Cd	—	2.12%	—	97.88%	0.05	0.05	1.0000
Cr	—	93.07%	6.93%	—	1.65	1.65	1.0000
Mn	20.67%	73.53%	5.80%	—	7.30	7.30	1.0000
Fe	40.46%	59.54%	—	—	78.36	78.36	1.0000
Ni	—	46.39%	53.61%	—	0.84	0.84	1.0000
Cu	15.64%	—	84.36%	—	1.03	1.03	1.0007
Zn	91.41%	—	8.59%	—	5.26	5.26	0.9996
Pb	26.15%	58.28%	—	15.57%	0.53	0.53	0.9999
As	6.37%	93.63%	—	—	3.33	3.33	0.9995

Note(s): — The value is small and can be ignored.

The Cr, Mn, Fe, Pb, and As in DLQ were primarily derived from the influx of geothermal water, with contributions of 93.07%, 73.53%, 59.54, 58.28%, and 93.63%, respectively. This establishes that YBJ geothermal water strongly contributes to the heavy metals of river water in DLQ. In addition, Mn, Fe, and Pb were also influenced by natural factors, with 20.67%, 40.46%, and 26.15%, respectively. The contribution of natural factors was 91.41% for Zn. Ni mainly came from the combined influence of mining activities and geothermal springs, with contributions of 53.61% and 46.39%, respectively. Cu mainly came from mining activities, with a contribution of 91.41%. In addition, the existence of Cd and Pb in 97.88% and 15.57% of the unidentified sources needs to be further explored. In conclusion, natural factors, geothermal springs, and mining activities had different degrees of contribution to the nine heavy metals.

4. Practical Applications

The preliminary results of this work will help to fill the current knowledge gaps regarding the water environment in small watersheds on the TP. It will also provide information for the integrated and refined management of water ecosystems. In addition, this work provides a basis for the development of water environment risk contingency plans. Regular surveys of water quality around geothermal fields can reduce health risks to the residents in the surrounding area. Therefore, their implementation would be useful managing the environmental safety of the water environment in the future.

5. Conclusions

This study focuses on the composition of the major ions and the sources of heavy metals in the DLQ. The water ion chemistry in the LR and DLQ rivers is mainly produced by the weathering of carbonate rocks. In addition, solutes in the rivers are also affected by the inflow of geothermal water. heavy metals in the DLQ rivers are mainly from natural, geothermal, and mining sources. The source of heavy metals to the river from the inflow of geothermal water cannot be ignored. In addition, increased precipitation and runoff due to climate change may have a diluting effect on the concentration of heavy metals such as As in rivers. The harsh environment of the highlands and the lack of meteorological and hydrological stations have led to a lack of long-term spatial and temporal observations of rivers. These factors limit the understanding of space-time variation of ion fractions and heavy metal pollution in small watershed rivers. Considering that the DLQ is used as a source of domestic drinking water, this study can provide a basis for adequate water quality protection and support further research on aquatic environmental evolution in small river basins in the Tibetan Plateau.

Author Contributions: W.X., methodology, writing, original draft preparation, and software; L.W., sample collection, methodology, and investigation; X.W. and W.M., sample collection and experiment; J.H., methodology and investigation; J.L. and X.L., conceptualization, investigation, data analysis, and editing manuscript. All authors have read and agreed to the published version of the manuscript.

Funding: This research was funded by the National Basic Research Program of China (2013CB956401) and National Natural Science Foundation of China (Nos. 41661144029 and 41672351).

Institutional Review Board Statement: Not applicable.

Informed Consent Statement: Not applicable.

Data Availability Statement: The datasets used or analyzed during the current study are available from the corresponding author on reasonable request.

Conflicts of Interest: The authors declare no conflict of interest.

References

1. Giri, S. Water quality prospective in Twenty First Century: Status of water quality in major river basins, contemporary strategies and impediments: A review. *Environ. Pollut.* **2021**, *271*, 116332. [CrossRef] [PubMed]
2. AWI. *2020 Mississippi River Watershed Report Card*; America's Watershed Initiative: St. Louis, MO, USA, 2011. Available online: https://americaswatershed.org/reportcard/ (accessed on 8 December 2020).
3. Duan, W.L.; He, B.; Chen, Y.; Zou, S.; Wang, Y.; Nover, D.; Chen, W.; Yang, G. Identification of long-term trends and seasonality in high-frequency water quality data from the Yangtze River basin, China. *PLoS ONE* **2018**, *13*, e0188889. [CrossRef] [PubMed]
4. Murty, M.; Kumar, S. Water pollution in India: An economic appraisal. *India Infrastruct. Rep.* **2011**, *19*, 285–298.
5. Huang, X.; Sillanpaa, M.; Gjessing, E.T.; Vogt, R.D. Water quality in the Tibetan Plateau: Major ions and trace elements in the headwaters of four major Asian rivers. *Sci. Total Environ.* **2009**, *407*, 6242–6254. [CrossRef]
6. Raymond, P.A.; Oh, N.H.; Turner, R.E.; Broussard, W. Anthropogenically enhanced fluxes of water and carbon from the Mississippi River. *Nature* **2008**, *451*, 449–452. [CrossRef]
7. Bera, T.; Kumar, V.; Sarkar, D.J.; Devi, M.S.; Behera, B.K.; Das, B.K. Pollution assessment and mapping of potentially toxic elements (PTE) distribution in urban wastewater fed natural wetland, Kolkata, India. *Environ. Sci. Pollut. R.* **2022**, *29*, 67801–67820. [CrossRef]
8. WHO. 1 in 3 People Globally Do Not Have Access to Safe Drinking Water. Available online: http://www.who.int/.2019 (accessed on 17 June 2019).
9. Qu, B.; Zhang, Y.L.; Kang, S.C.; Sillanpaa, M. Water quality in the Tibetan Plateau: Major ions and trace elements in rivers of the "Water Tower of Asia". *Sci. Total Environ.* **2019**, *649*, 571–581. [CrossRef]
10. Han, G.; Tang, Y.; Liu, M.; Van Zwieten, L.; Yang, X.M.; Yu, C.G.; Wang, H.L.; Song, Z.L. Carbon-nitrogen isotope coupling of soil organic matter in a karst region under land use change, Southwest China. *Agric. Ecosyst. Environ.* **2020**, *301*, 107027. [CrossRef]
11. Liu, M.; Han, G.L.; Zhang, Q.J.A. Effects of agricultural abandonment on soil aggregation, soil organic carbon storage and stabilization: Results from observation in a small karst catchment, Southwest China. *Agric. Ecosyst. Environ.* **2020**, *288*, 106719. [CrossRef]
12. Mao, G.; Zhao, Y.; Zhang, F.; Liu, J.; Huang, X. Spatiotemporal variability of heavy metals and identification of potential source tracers in the surface water of the Lhasa River basin. *Environ. Sci. Pollut. Res. Int.* **2019**, *26*, 7442–7452. [CrossRef]
13. Qin, H.H.; Gao, B.; He, L.; Hu, X.H.; Dong, L.; Sanjay, D.; Dong, A.; Sun, Z.X.; Wan, W. Hydrogeochemical Characteristics and Controlling Factors of the Lhasa River under the Influence of Anthropogenic Activities. *Water* **2019**, *11*, 948. [CrossRef]
14. Guo, Q.; Wang, Y.; Liu, W. B, As, and F contamination of river water due to wastewater discharge of the Yangbajing geothermal power plant, Tibet, China. *Environ. Geol.* **2007**, *56*, 197–205. [CrossRef]
15. Wang, G.; Zeng, C.; Zhang, F.; Zhang, Y.; Scott, C.; Yan, X. Traffic-related trace elements in soils along six highway segments on the Tibetan Plateau: Influence factors and spatial variation. *Sci. Total Environ.* **2017**, *581*, 811–821. [CrossRef] [PubMed]
16. Liu, M.; Zhao, L.Y.; Li, Q.Y.; Hu, Y.; Huang, H.W.; Zou, J.Y.; Gao, F.; Tao, J.X.; Zhang, Y.Z.; Xu, P. Distribution characteristics, enrichment patterns and health risk assessment of dissolved trace elements in river water in the source region of the Yangtze River. *J. Water Clim. Change* **2021**, *12*, 2288–2298. [CrossRef]

17. Huang, X.; Sillanpaa, M.; Gjessing, E.T.; Peraniemi, S.; Vogt, R.D. Environmental impact of mining activities on the surface water quality in Tibet: Gyama valley. *Sci. Total Environ.* **2010**, *408*, 4177–4184. [CrossRef]
18. Fuoco, I.; Figoli, A.; Criscuoli, A.; Brozzo, G.; De Rosa, R.; Gabriele, B.; Apollaro, C. Geochemical modeling of chromium release in natural waters and treatment by RO/NF membrane processes. *Chemosphere* **2020**, *254*, 126696. [CrossRef]
19. Fuoco, I.; Marini, L.; De Rosa, R.; Figoli, A.; Gabriele, B.; Apollaro, C. Use of reaction path modelling to investigate the evolution of water chemistry in shallow to deep crystalline aquifers with a special focus on fluoride. *Sci. Total Environ.* **2022**, *830*, 154566. [CrossRef]
20. Mohamed, A.; Asmoay, A.; Alshehri, F.; Abdelrady, A.; Othman, A. Hydro-geochemical applications and multivariate analysis to assess the water–rock interaction in arid environments. *Appl. Sci.* **2022**, *12*, 6340. [CrossRef]
21. Zhang, J.Y.; Liu, J.F.; Jin, J.L.; Ma, T.; Wang, G.Q.; Liu, H.W.; Min, X.; Wang, H.; Lin, J.; Bao, Z.X.; et al. Evolution and trend of water resources in Qinghai-Tibet Plateau. *Bull. Chin. Acad. Sci. Chin. Version* **2019**, *34*, 1264–1273. [CrossRef]
22. Huang, X.; Sillanpää, M.; Gjessing, E.T.; Peräniemi, S.; Vogt, R.D. Water quality in the southern Tibetan Plateau: Chemical evaluation of the Yarlung Tsangpo (Brahmaputra). *River Res. Appl.* **2011**, *27*, 113–121. [CrossRef]
23. Li, C.L.; Kang, S.C.; Chen, P.F.; Zhang, Q.G.; Mi, J.; Gao, S.P.; Sillanpää, M. Geothermal spring causes arsenic contamination in river waters of the southern Tibetan Plateau, China. *Environ. Earth Sci.* **2013**, *71*, 4143–4148. [CrossRef]
24. Zhu, M.Z.; Kuang, X.X.; Feng, Y.Q.; Hao, Y.L.; He, Q.L.; Zhou, H.; Chen, J.X.; Zou, Y.; Zheng, C. Hydrochemistry of the Lhasa River, Tibetan Plateau: Spatiotemporal Variations of Major Ions Compositions and Controlling Factors Using Multivariate Statistical Approaches. *Water* **2021**, *13*, 3660. [CrossRef]
25. Li, Z.; Liu, J.J.; Chen, H.L.; Li, Q.Q.; Yu, C.; Huang, X.; Guo, H.C. Water environment in the Tibetan Plateau: Heavy metal distribution analysis of surface sediments in the Yarlung Tsangpo River Basin. *Environ. Geochem. Health* **2020**, *42*, 2451–2469. [CrossRef]
26. Zhang, G.P.; Liu, C.Q.; Liu, H.; Jin, Z.S.; Han, G.L.; Li, L. Geochemistry of the Rehai and Ruidian geothermal waters, Yunnan Province, China. *Geothermics* **2008**, *37*, 73–83. [CrossRef]
27. Guo, Q.H.; Wang, Y.X.; Liu, W. Major hydrogeochemical processes in the two reservoirs of the Yangbajing geothermal field, Tibet, China. *J. Volcanol. Geoth. Res.* **2007**, *166*, 255–268. [CrossRef]
28. Apollaro, C.; Tripodi, V.; Vespasiano, G.; De Rosa, R.; Dotsika, E.; Fuoco, I.; Critelli, S.; Muto, F. Chemical, isotopic and geotectonic relations of the warm and cold waters of the Galatro and Antonimina thermal areas, southern Calabria, Italy. *Mar. Petrol. Geol.* **2019**, *109*, 469–483. [CrossRef]
29. Dittmar, T. Hydrochemical processes controlling arsenic and heavy metal contamination in the Elqui river system (Chile). *Sci. Total Environ.* **2004**, *325*, 193–207. [CrossRef]
30. Timperley, M.H.; Huser, B.A. Inflows of geothermal fluid chemicals to the Waikato River catchment, New Zealand. *New Zeal. J. Mar. Fresh* **1996**, *30*, 525–535. [CrossRef]
31. Demirel, Z.; Yildirim, N. Boron pollution due to geothermal wastewater discharge into the Buyuk Menderes River, Turkey. *Int. J. Environ. Pollut.* **2002**, *18*, 602–608. [CrossRef]
32. McCleskey, R.B.; Roth, D.A.; Mahony, D.; Nordstrom, D.K.; Kinsey, S. Sources, fate, and flux of geothermal solutes in the Yellowstone and Gardner Rivers, Yellowstone National Park, WY. *Appl. Geochem.* **2019**, *111*, 104458. [CrossRef]
33. Romero, L.; Alonso, H.; Campano, P.; Fanfani, L.; Farago, M. Arsenic enrichment in waters and sediments of the Rio Loa (Second Region, Chile). *Appl. Geochem.* **2003**, *18*, 1399–1416. [CrossRef]
34. Robinson, B.; Outred, H.; Brooks, R.; Kirkman, J. The distribution and fate of arsenic in the Waikato River system, North Island, New Zealand. *Chem. Spec. Bioavailab.* **1995**, *7*, 89–96. [CrossRef]
35. Sorey, M.L.; Colvard, E.M. Hydrologic investigations in the Mammoth corridor, Yellowstone National Park and vicinity, USA. *Geothermics* **1997**, *26*, 221–249. [CrossRef]
36. Wang, Y.C.; Li, L.; Wen, H.G.; Hao, Y.L. Geochemical evidence for the nonexistence of supercritical geothermal fluids at the Yangbajing geothermal field, southern Tibet. *J. Hydrol.* **2022**, *604*, 127243. [CrossRef]
37. Gad, M.; Saleh, A.H.; Hussein, H.; Farouk, M.; Elsayed, S. Appraisal of Surface Water Quality of Nile River Using Water Quality Indices, Spectral Signature and Multivariate Modeling. *Water* **2022**, *14*, 1131. [CrossRef]
38. Liu, L.L.; Tang, Z.; Kong, M.; Chen, X.; Zhou, C.C.; Huang, K.; Wang, Z.P. Tracing the potential pollution sources of the coastal water in Hong Kong with statistical models combining APCS-MLR. *J. Environ. Manag.* **2019**, *245*, 143–150. [CrossRef]
39. Thurston, G.D.; Spengler, J. A quantitative assessment of source contributions to inhalable particulate matter pollution in metropolitan Boston. *Atmos. Environ.* **1985**, *19*, 9–25. [CrossRef]
40. Zhang, T.H.; Huang, Q.Z. Research on geothermal wastewater pollution from Yangbajing Geothermal Test Power Plant in Tibet. *J. Environ. Sci.* **1997**, *17*, 252–255. (In Chinese)
41. Coetsiers, M.; Walraevens, K. Chemical characterization of the Neogene Aquifer, Belgium. *Hydrogeol. J.* **2006**, *14*, 1556–1568. [CrossRef]
42. Ji, D.; Zhao, P.J. Characteristics and genesis of the Yangbajing geothermal field, Tibet. In Proceedings of the World Geothermal Congress, Kyushu-Tohoku, Japan, 28 May–10 June 2000.
43. Pi, X.H.; Feng, L.; Li, W.F.; Liu, J.G.; Kuang, X.X.; Shi, K.; Qi, W.; Chen, D.; Tang, J. Chlorophyll-a concentrations in 82 large alpine lakes on the Tibetan Plateau during 2003–2017: Temporal-spatial variations and influencing factors. *Int. J. Digit. Earth* **2021**, *14*, 714–735. [CrossRef]

44. Liu, X.L.; Han, G.L.; Zeng, J.; Liu, J.K.; Li, X.; Boeckx, P. The effects of clean energy production and urbanization on sources and transformation processes of nitrate in a subtropical river system: Insights from the dual isotopes of nitrate and Bayesian model. *J. Clean. Prod.* **2021**, *325*, 129317. [CrossRef]
45. Liu, X.L.; Han, G.; Zeng, J.; Liu, M.; Li, X.Q.; Boeckx, P. Identifying the sources of nitrate contamination using a combined dual isotope, chemical and Bayesian model approach in a tropical agricultural river: Case study in the Mun River, Thailand. *Sci. Total Environ.* **2021**, *760*, 14393836. [CrossRef] [PubMed]
46. Liu, J.K.; Han, G.L. Tracing Riverine particulate black carbon sources in Xijiang river Basin: Insight from stable isotopic composition and Bayesian mixing model. *Water Res.* **2021**, *194*, 116932. [CrossRef] [PubMed]
47. Liu, M.; Zhang, Z.W.; Lin, L.; Zhao, L.Y.; Dong, L.; Jin, H.Y.; Zou, J.Y.; Li, R.; He, Y.J. The Content Level, Spatial and Temporal Distribution Characteristics, and Health-Risk Assessment of Trace Elements in Upper Lancang River (Changdu Section). *Water* **2022**, *14*, 1115. [CrossRef]
48. Li, H.; Li, J.; Liu, X.L.; Yang, X.; Zhang, W.; Wang, J.; Niu, Y.Q. Composition characteristics and source analysis of major ions in four small lake-watersheds on the Tibetan Plateau, China. *Huan Jing ke Xue* **2015**, *36*, 430–437. (In Chinese)
49. Wu, Z.Y. Study on the Pollution Control Scheme of Yangbajing Geothermal Power Plant Drainage. Master's Degree, Sichuan University, Chengdu, China, 2002. (In Chinese).
50. Li, S.L.; Xia, X.H.; Zhou, B.; Zhang, S.B.; Zhang, L.W.; Mou, X.L. Chemical balance of the Yellow River source region, the northeastern Qinghai-Tibetan Plateau: Insights about critical zone reactivity. *Appl. Geochem.* **2018**, *90*, 1–12. [CrossRef]
51. Zhao, L.Y.; Li, W.; Lin, L.; Guo, W.J.; Zhao, W.H.; Tang, X.Q.; Gong, D.D.; Li, Q.Y.; Xu, P. Field investigation on river hydrochemical characteristics and larval and juvenile fish in the source region of the Yangtze River. *Water* **2019**, *11*, 1342. [CrossRef]
52. Bundschuh, J.; Maity, J.P. Geothermal arsenic: Occurrence, mobility and environmental implications. *Renew. Sust. Energy Rev.* **2015**, *42*, 1214–1222. [CrossRef]
53. Spence, J.; Telmer, K. The role of sulfur in chemical weathering and atmospheric CO_2 fluxes: Evidence from major ions, $\delta^{13}C_{DIC}$, and $\delta^{34}S_{SO4}$ in rivers of the Canadian Cordillera. *Geochim. Cosmochim. Acta* **2005**, *69*, 5441–5458. [CrossRef]
54. Gibbs, R.J. Mechanisms controlling world water chemistry. *Science* **1970**, *170*, 1088–1090. [CrossRef]
55. Tian, Y.; Yu, C.Q.; Zha, X.J.; Gao, X.; Dai, E.F. Hydrochemical characteristics and controlling factors of natural water in the border areas of the Qinghai-Tibet Plateau. *J. Geogr. Sci.* **2019**, *29*, 1876–1894. [CrossRef]
56. Li, X.; Han, G.; Liu, M.; Liu, J.; Zhang, Q.; Qu, R. Potassium and its isotope behaviour during chemical weathering in a tropical catchment affected by evaporite dissolution. *Geochim. Cosmochim. Acta* **2022**, *316*, 105–121. [CrossRef]
57. Guo, Q.H.; Cao, Y.W.; Li, J.X.; Zhang, X.B.; Wang, Y.X. Natural attenuation of geothermal arsenic from Yangbajain power plant discharge in the Zangbo River, Tibet, China. *Appl. Geochem.* **2015**, *62*, 164–170. [CrossRef]
58. Fuoco, I.; De Rosa, R.; Barca, D.; Figoli, A.; Gabriele, B.; Apollaro, C. Arsenic polluted waters: Application of geochemical modelling as a tool to understand the release and fate of the pollutant in crystalline aquifers. *J. Environ. Manag.* **2022**, *301*, 113796. [CrossRef] [PubMed]
59. He, L. Hydrogeochemical Characteristics and Weathering Indicators in Lhasa River Basin. Master's Degree, East China University of Technology, Fuzhou, China, 2019. (In Chinese).
60. Li, S.L.; Liu, X.; Yue, F.J.; Yan, Z.F.; Wang, T.J.; Li, S.J.; Liu, C.Q. Nitorgen dynamics in the Critical Zones of China. *Prog. Phys. Geog.* **2022**, accepted.
61. Tian, Y.; Yu, C.Q.; Luo, K.L.; Zha, X.J.; Wu, J.S.; Zhang, X.Z.; Ni, R.X. Hydrochemical characteristics and element of natural waters in Tibet, China. *J. Geogr. Sci.* **2015**, *25*, 669–686. [CrossRef]
62. Gaillardet, J.; Viers, J.; Dupré, B. Trace elements in river waters. In *Treatise on Geochemistry*, 2nd ed.; Holland, H., Turekian, K., Eds.; Elsevier Science Ltd.: Amsterdam, The Netherlands, 2014; Volume 7, pp. 195–235. [CrossRef]
63. WHO. *Guidelines for Drinking-Water Quality*; WHO Chronicle; World Health Organization: Geneva, Switzerland, 2011.
64. *GB 3838-2002*; Chinese Standards for Surface Water Quality. Ministry of Public Health of China: Beijing, China, 2002. (In Chinese)
65. *GB 5749-2006*; Chinese Standards for Drinking Water Quality. Ministry of Public Health of China: Beijing, China, 2006. (In Chinese)
66. *EPA816-F-09-004*; National Primary Drinking Water Standards. National Secondary Drinking Water Standards. Environmental Protection Agency: Washington, DC, USA, 2009.
67. Lu, H.; Yan, Y.; Zhu, J.; Jin, T.; Liu, G.; Wu, G.; Stringer, L.C.; Dallimer, M. Spatiotemporal Water Yield Variations and Influencing Factors in the Lhasa River Basin, Tibetan Plateau. *Water* **2020**, *12*, 1498. [CrossRef]
68. Qiu, D.R.; Du, X.H.; Zha, X.L.J.; La, B.R.C.; Gu, S.Y.J.; Cui, J.Y. Analysis of Climatic Characteristics of Precipitation in Lhasa from 1980 to 2019. *J. Anhui Agric. Sci.* **2022**, *50*, 193–196. (In Chinese) [CrossRef]
69. Yu, T.T.; Gan, Y.Q.; Zhou, A.G.; Liu, C.F.; Liu, Y.D.; Li, X.Q.; Cai, H.S. Characteristics of oxygen and hydrogen isotope distribution of surface runoff in the Lhasa River basin. *Earth Sci.* **2010**, *35*, 873–878. (In Chinese)
70. Nimick, D.A.; Moore, J.N.; Dalby, C.E.; Savka, M.W. The fate of geothermal arsenic in the Madison and Missouri Rivers, Montana and Wyoming. *Water Resour. Res.* **1998**, *34*, 3051–3067. [CrossRef]
71. Qin, H.H.; Gao, B.; Huang, B.X.; Zhang, S.Q.; Dong, L.; Sun, Z.X. Distribution characteristics and pollution risk assessment of heavy metals in river water in Lhasa River Basin. *Nonferrous Met. Smelt. Part* **2020**, 79–86. (In Chinese)

72. Zwolsman, J.J.; van Bokhoven, A.J. Impact of summer droughts on water quality of the Rhine River—A preview of climate change? *Water Sci. Technol.* **2007**, *56*, 45–55. [CrossRef] [PubMed]
73. Zhang, H.; Wang, Z.F.; Zhang, Y.L.; Ding, M.J.; Li, L.H. Identification of traffic-related metals and the effects of different environments on their enrichment in roadside soils along the Qinghai-Tibet highway. *Sci. Total Environ.* **2015**, *521–522*, 160–172. [CrossRef] [PubMed]

water

MDPI

Article

Hydrochemical Characteristics and Risk Assessment of Tongzi River, Guizhou Province, Southwest China

Jiemei Lü [1,2] and Yanling An [1,2,*]

[1] The College of Resources and Environmental Engineering, Guizhou Institute of Technology, Guiyang 550003, China
[2] Key Laboratory of Karst Georesources and Environment, Ministry of Education, College of Resources and Environmental Engineering, Guizhou University, Guiyang 500025, China
* Correspondence: anyanling@git.edu.cn; Tel.: +86-150-8590-7470

Abstract: This study collected 31 water samples from the Tongzi River, Guizhou Province, Southwest China to conduct a risk assessment to understand the hydrochemical characteristics and major ion sources of irrigation and drinking water quality and their effects on human health. The results showed that ion abundance in the Tongzi River is Ca^{2+} (66%) > Mg^{2+} (24%) > K^{+}+Na^{+} (10%) for cations and HCO_3^- (75%) > SO_4^{2-}(21%) > Cl^- (4%) for anions. Additionally, the hydrochemical type of the water is Ca-Mg-HCO_3, controlled by carbonate weathering. Methods including ion ratios, principal component analysis (PCA), and correlation analysis (CA) were used to analyze the source of main ions in the river water. PC1, with the most significant variance (54.9%), decides the hydrochemical characteristics and is affected by the positive loadings of SO_4^{2-} (0.92), pH (0.85), Ca^{2+} (0.80), Cl^- (0.72), Na^+(0.66), NO_3^- (0.65), and K^+ (0.57). PC2 explains 19.2% of the total variance, with strong positive loadings of Na^+ (0.75), K^+ (0.63) and Cl^- (0.59). Mg^{2+} (0.84) and HCO_3^- (0.85) exhibits high loadings in PC3, explaining 9.3% of the variance. The results showed that intensive agricultural activities in the basin were the main source of nitrate NO_3^-, whereas SO_4^{2-} was mainly derived from mining activities. The lower concentrations of Na^+, K^+, and Cl^- were from coal combustion, domestic wastewater discharge, and agricultural fertilizer applications. The study area was mainly affected by carbonate rock weathering; natural processes (mainly the weathering of carbonate rocks) were still the main origin of Ca^{2+}, Mg^{2+}, and HCO_3^-. Moreover, the United States Salinity Laboratory (USSL) diagram and the Wilcox diagram showed that 100% of the samples fell in the C2S1 zones, and the water quality had good suitability for irrigation. The health risk assessment (HRA) results showed that $HQ_{NO_3^-}$ was much larger than HQ_{NH4+} and indicated that nitrate pollution dominated non-health hazards. About 6.5% of the samples in the tributaries represented an unacceptable risk for infants and children, and the HQ value for infants and children was always higher than that for adults. Additionally, the non-carcinogenic health risk of riverine ions for infants and children in the Tongzi River was very noteworthy, especially in the tributaries.

Keywords: hydrochemical characteristics; major ion sources; agricultural activities; risk assessment

check for updates

Citation: Lü, J.; An, Y. Hydrochemical Characteristics and Risk Assessment of Tongzi River, Guizhou Province, Southwest China. *Water* **2023**, *15*, 802. https://doi.org/10.3390/w15040802

Academic Editor: Tammo Steenhuis

Received: 28 November 2022
Revised: 15 February 2023
Accepted: 16 February 2023
Published: 18 February 2023

1. Introduction

River pollution generally comes from natural sources (weathering, salt domes, waterways, etc.) and anthropogenic sources (such as industrial wastewater, irrigation wastewater, domestic sewage, etc.) [1]. With the development of the economy and society, human activities have exacerbated river pollution [2,3], endangering the safety of agricultural irrigation water and drinking water, further posing a threat to human health [4,5]. For example, a high concentration of Na^+ in water will replace Ca^{2+} and Mg^{2+} in soil, reducing the permeability of the soil and its ability to form stable polymers, which is not conducive to plant growth [6,7]. Applying inorganic fertilizers and animal manure in agricultural areas has increased the nitrate content in water resources. The long-term drinking of this water can induce diseases, such as methemoglobinemia, blue baby syndrome, and cancer, which seriously threaten human health [8–10]. Therefore, polluted rivers have become an environmental and health hazard. It is necessary to accurately analyze the sources of pollutants in river water and evaluate the water quality [11], as this is of great significance for the management and protection of the river water environment. At present, current assessments of water pollution include irrigation water quality, a risk assessment, etc. Common methods include the USSL diagram, the Wilcox diagram, irrigation water quality [12–15], the water-quality index (WQI) [16,17], and the health risk assessment [18].

Hydrochemical characteristics of river water result from long-term interactions between water and environments [19]. By analyzing the hydrochemical characteristics of river water and the concentration of the main ions (Na^+, K^+, Mg^{2+}, Ca^{2+}, F^-, Cl^-, NO_3^-, SO_4^{2-}, and HCO_3^-), the main contribution sources and migration of ions can be identified [20,21], which help to determine the controlling factors of hydrochemical evolution and assist in tracking pollutants from natural and anthropogenic processes [22]. Methods such as main ion ratios, isotope fractionation [23], principal component analysis (PCA), correlation analysis (CA), hierarchical cluster analysis (HACA), and discriminant analysis (CDA) are widely used to identify the source of riverine ions [21,24–30], and multivariate statistical analyses of fluvial ions could better help in identification [31].

Typical carbonate landforms develop in the karst area of Southwest China, where relatively barren and thin soil is formed [32,33]. Due to the unique geological conditions and strong karstification, the river system is susceptible and fragile [34]; therefore, the hydrochemical characteristics of rivers are greatly affected by the geological background. In particular, rock weathering, gypsum dissolution, ion exchange, and other processes affect the composition of water chemical ions [35,36]. The Tongzi River is located in the largest tributary of the Chishui River (a tributary without a dam in the Yangtze River), which belongs to the typical karst basin in Southwest China. The agricultural activities in the basin are frequent, and the rivers provide residents with irrigation and drinking water along the coast. In recent years, human activities, including the watershed cultivation of slope farmland and deforestation, have led to soil erosion and nonpoint source pollution [37]. Therefore, this study systematically analyzes the fluvial ions of the Tongzi River in the typical karst area, aiming to achieve the following: (1) determine the ionic composition of the river water; (2) identify the source of the main ions in the river; and (3) explore the potential irrigation hazards and health risks of water quality.

2. Materials and Methods

2.1. Study Area

The Tongzi River is the longest tributary of the Chishui River, which is a tributary of the Yangtze River in Southwest China, and the basin area is 3348 square kilometers, spanning Tongzi, Zunyi, Renhuai, Xishui, and other counties (Figure 1). The mainstream has a total length of 122 km, and annual precipitation varies between 850 and 1200 mm, with a maximum rainfall of 1550 mm. Additionally, the annual precipitation is mainly concentrated between May and August, accounting for about 60% of the yearly rainfall; it is a typical rain-source mountain river. The urban construction in the Tongzi River basin accounts for a relatively small proportion of land. It is relatively concentrated in the county, and the Tianmen River Reservoir in the upper reaches is the primary drinking water source of the county. The middle and lower reaches flow through the mountainous karst area with steep slopes and high terrain, which consists mainly of cultivated land and forest land. The proportion of cultivated land in the basin reached 28%, which has a major impact on the ecological environment through soil erosion caused by slope farmland. There are many mining enterprises in the basin, mainly in Tongzi County and Xishui County, and some coal plant sites in the lower reaches of the mainstream are built along the river. As shown in Figure 1c, the exposed lithology in the Tongzi River basin is mainly carbonate rock, silicate rock, and migmatite, which are mainly carbonate rocks. Basin outcrop strata include

Sinian, Triassic, Quaternary, etc., with the upper reaches composed of Paleozoic Permian and Triassic strata, with a wide limestone distribution, as well as shale and coal measures.

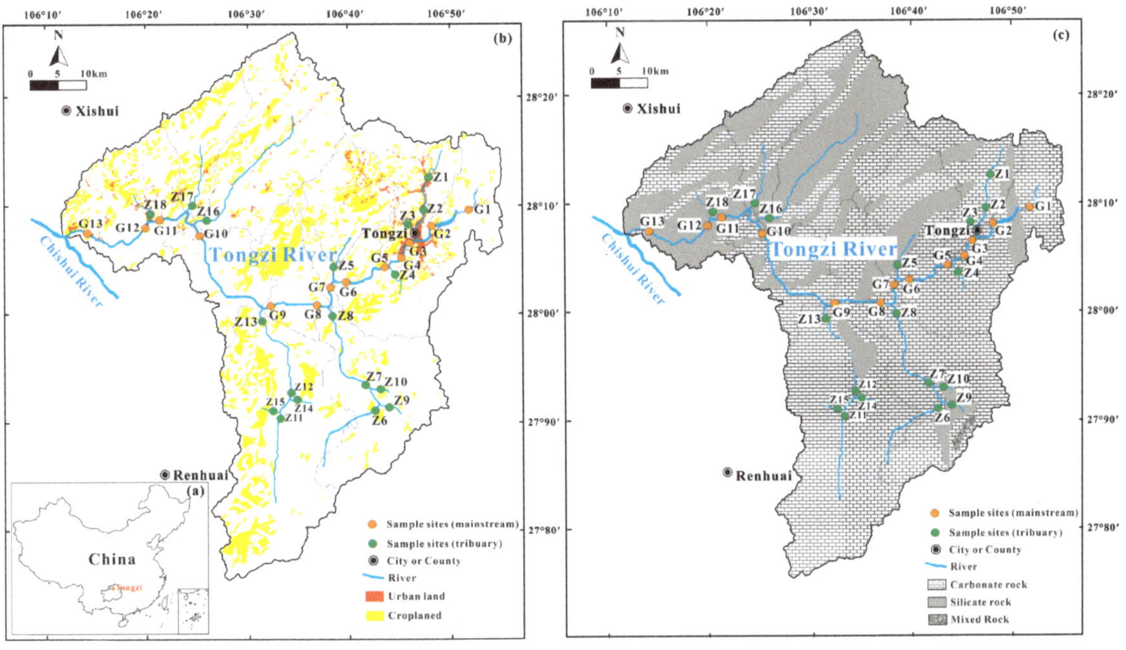

Figure 1. Location of Tongzi County in Southwest China (**a**), sampling maps of mainstream and tributaries of Tongzi River Basin under different land uses (**b**), the lithology distribution of Tongzi River watershed (**c**).

2.2. Sampling and Analysis

The implemented sampling campaign occurred during the basin's flood season in August of 2019, with a total of 31 samples obtained (13 samples (G1-G13) from the mainstream and 18 samples (Z1-Z18) from the tributaries) (Figure 1b). The sampling point information is shown in Figure 1. Water samples were collected at 0–20 cm below the surface using washed high-density polyethylene bottles. Temperature (T) and pH were measured in situ using WTW340i multiparameter water-quality meters (made in Germany, calibrated before sampling). On-site titration of HCO_3^- with 0.025 mol/L of HCl was carried out three times per water sample with an error of less than 5%. The collected river water samples were filtered on the same day (0.45 μm Millipore filter membrane), stored in a polyethylene bottle, and sealed in a refrigerator at 4 °C. Na^+, K^+, Ca^{2+}, Mg^{2+}, Cl^-, NO_3^-, F^-, and SO_4^{2-} were analyzed and determined via ion chromatography (ICS-1100, Thermo Fisher, Waltham, MA, USA, IonPac AG-19 anion exchange cartridge, and IonPac CS-12A cation exchange cartridge), and the test accuracy was better than ±5%.

2.3. Assessment Method

2.3.1. Irrigation Water Quality

The Tongzi River is the most crucial water resource for agricultural irrigation in the basin, and its water quality directly affects the growth of surrounding crops. Sodium adsorption ratio (SAR) and soluble sodium percentage (Na%) are indicators that reflect the alkali (sodium) damage of the irrigation water. A more considerable SAR value indicates that the stronger the soil adsorption of sodium ions, the more difficult it is for the vegetation roots to absorb water [38]. The larger the Na% value, the worse the permeability of the soil, thus affecting vegetation growth. For irrigation water quality, different irrigation water salinity and alkalinity levels affect soil quality attributes, thereby changing the farmland yield. To comprehensively evaluate the salinity and alkalinity hazards of irrigation water, the sodium adsorption rate and other indicators are commonly used. The soluble sodium percentage (Na%) and residual sodium carbonate (RSC) are calculated according to the ion equivalent concentration (meq/L) of river water [39].

$$SAR = \sqrt{2} \times Na^{+} / \left(Ca^{2+} + Mg^{2+}\right)^{1/2} \quad (1)$$

$$Na\% = Na^{+} / \left(Ca^{2+} + Mg^{2+} + Na^{+} + K^{+}\right) \times 100\% \quad (2)$$

$$RSC = (HCO_3^{-} - CO_3^{2-}) - (Ca^{2+} + Mg^{2+}) \quad (3)$$

2.3.2. Health Risk Assessment

The health risk assessment method is a quantitative analysis method for evaluating substances that are harmful to human health. Water pollutants come into contact with the human body mainly through drinking water and the skin, causing damage to human organs and threatening human health [40]. However, ingestion is identified as a more severe exposure pathway compared to the dermal route [41,42]. The average total nitrogen content in the Tongzi River basin was 2.85 mg/L [43], which is 1.4 times higher than the Chinese (GB3838-2002) V water-quality standard, and the main forms of nitrogen in natural rivers were NO_3^- and NH_4^+. Therefore, according to the USEPA risk guidelines (USEPA 2004), the exposure doses of direct intake ($ADD_{ingestion}$) were calculated for NO_3^- and NH_4^+ as follows:

$$ADD_{ingestion} = C_w \times IR \times EF \times ED/(BW \times AT) \quad (4)$$

where $ADD_{ingestion}$ is the average daily exposure dose ($mg{\cdot}kg^{-1}d^{-1}$) from drinking water; C_w is the measured concentration of nitrate in water (mg/L); IR is the ingestion rate (L/day); EF is the exposure frequency (days/year); ED is the exposure duration (years); BW is the body weight (kg); and AT is the average time for non-carcinogens to stay in the body (days). Table 1 shows the parameters of the health risk assessment model for the assessment of the risk NO_3^- and NH_4^+ pose to human health.

Table 1. Parameters employed for human health risk assessment.

Parameters	Unit	Infants	Children	Adults	Reference
EF	$d{\cdot}a^{-1}$	365	365	365	[44]
IR	$L{\cdot}d^{-1}$	0.8	1.5	2	[45]
AT	d	365 × ED	365 × ED	365×ED	[46]
ED	a	1	12	30	[46]
BW	kg	10	20	70	[45]

The hazard quotient (HQ) is usually used to evaluate non-carcinogenic health risks, which was defined as:

$$\text{Hazard Quotient (HQ)} = ADD_{ingestion}/RfD_{ingestion} \quad (5)$$

where $ADD_{ingestion}$ is the average daily exposure dose ($mg{\cdot}kg^{-1}d^{-1}$), and $DfR_{ingestion}$ is the reference dose of different ions (1.6 and 0.97 ppm/day for NO_3^- and NH_4^+) [47]. When HQ < 1, the human health risk caused by pollutants is permissible, whereas when HQ > 1, non-carcinogenic effects should be considered.

2.4. Data Analysis

Principal component analysis (PCA) and a correlation matrix were carried out using SPSS 21.0(IBM SPSS Statistics, Chicago, IL, USA) (PCA is a common method for exploring potential sources). A Piper diagram was created for the relationship between ion concentrations and physicochemical parameters in the Tongzi River, and Origin 2022(OriginLab, Northampton, UK) was used to edit the graphs.

3. Results and Discussion

3.1. Hydrochemical Compositions of Tongzi River

The chemical characteristics of river water demonstrate its environmental properties and functions. The water temperature (T), dissolved oxygen (DO), pH value, electrical conductivity (EC), main ion concentrations, and the composition of river water all reflect the basic hydrochemical properties and characteristics of river water. The statistical results of the major ion concentrations and the water parameters (T, EC, pH, and DO) during the study period are listed in Table 2. The Tongzi River water temperature ranged from 20.0 to 29.0 °C with a mean value of 23.6 °C. The EC values ranged from 330 to 622 μS/cm with a mean value of 436 μS/cm, whereas the pH values varied between 8.54 and 9.36 with a mean value of 9.13, meaning the river waters were slightly alkaline. The CO_3^{2-} and CO_2 in the aqueous solution were negligible with regards to the carbonate equilibrium and the main inorganic carbon species in the Tongzi River was HCO_3^-. The total cationic charges ($TZ^+ = Na^+ + K^+ + 2Mg^{2+} + 2Ca^{2+}$) in the river water ranged from 3.33 to 6.37 meq·L^{-1} (mean value of 4.55 meq·L^{-1}), and the total anionic charges ($TZ^- = Cl^- + NO_3^- + HCO_3^- + 2SO_4^{2-}$) in the river water ranged from 3.25 to 6.51 meq·L^{-1} (mean value of 4.56 meq·L^{-1}). The mean value of the normalized inorganic charge balance ($NICB = (TZ^+ - TZ^-)/TZ^+$) [48,49] was generally <4% for each sampling campaign (Table S1), suggesting that the anions and cations in the Tongzi River were basically balanced.

Table 2. Values of Major Ions and Some Hydrogeochemical Parameters in the Tongzi River.

	Min	Max	Mean	SD	Chinese Guideline	WHO Guideline
T (°C)	20.0	29.0	23.6	2.9		
EC (μS/cm)	330	622	436	61		
pH	8.54	9.36	9.13	0.17	6.5–8.5	6.5–8.5
DO (mg/L)	1.24	13.01	8.91	2.37		
Na^+(mmol/L)	0.06	0.58	0.20	0.11		
K^+ (mmol/L)	0.02	0.16	0.05	0.02		
Mg^{2+}(mmol/L)	0.31	1.09	0.56	0.22		
Ca^{2+}(mmol/L)	1.16	2.44	1.58	0.30		
F^- (mmol/L)	0.004	0.011	0.007	0.002	0.05	0.08
Cl^- (mmol/L)	0.03	0.51	0.14	0.10	7.05	7.05
NO_3^- (mmol/L)	0.08	0.47	0.20	0.07	1.428	3.570
SO_4^{2-} (mmol/L)	0.37	1.18	0.76	0.20	2.60	2.60
HCO_3^- (mmol/L)	2.01	3.66	2.69	0.43		
NH_4^+ (mg/L)	0.01	2.65	0.25	0.46	0.036	0.107
SAR	0.06	0.49	0.20	0.09		
Na%	2.56	16.57	8.33	0.03		
RSC	0.16	1.26	0.53	0.26		

Note: the unit of corresponding values in the Chinese guidelines and WHO guidelines are converted to mmol/L.

As shown in Figure 2, the mean proportion of cations in the Tongzi River water followed a decreasing order of Ca^{2+} (66%) > Mg^{2+} (24%) > K^+ + Na^+ (10%), and the dominant anion was HCO_3^- which, on average, accounted for 75% of the total anions, followed by SO_4^{2-} (21%) and Cl^- (4%). These findings are basically consistent with the hydrochemical characteristics of the Chishui River [47], where the dominant cations were found to be Ca^{2+} and Mg^{2+}, whereas the dominant anions were HCO_3^- and SO_4^{2-}; these data essentially comply with the component characteristics of major ions in the

Guizhou karst rivers, and the hydrochemical type of the Tongzi River water was Ca-Mg-HCO_3, controlled by carbonate weathering. Compared with the Changjiang River (urban area), the Jialing River (Chongqing area), and the Nanming River, the regions where carbonate rocks were distributed were all dominated by HCO_3^- and Ca^{2+}. The Tongzi River has a rich proportion of SO_4^{2-} (21%), which is obviously higher than in the Changjiang River [50,51] (16% and 10%), the Nanming River [31] (17%), and the Jialing River [26] (14%). Although the SO_4^{2-} content of the Tongzi River is higher than that of the above river, it is probably subjected to human activities and the dissolution of sulfide oxidation (e.g., pyrite) [47,52]. The Na^++K^+ and Cl^- contents are relatively low, and the proportions are obviously lower than in the Nanming River [31] (Na^++K^+:26%, Cl^-:10%) and the Jialing River [26] (Na^++K^+:28%, Cl^-:16%), and are much lower than in the Changjiang River [50,51] (Na^++K^+:31% and 42%, Cl^-: 13% and 24%), all of which are subject to the influence of different anthropogenic activities [26,31,47,50,51].

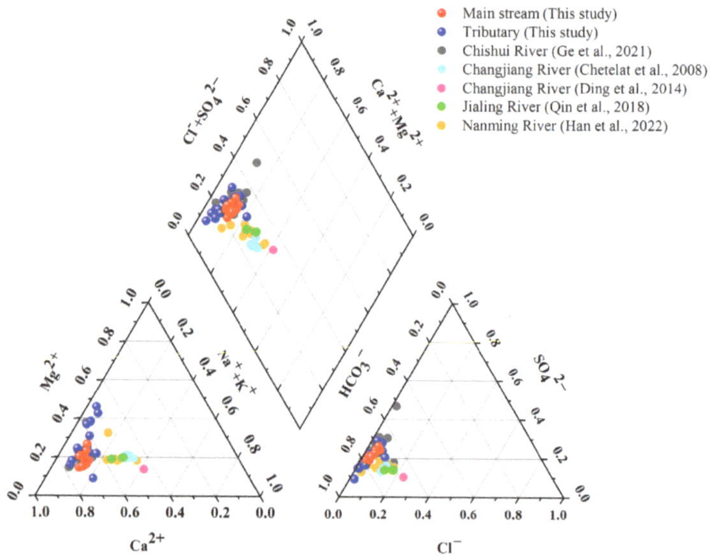

Figure 2. Piper diagrams representing the percentages of hydrochemical species in Tongzi River, Chishui River [47], Changjiang River [50,51], Jialing River [26], and Nanming River [31].

3.2. Sources of River Solutes

3.2.1. Correlation Matrix and PCA Analysis of Major Ions

Spearman's correlation among the major ionic components in the water allows for a systematic analysis of the interrelationships between the major ions. Based on their relationships, we can determine the source and identify the possible factors that affect the major ions [53]. Additionally, to further explore the source of the main ions in the Tongzi River basin, PCA was also used for analyses [54]. In general, KMO values below 0.5 are not desirable, whereas values from 0.5 to 0.7 are considered sufficient, and higher values (above 0.7) are outstanding [55]. The current study achieved a KMO value of 0.65; therefore, we used the correlation matrix and PCA to explore the source of major ions in the Tongzi River basin. For the Tongzi River, three principal components (PCs) with eigenvalues above 1 were retained, which explains 83.4% of the total variance (Table 3). PC1, with the most significant variance (54.9%), decides the hydrochemical characteristics and is affected by positive loadings of SO_4^{2-} (0.92), pH (0.85), Ca^{2+} (0.80), Cl^- (0.72), Na^+(0.66), NO_3^- (0.65), and K^+ (0.57). PC2 explains 19.2% of the total variance, with strong positive loadings of Na^+ (0.75), K^+ (0.63) and Cl^- (0.59). Mg^{2+} (0.84) and HCO_3^- (0.85) exhibited high loadings in PC3, explaining 9.3% of the variance.

Table 3. Varimax-rotated component matrix for water ions in the Tongzi River.

Variable	PC1	PC2	PC3	Communalities
EC	−0.16	−0.87	−0.06	0.78
pH	0.85	0.35	0.38	0.98
Na^+	0.66	0.63	−0.19	0.87
K^+	0.57	0.75	−0.11	0.89
Mg^{2+}	0.03	−0.44	0.84	0.90
Ca^{2+}	0.80	0.47	0.03	0.86
Cl^-	0.72	0.59	0.01	0.86
NO_3^-	0.65	0.38	0.05	0.56
SO_4^{2-}	0.92	0.04	−0.04	0.84
HCO_3^-	0.28	0.23	0.85	0.86
F^-	0.55	−0.04	−0.68	0.76
eigenvalues	6.04	2.11	1.03	
variance (%)	54.86	19.19	9.33	
cumulative (%)	54.86	74.05	83.38	

Note: extraction method—principal component analysis; rotation method—varimax with Kaiser normalization.

PC1 and PC2 exhibited high positive loadings (>0.50) of Na^+, K^+, and Cl^-, and there was a strong positive correlation among Na^+, K^+, and Cl^-. There is no obvious evaporite (KCl and NaCl) exposed in the river basin; therefore, in addition to rock-weathering sources, Na^+, K^+, and Cl^- in the river water also have other anthropogenic sources, such as industrial (coal combustion) [20,56] and domestic wastewater discharge [48] and agricultural fertilizer [24]. Moreover, there are some coal mines in the basin, many villages and towns along the basin, and a large proportion of agricultural land. Samples from the Tongzi River presented low concentrations of Cl^-, Na^+, and K^+, which means that Na^+, K^+, and Cl^- are mainly from coal combustion, rainfall [57], domestic wastewater discharge (NaCl), and agricultural fertilizer (KCl).

The results presented in Figure 3 show that SO_4^{2-} and Mg^{2+}, and especially HCO_3^- and Ca^{2+}, were the ions with a high concentration in the river water. SO_4^{2-} and Ca^{2+} had a significant correlation, but sulfate was also strongly correlated with other ions, such as Na^+, K^+, and Cl^-; the significance of SO_4^{2-} and anthropogenic input ions (Na^+, K^+, and Cl^-) is higher, inferring that they may be from the same source. SO_4^{2-} is derived from human activities and the dissolution of sulfide oxidation (e.g., pyrite) [47,52], which is consistent with the presence of SO_4^{2-} in the whole Chishui River [58]. HCO_3^-, Ca^{2+}, and Mg^{2+} are significantly correlated, indicating that carbonate weathering inputs account for their high proportion. In addition, Mg^{2+}, Ca^{2+}, and HCO_3^- mainly come from the dissolution of dolomite and limestone, which is consistent with the main rock types in the study area [31,59,60].

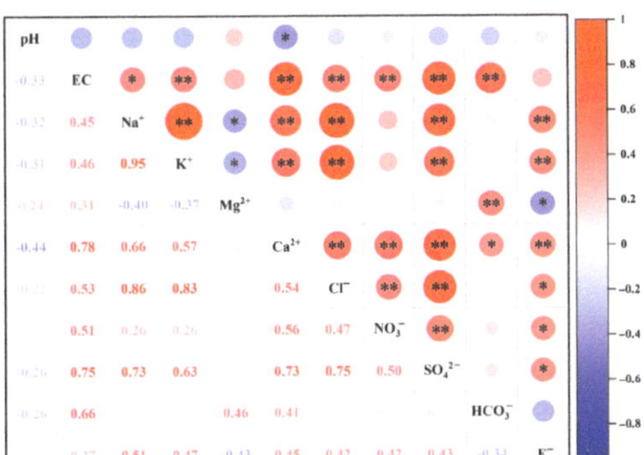

Figure 3. Spearman's correlation for the Tongzi River ions and other parameters; an asterisk (*) denotes $p < 0.05$; two asterisks (**) denote $p < 0.01$.

According to previous studies, nitrate mainly comes from anthropogenic sources, such as urban sewage, mobile emission sources (transportation) [61,62], industrial pollution [63], agricultural nitrogen fertilizers [64], and animal waste [65]. In this study, NO_3^- exhibited a much higher concentration (mean 12.7 mg/L) than that in the Xijiang River, where extensive agricultural activities exist (with a high average value of 6.4 mg/L) [65]. Furthermore, the high correlation between Cl^- and NO_3^- in PC1 indicates that their sources are consistent and are closely related to the use of agricultural nitrogen fertilizers [66]. Therefore, NO_3^- mainly comes from agricultural nitrogen fertilizers.

3.2.2. Mineral Dissolution

The Gibbs diagram [67] method defined three mechanisms (evaporation–crystallization, rock dominance, and atmospheric precipitation) that control world surface water chemistry. The relationship between TDS and $Na^+/(Na^++Ca^{2+})$ and $Cl^-/(Cl^-+HCO_3^-)$ can be used to judge the main control types of the main ions in river water and the influencing factors of hydrochemical changes in the river basin [67]. As shown in Figure 4, the $Na^+/(Na^++Ca^{2+})$ of most river samples in the Tongzi River is less than 0.6, and the $Cl^-/(Cl^- + HCO_3^-)$ is less than 0.2. All river sample points were from the rock-weathering control area and away from the atmospheric action zone, indicating that the ion composition of the river is mainly controlled by rock weathering, which is similar to that in the Yangtze River [26], the Xijiang River [68], the Wujiang River [69], and other large basins in China, as well as the Qingshui River [70], the Chishui River [47], and other karst regions.

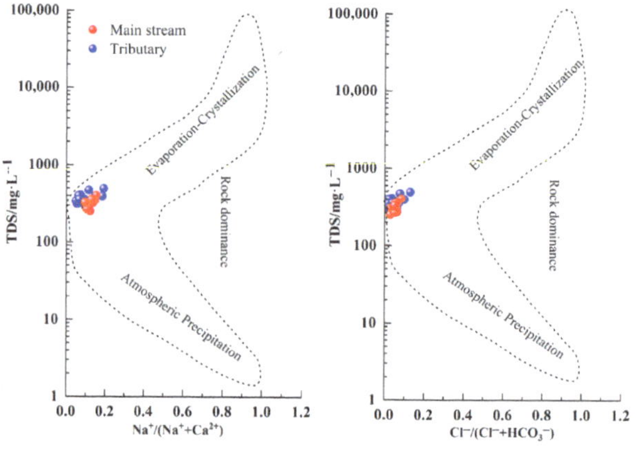

Figure 4. Gibbs plots of river water in the Tongzi River Basin [67].

For further analysis of rock-weathering types and to explore the source of the main ions generated by rock weathering in the region [71], end-member diagrams [72], including Mg^{2+}/Na^+ vs. Ca^{2+}/Na^+(Figure 5a) and HCO_3^-/Na^+ vs. Ca^{2+}/Na^+ (Figure 5b), can be used to further determine the types of rock-weathering sources [58]. As shown in Figure 5a, the samples from the mainstream and tributaries of the Tongzi River basin were mainly concentrated in the vicinity of the carbonate rock end-member and between the carbonate rock and silicate rock end-member. Still, most of the points are biased toward the carbonate rock area. Some fall in the carbonate rock-weathering area, indicating that the study area is more affected by carbonate rock than silicate rock.

As shown in Figure 5c, most water samples fall to the right of the 1:1 line, which may be affected, to a certain degree, by silicate rock weathering or the cation exchange process [73,74]. The ratios of Ca^{2+}+Mg^{2+} to HCO_3^- are used to characterize the dissolution of carbonate minerals' (such as calcite and dolomite) contribution to the chemical characteristics of river water [75]. Figure 5d shows that the water samples are mainly concentrated on the light side of the 1:1 line, indicating that the weathering of carbonate rocks is the main source of Ca^{2+}, Mg^{2+}, and HCO_3^- in the Tongzi River [76]. As shown in Figure 5e, all the water samples are mainly concentrated on the 1:1 line, which indicates that the dissolution of carbonate and sulfate minerals dominate the hydrochemical characteristics of the Tongzi River. In addition, the ratio between Na^++K^+–Cl^- and HCO_3^-+SO_4^{2-}–Ca^{2+}–Mg^{2+} is commonly utilized to identify the influence of cation exchange [46,58,77]. The ratios of most water samples were near the 1:1 line, indicating that the cation exchange process significantly affects water hydrochemistry [7]. Figure 5f shows that all water samples fall on the left side of the 1:1 line, revealing that cation exchange does not affect the Tongzi River.

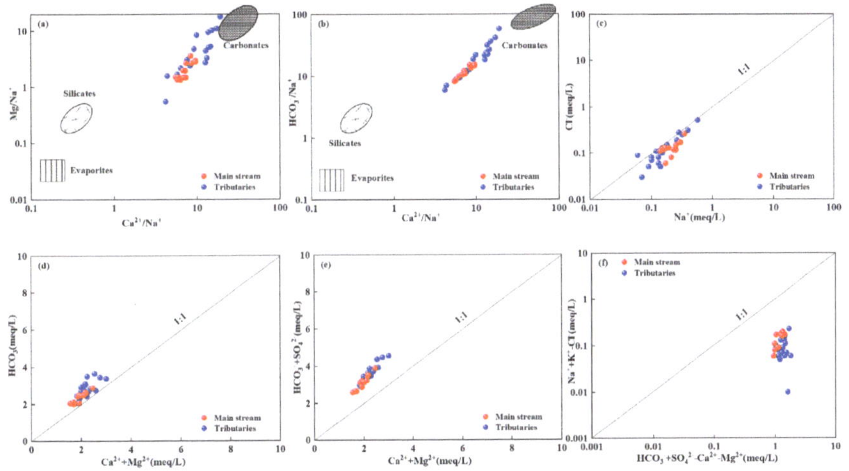

Figure 5. Plots of Ca^{2+}/Na^+ versus Mg^{2+}/Na^+ (**a**), Ca^{2+}/Na^+ versus HCO_3^-/Na^+ (**b**), Na^+ versus Cl^- (**c**), (Ca^{2+}+Mg^{2+}) versus HCO_3^- (**d**), (Ca^{2+}+Mg^{2+}) versus (HCO_3^-+SO_4^{2-}) (**e**), and (HCO_3^-+SO_4^{2-}−Ca^{2+}−Mg^{2+}) versus (Na^++K^+−Cl^-) (**f**) for the study area.

3.2.3. Anthropogenic Inputs

Human activities significantly impact the chemical characteristics of water bodies. In areas of intense human activities, anthropogenic inputs often influence rivers, with intensive human activities leading to elevated levels of Cl^-, NO_3^-, and SO_4^{2-} [78,79]. In addition, polluted water is generally considered to have high Cl^-/Na^+, NO_3^-/Na^+, and SO_4^{2-}/Na^+ ratios [46,73]. Therefore, the present study used the Cl^-/Na^+, NO_3^-/Na^+, and SO_4^{2-}/Na^+ ratios of the Tongzi River in comparison with those from other rivers to further evaluate the ion ratio characteristics of rivers under different human influences (Figure 6a,b). The results show that the Tongzi River has a higher ratio of NO_3^-/Na^+ than other urban rivers (Yangtze River, Jialing River, and Nanming River), and the water samples located close to agricultural activities are similar to those from the Chishui River, indicating that agricultural activity is an important source of NO_3^- in the Tongzi River. In addition, Figure 6b shows that the Tongzi River was also affected by industrial activities to a certain extent and is similar to the Chishui River, indicating that SO_4^{2-} is mainly derived from industrial activities.

3.3. Irrigation and Guideline-Based Water Quality

The Tongzi River is a typical plateau agricultural river. The proportion of cultivated land in the basin is 18%. The crops are mainly wheat, corn, and rice; therefore, Tongzi River water is an important water source for agricultural irrigation. Additionally, the irrigation water quality is almost equally important in ensuring food security and social and economic development [80]. The concentrations of inorganic ions in the Tongzi River are also in line with the range given by the Chinese guidelines and the WHO guidelines. The suitable pH value of drinking water is suggested to be between 6.5 and 8.5. Still, most water samples have unsuitable pH values because carbonate weathering causes karst rivers to have high pH [81].

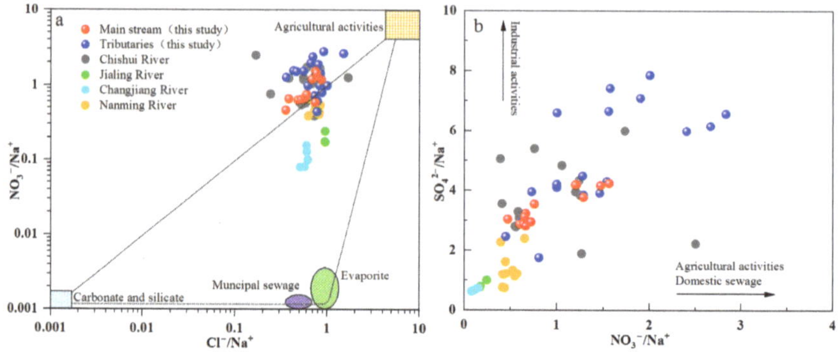

Figure 6. Plots of Cl^-/Na^+ versus NO_3^-/Na^+ (**a**) and NO_3^-/Na^+ versus SO_4^{2-}/Na^+ (**b**) for water samples of the Tongzi River, Chishui River [47], Changjiang River [50], Jialing River [26], and Nanming River [31].

The Na% ranged from 2.56 to 16.57%, with an average value of 8.21%, and the SAR varied from 0.06 to 0.49, with a mean value of 0.19. The United States Salinity Laboratory plot (Figure 7a) shows that 100% of the samples fell in the C2S1 zones, indicating a medium salinity hazard. In addition, according to the Wilcox diagram (Figure 7b), most samples were in the excellent to good section. In general, low sodium and moderate salinity hazards may exist in the Tongzi River, indicating a relatively suitable quality for irrigation without much harm to soil aggregates. In addition, the RSC value of the Tongzi River water sample exceeded 1.26 (Table S1), indicating the suitability of irrigation water.

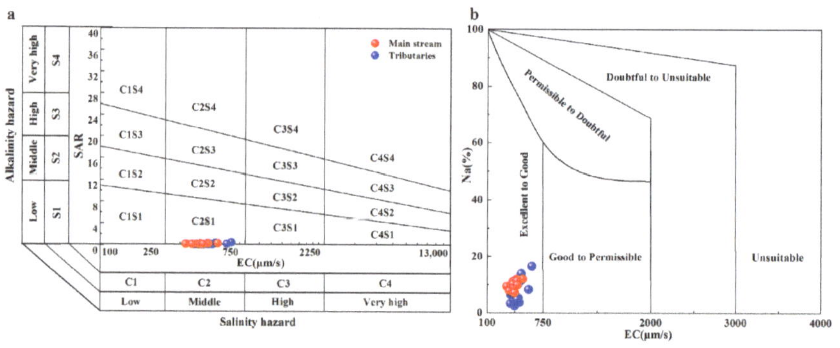

Figure 7. Salinity and alkalinity evaluation of irrigation water quality: (**a**) United States Salinity Laboratory (USSL) diagram and (**b**) Wilcox diagram.

3.4. Health Risk Assessment

A health risk assessment is a method used to assess possible harmful health effects in different groups of people [82]. The human body mainly absorbs toxic substances in the water body by drinking the water. The Tongzi River basin provides drinking water for nearby residents, especially those in the upstream Tianmen River Reservoir, as it is the main drinking water source in Tongzi County. NH_4^+ and NO_3^- in the water enter the human body mainly through the drinking of the water. Long-term consumption of water polluted by NH_4^+ and NO_3^- can lead to various diseases, such as leukemia, digestive system cancer, and blue baby syndrome [83]. Therefore, the present study used the model recommended by the US Environmental Protection Agency to evaluate the potential risks to human health posed by the non-point source pollution (NSP).

The non-carcinogenic health risks of the total HQ value ($HQ_T = HQ_{NO_3^-} + HQ_{NH_4^+}$) in the Tongzi River were calculated (Figure 8c). The results of the HRA showed that the $HQ_{NO_3^-}$ values for infants, children, and adults were 0.25–1.46, 0.23–1.37, and 0.09–0.52, with mean values of 0.63, 0.59, and 0.23 (Figure 8a), respectively. The HQ_{NH4+} values for infants, children, and adults were 0.001–0.219, 0.001–0.205, and 0–0.078, with mean values of 0.02, 0.019, and 0.007 (Figure 8b), respectively. This suggests that HQ_{NO3-} is much larger than HQ_{NH4+} and indicates that non-health hazards are dominated by nitrate pollution. For infants, children, and adults, the HQ_T of each ion and total HQ were higher than the hazard level (>1), indicating a higher health risk of the assessed ions. About 6.5% of the samples (Z3 and Z15) in the tributaries had an unacceptable risk for infants and children, and the HQ value for infants and children was always higher than that for adults, implying they face an increased risk from the riverine ions compared to adults. The non-carcinogenic health risk of riverine ions to infants and children in the Tongzi River is very noteworthy, especially in the tributaries.

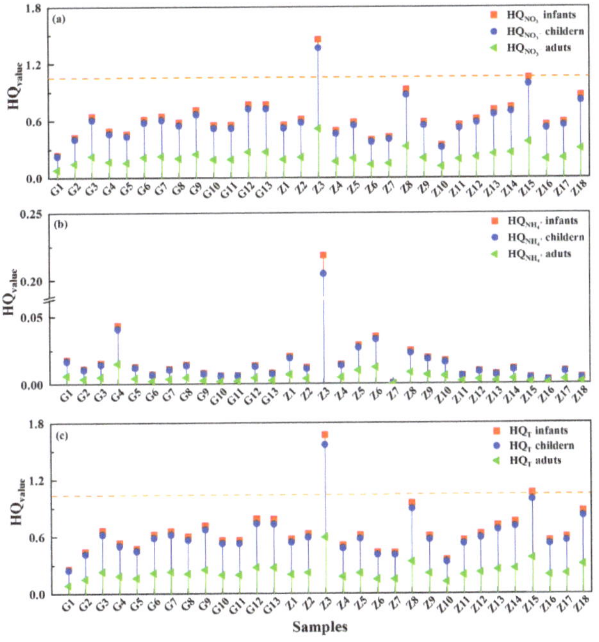

Figure 8. Non-carcinogenic health risk assessment in Tongzi River water. (**a**) HQ values of nitrate, (**b**) HQ values of ammonia, (**c**) total HQ value.

4. Conclusions

The present study collected 31 water samples from the Tongzi River to evaluate the hydrochemical characteristics, sources of river solutes, irrigation and guideline-based water quality, and risk to human health. The main conclusions of the present study are as follows: The ion abundance in the Tongzi River is Ca^{2+} (66%) > Mg^{2+} (24%) > $K^+ + Na^+$ (10%) for cations and HCO_3^- (75%) > SO_4^{2-} (21%) > Cl^- (4%) for anions. Additionally, the hydrochemical type of the water is Ca-Mg-HCO_3, which is mainly controlled by carbonate weathering; thus, the weathering of carbonate rock is still the main origin of Ca^{2+}, Mg^{2+}, and HCO_3^-. In a typical agricultural river, compared to urban river sources, intensive agricultural activities in the basin are the main source of nitrate NO_3^-, whereas SO_4^{2-} mainly derives from industrial activities (coal mine). The irrigation and guideline-based water-quality results showed that the concentrations of inorganic ions in the Tongzi River are also in line with the range given by the Chinese guidelines and the WHO guidelines, and that the irrigation water has good suitability. Moreover, the health risk assessment results showed that the amount of $HQ_{NO_3^-}$ is much larger than $HQ_{NH_4^+}$, and about 6.5% of the samples in the tributaries presented an unacceptable risk for infants and children, indicating that nitrate pollution dominates non-health hazards. Therefore, in the subsequent management of the water environment in the basin, the control of agricultural nonpoint source pollution should be strengthened to ensure the safety of water use in the basin.

Supplementary Materials: The following supporting information can be downloaded at: https://www.mdpi.com/article/10.3390/w15040802/s1, Table S1: The water parameters, major ion compositions, SAR, Na%, RSC, NICB of Tongzi River.

Author Contributions: J.L. methodology, formal analysis, investigation, data curation, and writing—original draft preparation. Y.A. writing—review and editing and funding acquisition. All authors have read and agreed to the published version of the manuscript.

Funding: This research was funded by the High-Level Talent Introduction Program for the Guizhou Institute of Technology (2019 [45]). The funder had no role in study design, data collection and analysis, decision to publish, or preparation of the manuscript.

Institutional Review Board Statement: Not applicable.

Informed Consent Statement: Not applicable.

Data Availability Statement: Not applicable.

Conflicts of Interest: The authors declare no conflict of interest.

References

1. Gnanachandrasamy, G.; Dushiyanthan, C.; Jeyavel Rajakumar, T.; Zhou, Y. Assessment of hydrogeochemical characteristics of groundwater in the lower Vellar river basin: Using Geographical Information System (GIS) and Water Quality Index (WQI). *Environ. Dev. Sustain.* **2020**, *22*, 759–789. [CrossRef]
2. Wang, J.; Jiang, Y.; Sun, J.; She, J.; Yin, M.; Fang, F.; Xiao, T.; Song, G.; Liu, J. Geochemical transfer of cadmium in river sediments near a lead-zinc smelter. *Ecotoxicological Environ. Saf.* **2020**, *196*, 110529. [CrossRef] [PubMed]
3. Zeng, J.; Han, G.; Zhang, S.; Liang, B.; Qu, R.; Liu, M.; Liu, J. Potentially toxic elements in cascade dams-influenced river originated from Tibetan Plateau. *Environ. Res.* **2022**, *208*, 112716. [CrossRef] [PubMed]
4. Liu, M.; Han, G.; Zhang, Q. Effects of agricultural abandonment on soil aggregation, soil organic carbon storage and stabilization: Results from observation in a small karst catchment, Southwest China. *Agric. Ecosyst. Environ.* **2020**, *288*, 106719. [CrossRef]
5. Chen, L.; Liu, J.-r.; Hu, W.-f.; Gao, J.; Yang, J.-y. Vanadium in soil-plant system: Source, fate, toxicity, and bioremediation. *J. Hazard. Mater.* **2021**, *405*, 124200. [CrossRef]
6. Yang, K.; Han, G.; Song, C.; Zhang, P. Stable H-O Isotopic Composition and Water Quality Assessment of Surface Water and Groundwater: A Case Study in the Dabie Mountains, Central China. *Int. J. Environ. Res. Public Health* **2019**, *16*, 4076. [CrossRef]
7. Lu, T.; Li, R.; Ferrer, A.S.N.; Xiong, S.; Zou, P.; Peng, H. Hydrochemical characteristics and quality assessment of shallow groundwater in Yangtze River Delta of eastern China. *Environ. Sci. Pollut. Res.* **2022**, *29*, 57215–57231. [CrossRef]
8. Ward, M.H.; Jones, R.R.; Brender, J.D.; De Kok, T.M.; Weyer, P.J.; Nolan, B.T.; Villanueva, C.M.; Van Breda, S.G. Drinking Water Nitrate and Human Health: An Updated Review. *Int. J. Environ. Res. Public Health* **2018**, *15*, 1557. [CrossRef]
9. Schullehner, J.; Hansen, B.; Thygesen, M.; Pedersen, C.B.; Sigsgaard, T. Nitrate in drinking water and colorectal cancer risk: A nationwide population-based cohort study. *Int. J. Cancer* **2018**, *143*, 73–79. [CrossRef]
10. Pennino, M.J.; Leibowitz, S.G.; Compton, J.E.; Hill, R.A.; Sabo, R.D. Patterns and predictions of drinking water nitrate violations across the conterminous United States. *Sci. Total Environ.* **2020**, *722*, 137661. [CrossRef]
11. Tsering, T.; Abdel Wahed, M.S.M.; Iftekhar, S.; Sillanpää, M. Major ion chemistry of the Teesta River in Sikkim Himalaya, India: Chemical weathering and assessment of water quality. *J. Hydrol. Reg. Stud.* **2019**, *24*, 100612. [CrossRef]
12. Zhang, B.; Song, X.; Zhang, Y.; Han, D.; Tang, C.; Yu, Y.; Ma, Y. Hydrochemical characteristics and water quality assessment of surface water and groundwater in Songnen plain, Northeast China. *Water Res.* **2012**, *46*, 2737–2748. [CrossRef] [PubMed]
13. Xia, C.; Liu, G.; Xia, H.; Jiang, F.; Meng, Y. Influence of saline intrusion on the wetland ecosystem revealed by isotopic and hydrochemical indicators in the Yellow River Delta, China. *Ecol. Indic.* **2021**, *133*, 108422. [CrossRef]
14. Zhou, Y.; Li, P.; Xue, L.; Dong, Z.; Li, D. Solute geochemistry and groundwater quality for drinking and irrigation purposes: A case study in Xinle City, North China. *Geochemistry* **2020**, *80*, 125609. [CrossRef]
15. Elumalai, V.; Nethononda, V.G.; Manivannan, V.; Rajmohan, N.; Li, P.; Elango, L. Groundwater quality assessment and application of multivariate statistical analysis in Luvuvhu catchment, Limpopo, South Africa. *J. Afr. Earth Sci.* **2020**, *171*, 103967. [CrossRef]

16. Wong, Y.J.; Shimizu, Y.; He, K.; Nik Sulaiman, N.M. Comparison among different ASEAN water quality indices for the assessment of the spatial variation of surface water quality in the Selangor river basin, Malaysia. *Environ. Monit. Assess.* **2020**, *192*, 644. [CrossRef]
17. Wong, Y.J.; Shimizu, Y.; Kamiya, A.; Maneechot, L.; Bharambe, K.P.; Fong, C.S.; Nik Sulaiman, N.M. Application of artificial intelligence methods for monsoonal river classification in Selangor river basin, Malaysia. *Environ. Monit. Assess.* **2021**, *193*, 438. [CrossRef]
18. Gao, Z.; Han, C.; Xu, Y.; Zhao, Z.; Luo, Z.; Liu, J. Assessment of the water quality of groundwater in Bohai Rim and the controlling factors—A case study of northern Shandong Peninsula, north China. *Environ. Pollut.* **2021**, *285*, 117482. [CrossRef]
19. Xia, C.; Liu, G.; Meng, Y.; Jiang, F. Reveal the threat of water quality risks in Yellow River Delta based on evidences from isotopic and hydrochemical analyses. *Mar. Pollut. Bull.* **2022**, *177*, 113532. [CrossRef]
20. Xu, Z.; Liu, C.-Q. Chemical weathering in the upper reaches of Xijiang River draining the Yunnan–Guizhou Plateau, Southwest China. *Chem. Geol.* **2007**, *239*, 83–95. [CrossRef]
21. Li, X.; Han, G.; Liu, M.; Liu, J.; Zhang, Q.; Qu, R. Potassium and its isotope behaviour during chemical weathering in a tropical catchment affected by evaporite dissolution. *Geochim. Et Cosmochim. Acta* **2022**, *316*, 105–121. [CrossRef]
22. Wei, M.; Duan, P.; Gao, P.; Guo, S.; Hu, Y.; Yao, L.; Li, M. Exploration and application of hydrochemical characteristics method for quantification of pollution sources in the Danjiangkou Reservoir area. *J. Hydrol.* **2020**, *590*, 125291. [CrossRef]
23. Apollaro, C.; Tripodi, V.; Vespasiano, G.; De Rosa, R.; Dotsika, E.; Fuoco, I.; Critelli, S.; Muto, F. Chemical, isotopic and geotectonic relations of the warm and cold waters of the Galatro and Antonimina thermal areas, southern Calabria, Italy. *Mar. Pet. Geol.* **2019**, *109*, 469–483. [CrossRef]
24. Zhang, S.; Han, G.; Zeng, J.; Malem, F. Source tracing and chemical weathering implications of strontium in agricultural basin in Thailand during flood season: A combined hydrochemical approach and strontium isotope. *Environ. Res.* **2022**, *212*, 113330. [CrossRef] [PubMed]
25. Han, G.; Liu, C.-Q. Strontium isotope and major ion chemistry of the rainwaters from Guiyang, Guizhou Province, China. *Sci. Total Environ.* **2006**, *364*, 165–174. [CrossRef] [PubMed]
26. Qin, T.; Yang, P.; Groves, C.; Chen, F.; Xie, G.; Zhan, Z. Natural and anthropogenic factors affecting geochemistry of the Jialing and Yangtze Rivers in urban Chongqing, SW China. *Appl. Geochem.* **2018**, *98*, 448–458. [CrossRef]
27. Ghaemi, Z.; Noshadi, M. Surface water quality analysis using multivariate statistical techniques: A case study of Fars Province rivers, Iran. *Environ. Monit. Assess.* **2022**, *194*, 178. [CrossRef] [PubMed]
28. Wang, J.; Liu, G.; Liu, H.; Lam, P.K.S. Multivariate statistical evaluation of dissolved trace elements and a water quality assessment in the middle reaches of Huaihe River, Anhui, China. *Sci. Total Environ.* **2017**, *583*, 421–431. [CrossRef] [PubMed]
29. Zhan, S.; Wu, J.; Wang, J.; Jing, M. Distribution characteristics, sources identification and risk assessment of n-alkanes and heavy metals in surface sediments, Tajikistan, Central Asia. *Sci. Total Environ.* **2020**, *709*, 136278. [CrossRef]
30. Zhou, J.; Wu, Q.; Gao, S.; Zhang, X.; Wang, Z.; Wu, P.; Zeng, J. Coupled controls of the infiltration of rivers, urban activities and carbonate on trace elements in a karst groundwater system from Guiyang, Southwest China. *Ecotoxicol. Environ. Saf.* **2023**, *249*, 114424. [CrossRef]
31. Han, R.; Xu, Z. Riverine Hydrochemical Characteristics of a Typical Karst Urban Watershed: Major Ion Compositions, Sources, Assessment, and Historical Evolution. *ACS Earth Space Chem.* **2022**, *6*, 1495–1505. [CrossRef]
32. Zeng, J.; Yue, F.-J.; Wang, Z.-J.; Wu, Q.; Qin, C.-Q.; Li, S.-L. Quantifying depression trapping effect on rainwater chemical composition during the rainy season in karst agricultural area, southwestern China. *Atmos. Environ.* **2019**, *218*, 116998. [CrossRef]
33. Han, G.; Tang, Y.; Liu, M.; Van Zwieten, L.; Yang, X.; Yu, C.; Wang, H.; Song, Z. Carbon-nitrogen isotope coupling of soil organic matter in a karst region under land use change, Southwest China. *Agric. Ecosyst. Environ.* **2020**, *301*, 107027. [CrossRef]
34. Yue, F.-J.; Waldron, S.; Li, S.-L.; Wang, Z.-J.; Zeng, J.; Xu, S.; Zhang, Z.-C.; Oliver, D.M. Land use interacts with changes in catchment hydrology to generate chronic nitrate pollution in karst waters and strong seasonality in excess nitrate export. *Sci. Total Environ.* **2019**, *696*, 134062. [CrossRef]
35. Karunanidhi, D.; Subramani, T.; Roy, P.D.; Li, H. Impact of groundwater contamination on human health. *Environ. Geochem. Health* **2021**, *43*, 643–647. [CrossRef] [PubMed]
36. Fuoco, I.; Marini, L.; De Rosa, R.; Figoli, A.; Gabriele, B.; Apollaro, C. Use of reaction path modelling to investigate the evolution of water chemistry in shallow to deep crystalline aquifers with a special focus on fluoride. *Sci. Total Environ.* **2022**, *830*, 154566. [CrossRef] [PubMed]
37. Tian, Y.; Wang, S.; Bai, X.; Zhang, q.; Tao, j.; Zhang, Y.; Liang, M.; Zhou, G.; Lao, Y. Response og runoff of climate and human activities in Tongzi River Basin. *Res. Soil Water Conserv.* **2020**, *27*, 76–82.
38. Singh, V.; Singh, U. Assessment of groundwater quality of parts of Gwalior (India) for agricultural purposes. *Indian J. Sci. Technol.* **2008**, *1*, 1–5. [CrossRef]
39. Santacruz de León, G.; Ramos Leal, J.A.; Moran Ramírez, J.; López Álvarez, B.; Santacruz de León, E.E. Quality indices of groundwater for agricultural use in the Soconusco, Chiapas, Mexico. *Earth Sci. Res. J.* **2017**, *21*, 117–127. [CrossRef]
40. Xia, F.; Niu, X.; Qu, L.; Dahlgren, R.A.; Zhang, M. Integrated source-risk and uncertainty assessment for metals contamination in sediments of an urban river system in eastern China. *Catena* **2021**, *203*, 105277. [CrossRef]
41. Adimalla, N.; Li, P. Occurrence, health risks, and geochemical mechanisms of fluoride and nitrate in groundwater of the rock-dominant semi-arid region, Telangana State, India. *Hum. Ecol. Risk Assess. Int. J.* **2019**, *25*, 81–103. [CrossRef]

42. Adimalla, N.; Qian, H.; Nandan, M.J. Groundwater chemistry integrating the pollution index of groundwater and evaluation of potential human health risk: A case study from hard rock terrain of south India. *Ecotoxicol. Environ. Saf.* **2020**, *206*, 111217. [CrossRef] [PubMed]
43. Zhou, X.; An, Y.; Wu, Q.; Lv, J.; Gao, S.; Li, F. Response of Spatial Distribution of Nitrogen and Phosphorus to Human Activitues in Mountainous Rivers—A Case Study of Tongzi River, a first-grade Tributary fo Chishui River. *Res. Soil Water Conserv.* **2021**, *28*, 179–185.
44. Chen, J.; Wu, H.; Qian, H. Groundwater Nitrate Contamination and Associated Health Risk for the Rural Communities in an Agricultural Area of Ningxia, Northwest China. *Expo. Health* **2016**, *8*, 349–359. [CrossRef]
45. Qasemi, M.; Afsharnia, M.; Zarei, A.; Farhang, M.; Allahdadi, M. Non-carcinogenic risk assessment to human health due to intake of fluoride in the groundwater in rural areas of Gonabad and Bajestan, Iran: A case study. *Hum. Ecol. Risk Assess. Int. J.* **2019**, *25*, 1222–1233. [CrossRef]
46. Liu, J.; Peng, Y.; Li, C.; Gao, Z.; Chen, S. Characterization of the hydrochemistry of water resources of the Weibei Plain, Northern China, as well as an assessment of the risk of high groundwater nitrate levels to human health. *Environ. Pollut.* **2021**, *268*, 115947. [CrossRef] [PubMed]
47. Ge, X.; Wu, Q.; Wang, Z.; Gao, S.; Wang, T. Sulfur isotope and stoichiometry–based source identification of major ions and risk assessment in Chishui River Basin, Southwest China. *Water* **2021**, *13*, 1231. [CrossRef]
48. Das, A.; Krishnaswami, S.; Sarin, M.; Pande, K. Chemical weathering in the Krishna Basin and Western Ghats of the Deccan Traps, India: Rates of basalt weathering and their controls. *Geochim. Et Cosmochim. Acta* **2005**, *69*, 2067–2084. [CrossRef]
49. Huh, Y.; Tsoi, M.; Zaitsev, A.; Edmond, J. The fluvial geochemistry of the rivers of Eastern Siberia: I. Tributaries of the Lena River draining the sedimentary platform of the Siberian Craton. *Geochim. Et Cosmochim. Acta* **1998**, *62*, 1657–1676. [CrossRef]
50. Chetelat, B.; Liu, C.-Q.; Zhao, Z.; Wang, Q.; Li, S.; Li, J.; Wang, B. Geochemistry of the dissolved load of the Changjiang Basin rivers: Anthropogenic impacts and chemical weathering. *Geochim. Et Cosmochim. Acta* **2008**, *72*, 4254–4277. [CrossRef]
51. Ding, T.; Gao, J.; Tian, S.; Shi, G.; Chen, F.; Wang, C.; Luo, X.; Han, D. Chemical and isotopic characteristics of the water and suspended particulate materials in the Yangtze River and their geological and environmental implications. *Acta Geol. Sin. Engl. Ed.* **2014**, *88*, 276–360. [CrossRef]
52. Fuoco, I.; De Rosa, R.; Barca, D.; Figoli, A.; Gabriele, B.; Apollaro, C. Arsenic polluted waters: Application of geochemical modelling as a tool to understand the release and fate of the pollutant in crystalline aquifers. *J. Environ. Manag.* **2022**, *301*, 113796. [CrossRef] [PubMed]
53. Zongxing, L.; Qi, F.; Wei, L.; Tingting, W.; Aifang, C.; Yan, G.; Xiaoyan, G.; Yanhui, P.; Jianguo, L.; Rui, G. Study on the contribution of cryosphere to runoff in the cold alpine basin: A case study of Hulugou River Basin in the Qilian Mountains. *Glob. Planet. Chang.* **2014**, *122*, 345–361. [CrossRef]
54. Yidana, S.M.; Ophori, D.; Banoeng-Yakubo, B. A multivariate statistical analysis of surface water chemistry data—The Ankobra Basin, Ghana. *J. Environ. Manag.* **2008**, *86*, 80–87. [CrossRef] [PubMed]
55. Ustaoğlu, F.; Tepe, Y.; Taş, B. Assessment of stream quality and health risk in a subtropical Turkey river system: A combined approach using statistical analysis and water quality index. *Ecol. Indic.* **2020**, *113*, 105815. [CrossRef]
56. Burke, A.; Present, T.M.; Paris, G.; Rae, E.C.M.; Sandilands, B.H.; Gaillardet, J.; Peucker-Ehrenbrink, B.; Fischer, W.W.; McClelland, J.W.; Spencer, R.G.M.; et al. Sulfur isotopes in rivers: Insights into global weathering budgets, pyrite oxidation, and the modern sulfur cycle. *Earth Planet. Sci. Lett.* **2018**, *496*, 168–177. [CrossRef]
57. Zeng, J.; Han, G. Rainwater Chemistry Reveals Air Pollution in a Karst Forest: Temporal Variations, Source Apportionment, and Implications for the Forest. *Atmosphere* **2020**, *11*, 1315. [CrossRef]
58. Liu, J.; Wang, F.; Cai, W.; Wang, Z.; Li, C. Numerical investigation on the effects of geological parameters and layered subsurface on the thermal performance of medium-deep borehole heat exchanger. *Renew. Energy* **2020**, *149*, 384–399. [CrossRef]
59. Guo, F.; Jiang, G.; Yuan, D.; Polk, J.S. Evolution of major environmental geological problems in karst areas of Southwestern China. *Environ. Earth Sci.* **2013**, *69*, 2427–2435. [CrossRef]
60. Liu, Q.; Gu, Z.; Lu, Y.; Xiao, S.; Li, G. Weathering processes of the dolomite in Shibing (Guizhou) and formation of collapse and stone peaks. *Environ. Earth Sci.* **2015**, *74*, 1823–1831. [CrossRef]
61. Zeng, J.; Han, G.; Zhang, S.; Xiao, X.; Li, Y.; Gao, X.; Wang, D.; Qu, R. Rainwater chemical evolution driven by extreme rainfall in megacity: Implication for the urban air pollution source identification. *J. Clean. Prod.* **2022**, *372*, 133732. [CrossRef]
62. Zeng, J.; Han, G.; Zhang, S.; Qu, R. Nitrate dynamics and source identification of rainwater in Beijing during rainy season: Insight from dual isotopes and Bayesian model. *Sci. Total Environ.* **2023**, *856*, 159234. [CrossRef] [PubMed]
63. Grabb, K.C.; Ding, S.; Ning, X.; Liu, S.M.; Qian, B. Characterizing the impact of Three Gorges Dam on the Changjiang (Yangtze River): A story of nitrogen biogeochemical cycling through the lens of nitrogen stable isotopes. *Environ. Res.* **2021**, *195*, 110759. [CrossRef] [PubMed]
64. Xu, S.; Li, S.; Su, J.; Yue, F.; Zhong, J.; Chen, S. Oxidation of pyrite and reducing nitrogen fertilizer enhanced the carbon cycle by driving terrestrial chemical weathering. *Sci. Total Environ.* **2021**, *768*, 144343. [CrossRef] [PubMed]
65. Wang, B.; Lee, X.-Q.; Yuan, H.-L.; Zhou, H.; Cheng, H.-G.; Cheng, J.-Z.; Zhou, Z.-H.; Xing, Y.; Fang, B.; Zhang, L.-K.; et al. Distinct patterns of chemical weathering in the drainage basins of the Huanghe and Xijiang River, China: Evidence from chemical and Sr-isotopic compositions. *J. Asian Earth Sci.* **2012**, *59*, 219–230. [CrossRef]

66. Hui, T.; Jizhong, D.; Shimin, M.; Zhuang, K.; Yan, G. Application of water quality index and multivariate statistical analysis in the hydrogeochemical assessment of shallow groundwater in Hailun, northeast China. *Hum. Ecol. Risk Assess. Int. J.* **2021**, *27*, 651–667. [CrossRef]
67. Gibbs, R.J. Mechanisms controlling world water chemistry. *Science* **1970**, *170*, 1088–1090. [CrossRef] [PubMed]
68. Yang, R.; Sun, H.; Chen, B.; Yang, M.; Zeng, Q.; Zeng, C.; Huang, J.; Luo, H.; Lin, D. Temporal variations in riverine hydrochemistry and estimation of the carbon sink produced by coupled carbonate weathering with aquatic photosynthesis on land: An example from the Xijiang River, a large subtropical karst-dominated river in China. *Environ. Sci. Pollut. Res.* **2020**, *27*, 13142–13154. [CrossRef]
69. Han, Q.; Wang, B.; Liu, C.-Q.; Wang, F.; Peng, X.; Liu, X.-L. Carbon biogeochemical cycle is enhanced by damming in a karst river. *Sci. Total Environ.* **2018**, *616–617*, 1181–1189. [CrossRef]
70. Lü, J.; An, Y.; Wu, Q.; Zhou, S.; Wu, Y. Chemical characteristics and CO_2 consumption of the Qingshuijiang River Basin, Guizhou Province, Southwestern China. *Geochem. J.* **2018**, *52*, 441–456. [CrossRef]
71. Shen, B.; Wu, J.; Zhan, S.; Jin, M.; Saparov, A.S.; Abuduwaili, J. Spatial variations and controls on the hydrochemistry of surface waters across the Ili-Balkhash Basin, arid Central Asia. *J. Hydrol.* **2021**, *600*, 126565. [CrossRef]
72. Devic, G.; Djordjevic, D.; Sakan, S. Natural and anthropogenic factors affecting the groundwater quality in Serbia. *Sci. Total Environ.* **2014**, *468–469*, 933–942. [CrossRef] [PubMed]
73. Liu, J.; Peng, Y.; Li, C.; Gao, Z.; Chen, S. An investigation into the hydrochemistry, quality and risk to human health of groundwater in the central region of Shandong Province, North China. *J. Clean. Prod.* **2021**, *282*, 125416. [CrossRef]
74. Li, P.; Song, X.; Wang, J.; Zhou, X.; Li, J.; Lin, F.; Hu, Z.; Zhang, X.; Cui, H.; Wang, W.; et al. Reduced sensitivity to neutral feedback versus negative feedback in subjects with mild depression: Evidence from event-related potentials study. *Brain Cogn.* **2015**, *100*, 15–20. [CrossRef] [PubMed]
75. Liu, J.; Gao, Z.; Zhang, Y.; Sun, Z.; Sun, T.; Fan, H.; Wu, B.; Li, M.; Qian, L. Hydrochemical evaluation of groundwater quality and human health risk assessment of nitrate in the largest peninsula of China based on high-density sampling: A case study of Weifang. *J. Clean. Prod.* **2021**, *322*, 129164. [CrossRef]
76. Dun, Y.; Ling, J.; Wang, R.; Wei, J.; Zhou, Q.; Cao, Y.; Zhang, Y.; Xuan, Y. Hydrochemical Evolution and Nitrogen Behaviors in Coastal Groundwater Suffered From Seawater Intrusion and Anthropogenic Inputs. *Front. Mar. Sci.* **2022**, *9*, 945330. [CrossRef]
77. Xiao, J.; Jin, Z.D.; Wang, J.; Zhang, F. Hydrochemical characteristics, controlling factors and solute sources of groundwater within the Tarim River Basin in the extreme arid region, NW Tibetan Plateau. *Quat. Int.* **2015**, *380–381*, 237–246. [CrossRef]
78. Widory, D.; Petelet-Giraud, E.; Négrel, P.; Ladouche, B. Tracking the Sources of Nitrate in Groundwater Using Coupled Nitrogen and Boron Isotopes: A Synthesis. *Environ. Sci. Technol.* **2005**, *39*, 539–548. [CrossRef]
79. Fan, B.-L.; Zhao, Z.-Q.; Tao, F.-X.; Liu, B.-J.; Tao, Z.-H.; Gao, S.; Zhang, L.-H. Characteristics of carbonate, evaporite and silicate weathering in Huanghe River basin: A comparison among the upstream, midstream and downstream. *J. Asian Earth Sci.* **2014**, *96*, 17–26. [CrossRef]
80. Nijesh, P.; Akpataku, K.V.; Patel, A.; Rai, P.; Rai, S.P. Spatial variability of hydrochemical characteristics and appraisal of water quality in stressed phreatic aquifer of Upper Ganga Plain, Uttar Pradesh, India. *Environ. Earth Sci.* **2021**, *80*, 185. [CrossRef]
81. Liu, J.; Han, G. Controlling factors of riverine CO_2 partial pressure and CO_2 outgassing in a large karst river under base flow condition. *J. Hydrol.* **2021**, *593*, 125638. [CrossRef]
82. Zhang, Q.; Xu, P.; Qian, H.; Yang, F. Hydrogeochemistry and fluoride contamination in Jiaokou Irrigation District, Central China: Assessment based on multivariate statistical approach and human health risk. *Sci. Total Environ.* **2020**, *741*, 140460. [CrossRef] [PubMed]
83. Lundberg, J.O.; Weitzberg, E.; Cole, J.A.; Benjamin, N. Nitrate, bacteria and human health. *Nat. Rev. Microbiol.* **2004**, *2*, 593–602. [CrossRef] [PubMed]

water

Article

Pollutants Source Assessment and Load Calculation in Baiyangdian Lake Using Multi-Model Statistical Analysis

Guangwei Wang [1,2], Cuicui Lv [3], Congke Gu [4], Yang Yu [2], Zhenglun Yang [1], Zhixiong Zhang [4] and Changyuan Tang [1,*]

1 Graduate School of Horticulture, Chiba University, 648 Matsudo, Matsudo-City 271-8510, Japan
2 Guangzhou South Surveying & Mapping Technology Co., Ltd., South Geo-Information Industrial Park, No. 39 Si Cheng Rd, Guangzhou 510640, China
3 Xiong'an Institute of Innovation, Chinese Academy of Sciences, Xiong'an 071700, China
4 Key Laboratory of Agricultural Water Resources, Hebei Laboratory of Agricultural Water-Saving, Center for Agricultural Resources Research, Institute of Genetics and Developmental Biology, Chinese Academy of Sciences, Shijiazhuang 050021, China
* Correspondence: cytang@faculty.chiba-u.jp

Abstract: Baiyangdian lake, the largest fresh lake on the Haihe Basin in North China, has attracted wide attention on account of the distinguished ecological water bodies in Xiong'an New Area. Although remarkable achievements have been made in pollution control in Baiyangdian lake, the problem facing the overall water environment remains serious. The complex pollutant sources, drastic pollutant flux changes, and the unclear impact of the role of pollutants pose great challenges to the water quality, water environment management, and long-term restoration of the ecological environment. Here, the potential pollution sources, their contribution ratio, and the influence of the pollution load of different sources on the water quality in the priority source areas of Baiyangdian lake are discussed in detail based on collected and existing monitoring data. It is proven that the primary pollution sources of for nitrogen, phosphorus, and organic pollutants are from agricultural and rural non-point source pollution; the load contribution rates exceed 50%, of which the contribution rate to the total phosphorus load reaches 73.37%. The total load contribution of runoff to the three pollutants was small, although the contribution of soil erosion to total nitrogen was 22.95%. The contribution of point source pollution to COD was high, with a rate of 22.33%. In order to ensure the environmental quality of Baiyangdian lake, it is obligatory to strengthen the control of agricultural and rural pollution discharge and to standardize the pollution discharge of livestock and poultry breeding. This study provides a helpful support for protecting the water ecology of the national Xiong'an New Area.

Keywords: Baiyangdian lake; pollutants source; load estimation; export coefficient model; PLOAD model

Citation: Wang, G.; Lv, C.; Gu, C.; Yu, Y.; Yang, Z.; Zhang, Z.; Tang, C. Pollutants Source Assessment and Load Calculation in Baiyangdian Lake Using Multi-Model Statistical Analysis. *Water* **2022**, *14*, 3386. https://doi.org/10.3390/w14213386

Academic Editors: Guilin Han and Xiaolong Liu

Received: 23 September 2022
Accepted: 21 October 2022
Published: 26 October 2022

1. Introduction

With the rapid development of social economy, the excessive enrichment of nutrient salts such as nitrogen, phosphorus, and organic pollutants can cause serious eutrophication problems for many large lakes [1,2]. Abundant nutrients affect the growth of algae in the water, which in turn leads to the death of fish and plankton, cyanobacterial bloom outbreaks, and biodiversity decline, causing water pollution and seriously threatening the health of aquatic ecosystems [3–6]. Therefore, it is of great significance to identify the distribution and sources of pollution in critical lake regions. However, due to the spatial extensiveness and temporal uncertainty of non-point source pollution, it is hard to obtain accurate information on pollutants in order to control and manage the aquatic environment of large lakes [7].

Baiyangdian lake, located in the middle of Haihe basin, is a typical shallow macrophytic lake in North China that was chosen as the core scope of the new-built Xiong'an

New Area. Baiyangdian lake has irreplaceable ecological functions in regulating the climate of North China, replenishing surrounding groundwater, and maintaining the ecology and biodiversity of Xiong'an New Area [8]. However, the location of its slow drainage area and human activities have led to degrading environment and vulnerable ecology around Baiyangdian lake, which seriously restricts new ecological city construction and coordinated economic development [9]. It has been proven that 22 antibiotics, including eight quinolones, nine sulfonamides, and five macrolides, were present and distributed in water, sediments, and biota samples from Baiyangdian lake, which is largely due to anthropogenic activities [10]. Yang et al., (2020) found that 30 PPCPs were frequently detected in 2019, of which sulfamethoxazole, ofloxacin, anhydro-erythromycin, carbamazepine, and caffeine were appraised as the dominant PPCPs [11]. The existence of non-point source pollution led the dissolved oxygen concentration to decrease by more than 1 mg/L [12]. Yang et al., (2022) estimated the ecological thresholds for Baiyangdian lake by applying the PCLake + model, which showed that nitrogen and phosphorus loads exceeded the threshold in 2012 and that the ecological thresholds would be exceeded in 2050 with limited environmental policies on human waste [13].

Clarifying the source of pollution and the contribution ratio is the basis of water environment treatment, remediation, and management, while the pollutant sources of Baiyangdian lake are sophisticated and mutative [14]. Baiyangdian lake flows through densely populated and industrial areas, and there are more than 200 key sewage wastewater treatment plants with great impact on surface water quality, among which many sewage enterprises are distributed in Mancheng, Gaoyang and Lixian counties. For historical reasons, Baiyangdian lake has been a gathering area of upstream sewage discharge for a long time, and the water sediment has accumulated a large number of historical pollutants which have been released to the water body for a long time through the boundary of water and sediment [9]. Non-point source pollution is the main source of pollutants in the surface water of Baiyangdian lake basin [4]. The non-point source pollution in the basin is essential to calculation of the surface water pollution load input into Baiyangdian lake.

Non-point source pollution model can simulate the migration and transformation process of pollutants from the source to the water body under different scenarios and different driving forces and can predict the non-point source pollutant load and its changes [15]. According to whether the formation process of the non-point source pollution is considered in the model build, non-point source pollution models can be categorized as mechanical models and empirically-based lumped models [16–18]. The mechanism models, such as SWAT [16,19], AnnAGNPS [20], and SPARROW [21], use mathematical methods to simulate the migration and formation of pollutants from non-point source pollution [22]. Such models, however, require a large amount of data to verify the model adaptation in the current study area and to calibrate and validate the parameter, which restricts their promotion [23]. The lumped models are known as "black box" models. These models construct the empirical relationship between the non-point source pollution load and influencing factors through regression analysis [24]. The lumped models use multiple lumped parameters which use monitoring data on hydrology, water quality, rainfall, etc., to describe the nutrient dynamics of the basin. Although the lumped models ignore the complex process of non-point source pollution migration and transformation, their advantages lie in simple operation, low data requirements, and strong robustness [25]. Among the lumped models of non-point source pollution, the Export Coefficient Model (ECM) and Revised Universal Soil Loss Equation (RUSLE) are widely used around the world because of their clear physical meaning and concise expression. They are simple methods for quantitatively identifying non-point source pollution load at the basin scale, especially for regions with significant spatial heterogeneity [15,24]. Lumped models are thus able to illustrate the non-point source pollution load in the Baiyangdian lake basin, which is characterized by intense human activities, complex land use, and geomorphic diversity. Existing studies have mainly focused on the analysis of nitrogen and phosphorus pollution in local areas, and little research has been carried out on the estimation and source analysis of nitrogen and phosphorus loads in the entire Baiyangdian lake basin.

In order to determine the influence of human activities and the migration of pollutants caused by natural environmental changes, we use lumped model statistical analysis methods to analyze the potential pollution sources, pollution load, and the proportion of pollution contribution from Baiyangdian lake based on the collected monitoring data, aiming to quantify the impact of pollution sources and load on the water quality of Baiyangdian lake. To improves the accuracy and reliability of the overall calculation, the RUSLE and PLOAD models are used to calculate the pollution load of surface runoff and ECM is used to estimate the agricultural pollution load alone. The LOADEST model is used to calculate the pollutant flux into the lake and compare it with the non-point source load model results to verify the accuracy of the overall pollution load calculation.

2. Study Area and Methods

2.1. Study Area and Basin Division

Geohydrology: Baiyangdian Lake Basin is located in the middle of the Haihe River Basin (113°40′~116°48′ E, 38°10′~40°03′ N), and has an area of about 31200 km^2. The Baiyangdian lake basin belongs to the Daqing River system and is surrounded by Yongding River on its northern part, the Taihang Mountains on its western part, the Ziya River Basin on its southern part, and Bohai Bay to the east. Baiyangdian lake is located in the catchment area of the lower reaches of the basin, with a total area of about 366 km^2 (39.5 km long from east to west and 28.5 km wide from north to south, and with an average annual water resource of 41.24 million cubic meters. There are nine rivers that flow into Baiyangdian lake: Lilong River, Xiaoyi River, Tang River, Fu River, Cao River, Bao River, Qingshui River, Ping River, and Baigouyin River. The lake is known in history as the "End of the Nine Rivers". The daily flow data of key sections of the rivers to Baiyangdian show that only Fu River and Xiaoyi River had long-term flow; Baigouyin River and Bao River are seasonal rivers which only flow into the lake during the rainy season and water supplement period, while the rest of the rivers have been cut off or diverted for more than two years.

Climate and Meteorology: Baiyangdian lake basin is situated in a continental monsoon climate zone. The average annual temperature is between 7~12 °C, the annual average rainfall is 554 mm, and the average annual evaporation is 1637 mm. The annual precipitation is unevenly distributed, with 89.7% concentrated between May and October.

Soil and vegetation: There are 14 soil types and 31 subtypes of soil in Baiyangdian Basin, mainly including brown soil, chestnut brown soil, calcareous stony soil, coarse bone soil, aquiferous soil and paddy soil, etc., as shown in Figure S1. The main land use types are paddy fields, drylands, woodlands, grasslands, orchards, and wetlands. Common woody plants in vegetation include poplars, willows, acacia, apples, pears, etc. Shrubs include purple locust, ceremony, mulberry, etc. Grasses consist mainly of poaceae, asteraceae, and legumes. Aquatic vegetation mainly includes phytoplankton and aquatic vascular plants such as algae, lotus root, reeds, and barnyard.

In this research, the scope of data collection covers the whole Baiyangdian lake Basin, shown in Figure 1. Because not all the rivers within the basin flow into Baiyangdian lake, we screened the four rivers (Fu River, Xiaoyi River, Baigouyin River, and Bao River) that had flow into Baiyangdian lake in 2020 as the sub-basin (shown in Figure 2) through the ArcSWAT model while analyzing the pollutants load and source in Baiyangdian lake.

Figure 1. The research area of Baiyangdian lake basin.

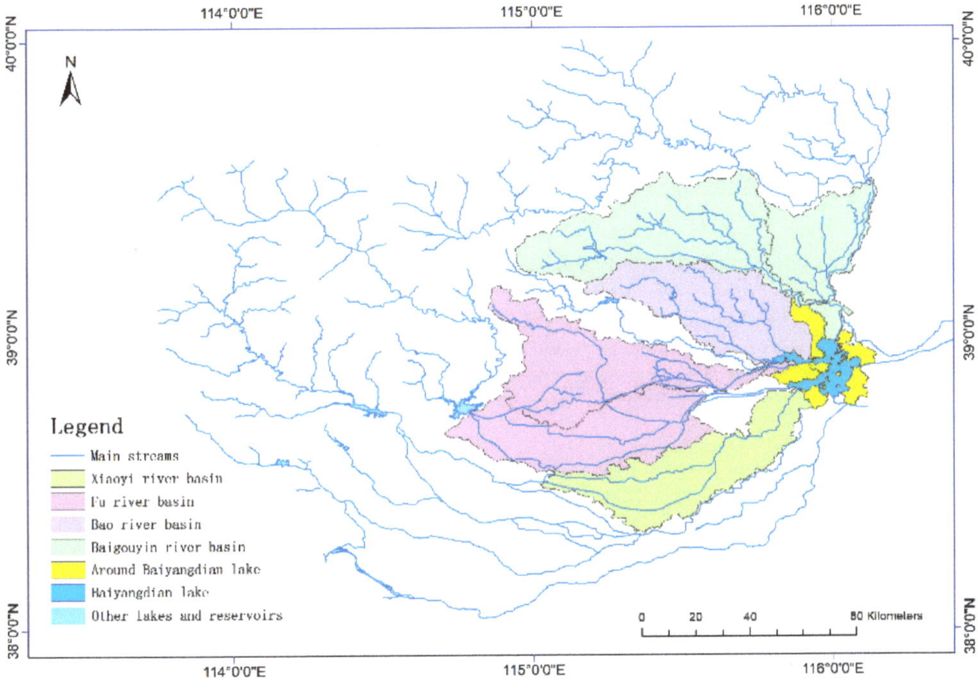

Figure 2. The sub-basin division of Baiyangdian lake basin.

2.2. Data Collection and Information

We collected more than 30 GB of data from social and economic statistical yearbooks, government affairs information, internal statistical reports, and the related literature to screen the statistical data needed to calculate source analysis. The names and sources of the data are shown in Table S1.

Because the data for 2020 were relatively complete, our study used the whole year of 2020 as the research period to explore the load of total nitrogen, total phosphorus, and chemical oxygen demand (COD) flowing into Baiyangdian lake through rivers in the Baiyangdian lake basin.

2.3. Export Coefficient Modelling (ECM)

The Export Coefficient Modelling (ECM) approach has been broadly recommended based on its convenient theory foundations, notably, that the sum pollutants of drainage basin cause non-point source pollution load [26,27]. In this model approach, livestock and poultry breeding and rural life factors such as livestock quantity and distribution status, land use type, and rural residents are considered to obtain differentiated output coefficients in order to analyze the agricultural non-point source pollution load and source analysis. Based on the ECM approach, the non-point source pollution load of rural life, livestock and poultry breeding, and agricultural planting in 2020 in Baiyangdian lake basin was calculated and the final non-point source load of agriculture and rural areas into Baiyangdian lake was corrected according to the relationship between the water quantity data of the hydrological station in the upper reaches of the Baiyangdian lake basin and the downstream water quantity data.

The ECM equation is

$$L_j = \sum_{i=1}^{m} E_{ij} \times A_i \tag{1}$$

where L_j is the average export coefficient of pollutants, kg/(km^2 a); i is the land use type with a total of m species; E_{ij} is the export coefficient of pollutants j in land use type i (kg/km^2), the excretion coefficient of each livestock and poultry of type i (kg/a), or the export coefficient per person (kg/a); and A_i is the area of land use type i in the river basin (km^2), the livestock quantity, or the population quantity. The acquisition of all these data were calculated and corrected according to the relevant data in Table S1 in the Supplementary Materials.

2.4. Revised Universal Soil Loss Equation (RUSLE)

The Revised Universal Soil Loss Equation (RUSLE) is used for loading of adsorbed contaminants such as nitrogen, phosphorus, and COD released into water bodies through soil erosion [28,29]. RUSLE is an empirical soil erosion model designed on the basis of the common soil erosion equation, as shown in Equation (2). In this model, watershed characteristics and hydrometeorological conditions can be used to predict soil erosion rates, allowing the spatial heterogeneity of soil erosion processes to be obtained.

$$W_{ads} = r_{ads} \times \left(\sum_{u=1}^{U} X_u \times A_u \times C_s \times t \right) \times 10^{-3} \tag{2}$$

$$X_u = R \times K_u \times LS \times S_u \times M_u \tag{3}$$

where W_{ads} is the adsorption pollution load; t/a; r_{ads} is the sediment transport ratio; u is the land use type; X_u is the annual soil erosion modulus, indicating the average soil loss caused by erosion, which is calculated by Equation (3), in t/hm^2; A_u is the area of land use type u, km^2; C_s is the background concentration of pollutants in the soil, g/kg; t is the pollutant enrichment rate; R is the driving factor for rainfall erosion; K_u is the soil erosion factor; LS is the topographic factor; S_u is the soil and water conservation factor; and M_u is the vegetation cover and management factor.

The driving factor R for rainfall erosion in Equation (3) quantifies the impact of rainfall on the soil and reflects the probability of erosion associated with precipitation events; R can be estimated by the following formula [30].

$$R = \sum_{i=1}^{12} 1.735 \times 10^{1.5\log \frac{P_i^2}{P_u} - 0.8188} \tag{4}$$

where P_u is the average annual precipitation and P_i is the average monthly precipitation.

The soil erosion factor K_u in Equation (3) is the rate of soil erosion measured on standard plots per unit rainwater erosion index, usually calculated by the soil characteristic attribute properties:

$$\begin{gathered} K = \left\{0.2 + 0.3exp\left[-0.0256S_d\left(1 - \frac{S_i}{100}\right)\right]\right\} \times \\ \left[\frac{S_i}{C_i + S_i}\right]^{0.3} \times \left\{1.0 - \frac{0.25C_0}{C_0 + exp(3.72 - 2.95C_0)}\right\} \times \\ \left[1.0 - 0.7\left(1 - \frac{S_d}{100}\right)\right] / \left\{1 - \frac{S_d}{100} + exp\left[-5.51 + 22.9\left(1 - \frac{S_d}{100}\right)\right]\right\} \end{gathered} \tag{5}$$

where S_d is the sand content%, S_i is the silt content%, C_i is the clay content%, and C_0 is the organic carbon content%.

The topographic factor LS in Equation (3) reflects the effect of terrain on soil erosion. Among them, the slope length factor L represents the effect of slope length on erosion and the slope factor S reflects the influence of slope on erosion. LS can be calculated by the following equation:

$$LS = \left(\frac{Q_A \times L_c}{22.13}\right)^{0.4} \times \left(\frac{\sin\theta}{0.0896}\right)^{1.3} \tag{6}$$

where Q_A is the flow accumulation, L_c is the cell size, and θ is the slope.

The vegetation cover and management factor M_u in equation (3) is used to reflect the influence of vegetation and human management behavior on the soil erosion rate, including crop planting and farmland management in agricultural areas and the inhibition of soil erosion by vegetation cover in forest areas. The value of this factor changes with the change of plant growth caused by seasonal factors. M_u can be estimated by

$$M = \begin{cases} 1 & m = 0 \\ 0.658 - 0.3436\lg m & 0 < m < 78.3\% \\ 0 & m > 78.3\% \end{cases} \tag{7}$$

where m is the average annual vegetation cover.

According to the RUSLE model, the parameters of the study area were calculated and a series of raster data were obtained, as shown in Figure 3. Other sources used for parameters are recorded in Table S2.

2.5. PLOAD Model

The PLOAD model is a simplified GIS-based model for calculating non-point source pollution loads for various sub-basins and land uses based on annual precipitation, land use, and best watershed management measures [31,32]. In this research, the PLOAD model was adopted to calculate the pollution load of dissolved pollutants generated by rainfall. The equation is as follows:

$$W_{RJ} = \lambda_{RJ} \times \sum_{u=1}^{U} \left(P \times P_J \times R_{Vu} \times A_u\right) \times C_u \times 10^{-5} \tag{8}$$

where W_{RJ} is the load of dissolved pollutants, t/a; λ_{RJ} is the river entry coefficient; P is the rainfall, mm/a; P_J is the runoff coefficient correction; R_{Vu} is the average runoff coefficient of the land use type u; A_u is the area of land use type u, hm^2; and C_u is the average output concentration of pollutants in rainfall event under the land use type u, mg/L.

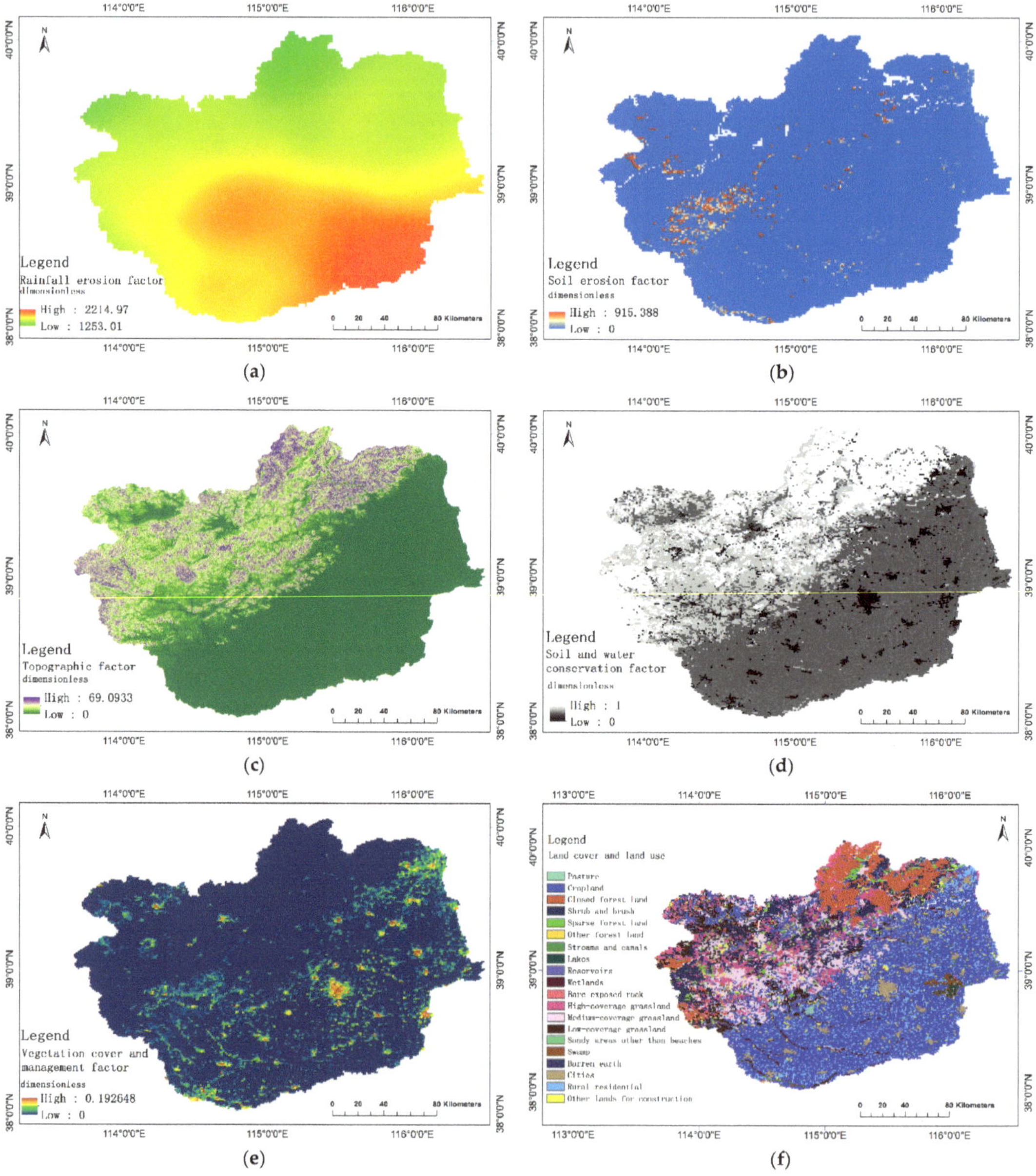

Figure 3. Calculation results of parameters in RUSLE model in the study area: (**a**) Rainfall erosion, (**b**) Soil erosion factor, (**c**) Topographic factor, (**d**) Soil and water conservation factor, (**e**) Vegetation cover factor, (**f**) Land use type.

The runoff coefficient was calculated using the Runoff Curve Number (SCS-CN) model [33]. This model can reflect the effects of different soil types, different land use patterns, and previous soil moisture content on rainfall runoff. The basic equations of the model are as follows:

$$\begin{cases} Q = \dfrac{(P-0.2S)^2}{P+0.8S}, & P \geq 0.2S \\ Q = 0, & P < 0.2S \end{cases} \tag{9}$$

$$S = \frac{25400}{CN} - 254 \tag{10}$$

where Q is the ground runoff, mm; P is the rainfall of a rainfall event, mm; S is the potential water storage capacity of the surface, mm; CN is the comprehensive parameter of the basin characteristics, which is related to soil characteristics, pre-humidity and land use. The final runoff coefficient is the ratio of Q to P. The rainfall flow coefficient correlation parameters in the study area are shown in Figure 4. The pollution concentration, river entry coefficient, and runoff correction factors were adopted according to Li et al., (2021), Li (2007), and Cui (2011) [4,34,35].

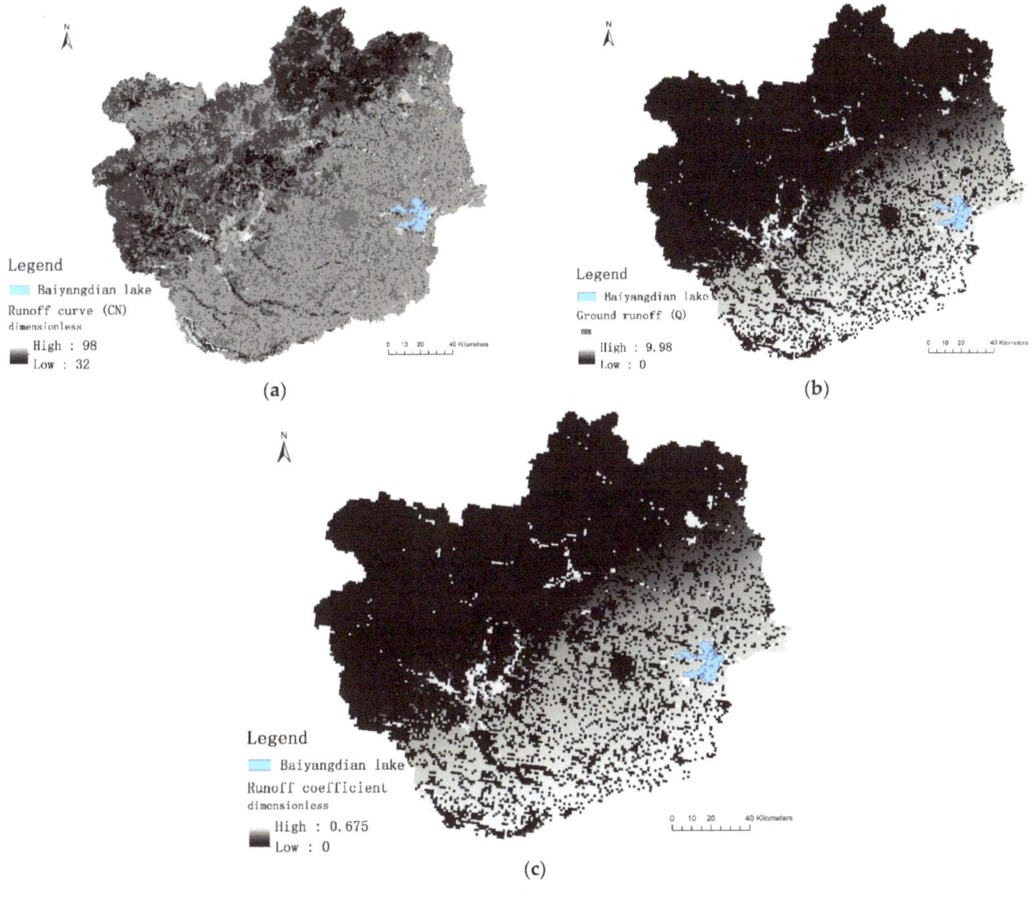

Figure 4. Coefficient parameters of rainfall in the study area: (**a**) Runoff curves (CN), (**b**) Ground runoff (Q), (**c**) Runoff coefficient.

2.6. Point Source Load

Point source pollution includes sewage from industrial enterprises and sewage treatment plants. Industrial wastewater pollution load is calculated by the following Equation (11):

$$W_P = \lambda_P \times \left(\sum_{k=1}^{K} \sum_{n=1}^{N} \left(Q_{IW,k,n} \times C_{IW,k,n} \times 10^{-6} \right) + p_{LW} \times P_{LW} \times (1 - f_{LW}) \times 10^{-3} \right) \quad (11)$$

where W_P is the point source pollution load, t/a; λ_P is the point source pollution into the river coefficient; K and N are the number of industrial pollution source industry types and the number of industrial enterprises, respectively; $Q_{IW,k,n}$ is the sewage discharge of enterprise n in industry k, m^3; $C_{IW,k,n}$ is the corresponding average emission concentration of pollutants, mg/L; p_{LW} is the yield factor of per capita pollution, kg/(per person·a); P_{LW} is the catchment area population; and f_{LW} is the sewage treatment rate. The corresponding collected data sources are shown in Table S1 in the Supplementary Materials.

2.7. LOADEST Model

LOADEST is a useful pollutant regression model based on various water quality data [36,37]. Three methods can be used to estimate parameter valuation, including (1) Minimum likelihood estimation, MLE [38]; (2) Adjusted maximum likelihood estimation, AMLE [39]; and (3) Least absolute deviation, LAD [40]. Different valuation methods are used on the basis of whether the residuals are normally distributed or whether there are censored data.

Factors such as the characteristics of pollution sources and the proportion of point sources and non-point sources carry an impact on the long-term flux calculation. When the pollutants are mainly derived from the point source, the change in runoff during the year does not change the pollutant flux of the cross-section; rather, the instantaneous concentration of the pollutants changes negatively with the flow rate. When the pollutants are mainly derived from non-point sources, such as soil erosion or endogenous release, the generation of pollutants is proportional to the runoff. When the contaminants originate from a mix of point and non-point sources, the relationship between runoff and pollutant flux is complicated, and positive correlation, negative correlation, and basically uncorrelated situations may occur.

For different cases, the general flux can be calculated using the following three equations:

$$W_1 = K \sum_{i=1}^{n} \frac{C_i Q_i}{n} \quad (12)$$

$$W_2 = K \sum_{i=1}^{n} C_i \overline{Q_P} \quad (13)$$

$$W_3 = K \frac{\sum_{i=1}^{n} C_i Q_i}{\sum_{i=1}^{n} Q_i} \overline{Q_r} \quad (14)$$

where K is the time conversion coefficient calculated according to the monitoring time frequency and flux period, C_i is the instantaneous monitoring concentration of pollutants, Q_i is the instantaneous flow of water, $\overline{Q_r}$ represents the mean data over the desired period, $\overline{Q_P}$ indicates the mean flow data of C_i, and n represents the amount of data.

Equation (12) weakens the time change of runoff, Equation (13) emphasizes the role of the corresponding runoff in each pollutant concentration measurement, and Equation (14) emphasizes the average time change of the total runoff and weakens the sensitivity to runoff change. Different formulas should be selected for the flux calculation for different situations.

3. Results

3.1. Non-Point Source Load from Agriculture and Rural Areas

Generally speaking, non-point sources from agriculture and rural areas include rural life, livestock and poultry farming, and agricultural planting. There are dense population and intensive life and production activities surrounding the Baiyangdian lake district, with a permanent population of about 240,000 in 2017 and an arable land area of about 12,000 hectares in 2018, which is the largest type of land use in the area [41,42]. In this study, the pollutant sources from agriculture and rural areas were first calculated.

Non-point source pollution generated by rural life is mainly derived from domestic garbage, domestic water, and human excrement [6]. The similar living habits of residents mean that similar contents of nitrogen and phosphorus elements are commonly ingested from food; thus, the rural population is the dominant factor that determine the intensity of non-point source pollution in rural life. For areas with dense rural populations, the corresponding pollution intensity is much higher. The non-point source pollution load of rural life calculated based on rural life information is shown in Figure S2.

We obtained the non-point source load of livestock and poultry farming according to the number of large livestock, pigs, sheep, and poultry in various districts and counties of Baiyangdian lake basin, in which the feeding method, fecal and urine treatment method, and proportion of breeding facility were considered in the calculation, as shown in Figure S3. Livestock and poultry are important sources of non-point source pollution in agriculture which emit a large amount of nitrogen, phosphorus, and organic pollutants into the environment through feces and urine during growth [4]. The environmental pollution problems caused by livestock and poultry manure and urine exist in most of North China.

Non-point source pollution from agricultural planting is a major component of non-point source pollution in agriculture and rural areas. The excessive use of chemical fertilizers and pesticides causes a large number of agricultural chemicals to remain on the surface of cultivated land, and the surface runoff generated by rainfall enters the natural water body, leading to non-point source pollution [43]. Planting and crop management methods were dominant determinants of the pollutant load from planting non-point sources. The non-point source pollution intensity of agricultural planting in each district and county of Baiyangdian lake basin was calculated through land use data to obtain the planting area of grain, vegetables, orchards, and other categories based on the correction to obtain the output coefficient, as shown in Figure S4.

The total load of non-point source pollution in rural areas obtained by the three types of rural life, livestock and poultry farming, and agricultural planting is shown in Figure 5.

Overall, Dingzhou City, Tang County and Yi County were the three areas with the highest non-point source loads in agriculture and rural areas in Baiyangdian River Basin, which accounted 19.6% for nitrogen, 18.5% for phosphorus and 26.9% for COD of the total agricultural and rural non-point source load. According to the Baoding Economy Statistical Yearbook (2020), the total area of crop planting in Dingzhou City in 2019 was as high as 159,000 hectares, of which 117,000 hectares were planted with grain and 18,000 hectares were planted with vegetables. In 2019, 6.53 million poultry were output, including 870,000 head of pigs. In addition, the rural population of Dingzhou City ranked fourth among all the cities and counties of Baiyangdian river basin, leading to the highest pollution output. Similarly, Tang County and Yi County had large-scale farming and planting activities, resulting in high non-point pollution sources.

However, Baiyangdian lake basin included the catchment area of all rivers naturally exported into Baiyangdian lake. In fact, many of the rivers were cut off or artificially diverted and no longer transmit water to Baiyangdian lake, meaning that the pollution load generated by the whole basin cannot be equated with the sediment load. Thus, we divided Baiyangdian lake basin into sub-basins based on the four rivers (Fu River, Xiaoyi River, Baigouyin River, Bao River) that transport water into the lake [44]. The corresponding potential sedimentation of non-point source pollution in agriculture and rural areas was rectified, and the resulting total nitrogen load, total phosphorus load, and total COD load were 1073.82 tons, 102.72, tons and 3349.33 tons, respectively (see Table 1). Among them, the contribution rate of livestock and poultry breeding to COD exceeds 70%, the contribution rate of agricultural planting to total phosphorus reaches 73.12%, and the contribution rate of the three types of non-point sources to total nitrogen is basically equal, of which the contribution rate of rural life is slightly higher at 38.48%.

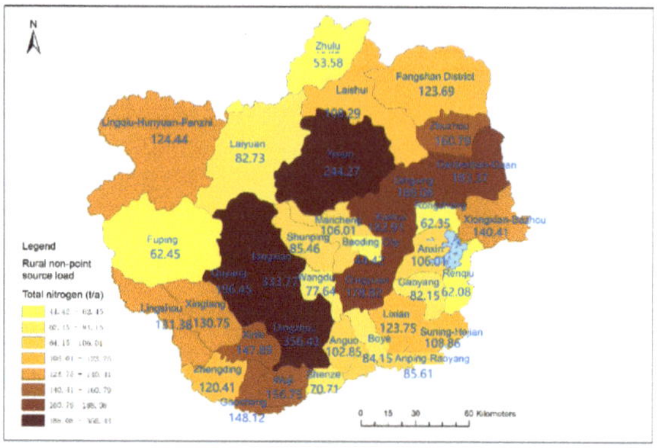

(**a**)

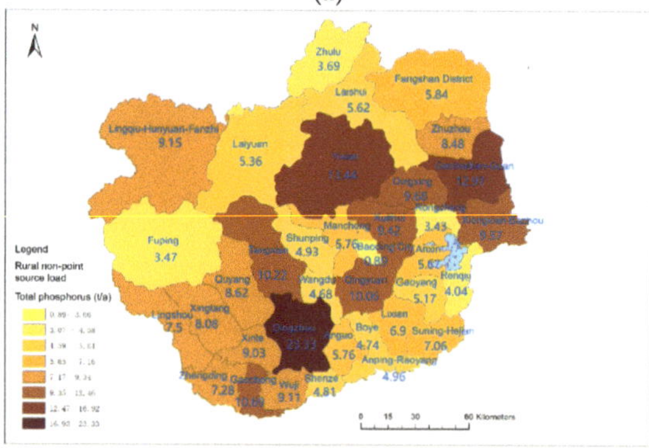

(**b**)

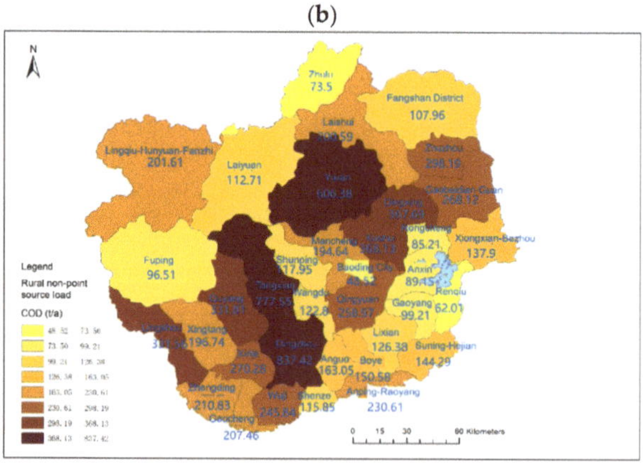

(**c**)

Figure 5. The total load of non-point source pollution in agriculture and rural areas in Baiyangdian lake: (**a**) Total nitrogen load, (**b**) Total phosphorus load, (**c**) COD load.

Table 1. Non-point source loads in agricultural and rural areas.

Pollutant	Amount (t)	Percentage(%)		
		Rural Life	Livestock and Poultry Breeding	Agricultural Cultivation
Total nitrogen	1073.82	38.48	29.59	31.93
Total phosphorus	102.72	7.93	18.95	73.12
COD	3349.33	18.15	70.16	11.69

It can be seen that livestock and poultry breeding have a significant impact on COD pollutants in downstream water bodies; thus, efficient treatment of aquaculture pollution, especially emphasizing the management of the small-scale decentralized breeding, should be considered to control the release of organic pollutants into Baiyangdian lake. In addition, the application of excessive agrochemicals in agricultural planting causes a large amount of nutrients to remain on the farmland soil surface, which flow into natural water bodies by rainfall, resulting in eutrophication. Therefore, adopting scientific fertilization methods and intensive farmland management methods are advisable measures to reduce non-point source pollution in agricultural planting.

3.2. *Pollutant Loads from Surface Runoff Erosion*

Precipitation causes surface water-soluble pollutants to migrate to water bodies and causes soil particles to enter the water body by eroding the topsoil [28]. Pollutants such as nitrogen, phosphorus, and organic matter adsorbed on the surface of soil particles have become important sources of non-point source pollution in water bodies. Due to long-term cultivation, the topsoil of farmland in the Baiyangdian lake basin is usually polluted by adsorbed organic matter and inorganic nitrogen, phosphorus. Outdoor farmland soil without mulching is easily washed away by precipitation, causing topsoil erosion. Therefore, the pollution load caused by soil erosion should be considered when calculating the non-point source pollution caused by rainfall in the Baiyangdian lake basin.

3.2.1. Adsorbed Pollutant Loads

In order to distinguish between the dissolved pollutants from rainfall runoff and the particulate pollutants carried by soil erosion, the total nitrogen and phosphorus content of the topsoil in the soil database was used to subtract the adsorbed nitrogen and adsorbed phosphorus content that was easily soluble in water, and the difference obtained was used as the particle pollutant content. For the soil COD pollutant background value, the soil organic matter content was used directly as the particle contaminant content.

On the basis of the calculation of each parameter by Equation (2), the adsorption pollutant load generated by soil erosion in Baiyangdian lake is shown in Figure 6.

It can be seen that the non-point source pollution load caused by soil erosion in the whole basin is generally low. The high load areas are concentrated in the Taihang Mountains, especially the upper reaches of Lilong River and Tang River. The difference between the loads of the three pollutants is mainly due to their different contents in the surface soil, generally in the order of total nitrogen > total phosphorus > soil organic matter. The non-point source pollution load from soil erosion was rectified based on the sub-basins of the four main rivers (Fu River, Xiaoyi River, Baigouyin River, Bao River), showing that the total adsorption load generated by soil erosion was 94.1 tons for nitrogen, 20.8 tons for phosphorus, and 1262.9 tons for COD, as shown in Table 2.

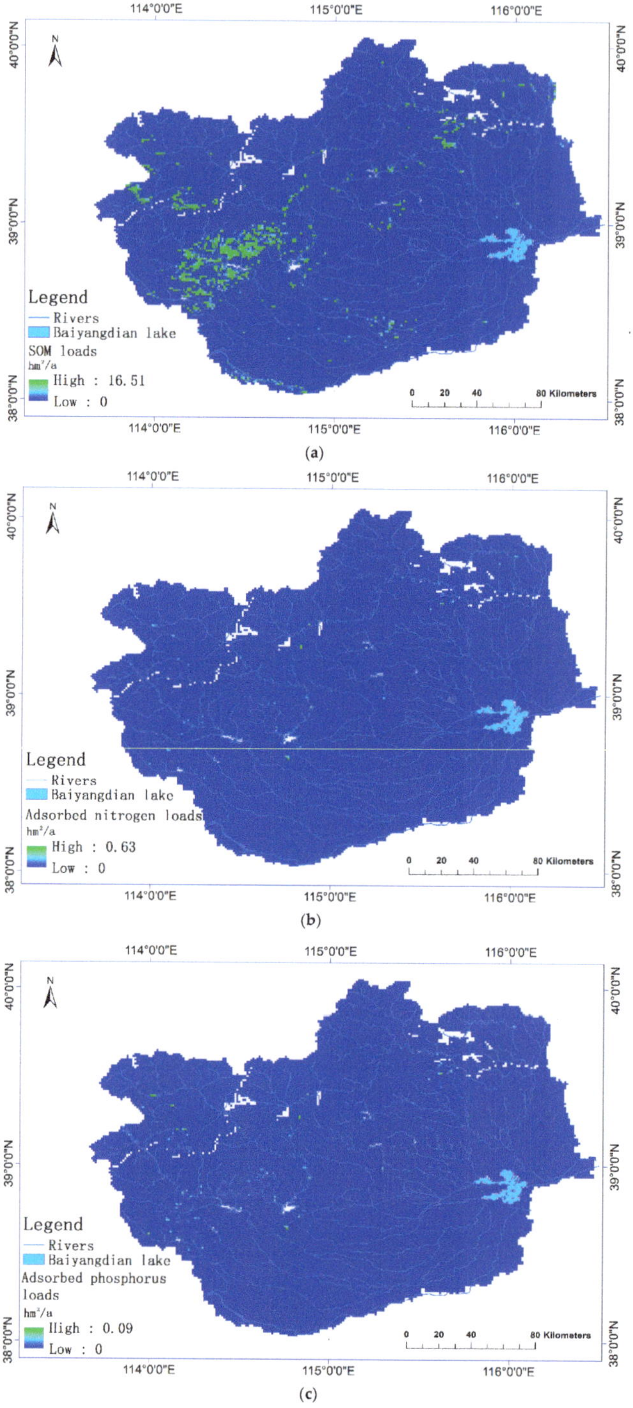

Figure 6. Adsorption pollutant loads from soil erosion in Baiyangdian lake: ((**a**) Soil organic matter, (**b**) Adsorbed nitrogen, (**c**) Adsorbed phosphorus.

Table 2. The load of adsorbed pollutants generated by soil erosion.

River into Baiyangdian Lake	Load of Adsorbed Pollutants Generated by Soil Erosion (t/a)		
	COD	**Total Nitrogen**	**Total Phosphorus**
Bao River	19.9	1.0	0.6
Fu River	585.9	27.5	15.3
Baigouyin River	495.4	56.4	2.6
Xiaoyi River	161.7	9.2	2.2
Total	1262.9	94.1	20.8

3.2.2. Dissolved Pollution Loads

Surface runoff is the driving force of most non-point source pollution, and the differential generation and migration of non-point source loads in agriculture and rural areas are caused by rainfall. Thus, the non-point source loads in agriculture and rural areas calculated using ECM model are essentially dissolved pollutant pollution loads caused by surface runoff. Here, the calculation of the dissolved pollutant load uses the pollution load generated by other land uses with surface pollution by rainfall, with the exception of land for agriculture planting. Combined with the average yield of rainfall and the concentration of pollutants in different regional production streams, the distribution of the dissolved pollutant load caused by rain erosion in the unit area of land use excepting agricultural planting land was finally calculated; see Figure 7.

The non-point source pollution load of the rainfall in the four inflow river basins was obtained (see Table 3) according to the sub-basin division in Figure 2. It was proven that the pollution load of rainfall showed less impact on the pollutant load than that from agriculture and rural areas.

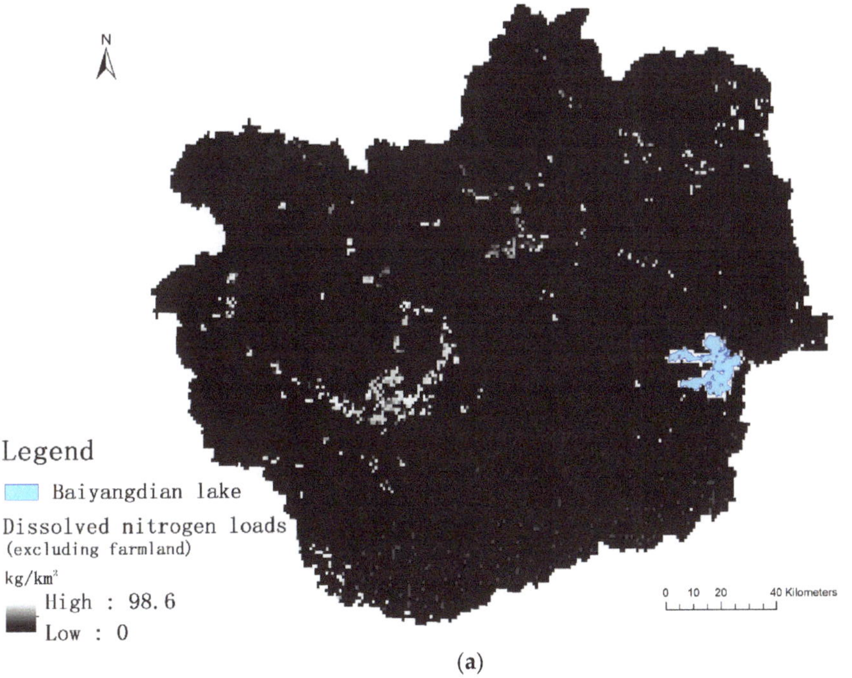

(a)

Figure 7. *Cont.*

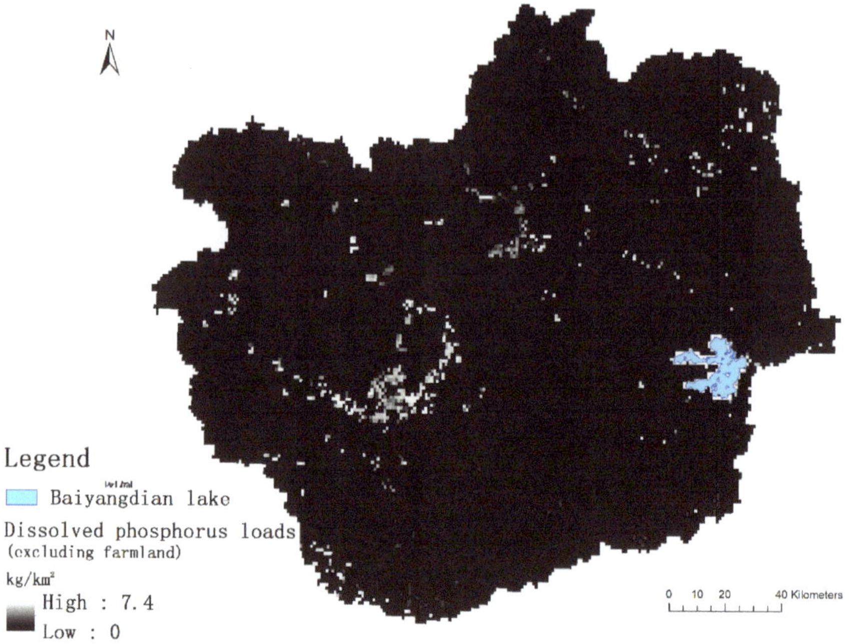

(b)

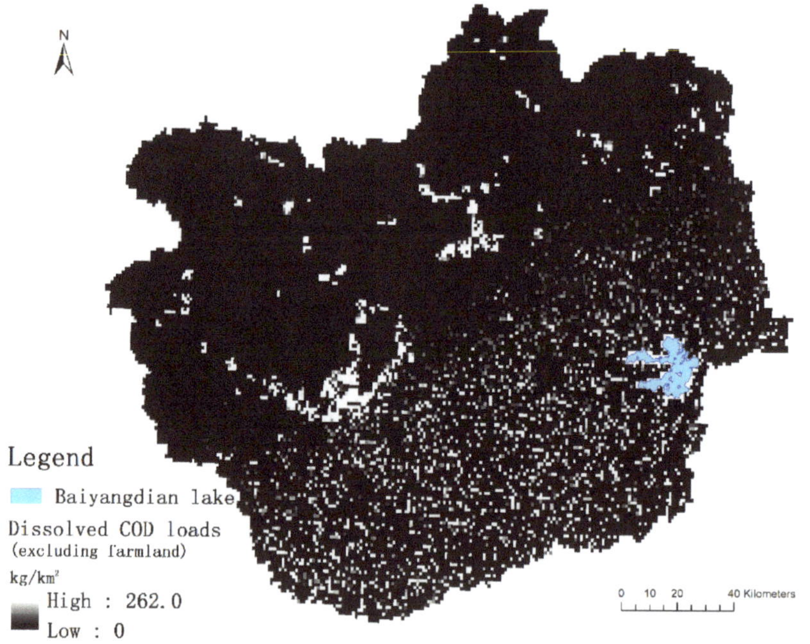

(c)

Figure 7. Dissolved pollutant load of rainfall per unit: (**a**) Total nitrogen load, (**b**) Total phosphorus load, (**c**) COD load.

Table 3. Non-point source pollution load of rainfall.

Pollutant	Fu River	Baigouyin River	Bao River	Xiaoyi River	Surrounding	Total
Total nitrogen (t/a)	38.80	37.88	1.72	27.20	2.31	107.91
Total phosphorus(t/a)	1.34	1.10	0.04	0.93	0.05	3.47
COD (t/a)	201.74	199.43	8.54	144.14	11.09	564.94

3.3. Point Source Pollutant

Point source load is completely based on the survey statistics of industrial enterprises and sewage treatment plants in Baiyangdian lake shown in Table S1. Through the comprehensive collation of the environmental disclosure information of the key sewage discharge units in 2021 by the Baoding Municipal Bureau of Ecology and Environment, there were a total of 37 key enterprises in Dingxing County, 20 key enterprises in Anguo City, 17 key enterprises in Qingyuan District, 36 key enterprises in Boye County, 12 key enterprises in Laiyuan County, 69 key enterprises in Lixian County, 37 key enterprises in Lianchi District, 133 key enterprises in Mancheng, 30 key enterprises in Yixian County, 19 key enterprises in Gaobeidian, 20 key enterprises in the high-tech zone, 38 key enterprises in Jingxiu District, 29 key enterprises in Tang County, 61 key enterprises in Xushui District, 69 key enterprises in Zhuozhou City, 27 key enterprises in Laishui County, 14 key enterprises in Wangdu County, 66 key enterprises in Gaoyang County, and 26 key enterprises in Quyang County. After excluding unclear data, incorrect reporting, and production stoppages, the remaining 682 items of enterprise information were sorted out, of which about one third of the enterprises had no obvious sewage discharge. Thus, 254 pieces of point source discharge information were finally obtained. The calculated total amount of pollutant discharge from point sources in 2020 was calculated as 271.01 tons of total nitrogen, 21.49 tons of total phosphorus, and 1302.56 tons of COD. As the point source pollution loads were mainly based on collected data, we should note that the reliability and accuracy of released data must have a great impact on the calculation results.

3.4. Flux of Pollutants

Based on the collected useful data, the total flux of nitrogen, phosphorus, and organic pollutants of the four rivers (Fu River, Xiaoyi River, Baigouyin River and Bao River) flowing into Baiyangdian lake in 2020 was calculated and summarized as shown in Table 4. It can be seen that Fu River delivered the largest amount of water into Baiyangdian lake, close to the sum of the other three rivers, as Fu River was filled with almost 87% upstream replenishment. Baigouyin River delivered 0.8 billion cubic meters of water to Baiyangdian lake in 2020, of which 35% came from upstream replenishment. Correspondingly, Fu River carried the most pollutants into Baiyangdian lake, with a total amount of 2813.3 tons of COD, 844.8 tons of total nitrogen, and 27.6 tons of total phosphorus, of which the input of COD, total nitrogen, and total phosphorus accounts for as much as 55.35%, 66.39%, and 75.64%, respectively, as shown in Table 4 and Figure 8. From the perspective of the water volume proportion, Fu River input 47.38% of the total water entering the Baiyangdian lake in 2020. It can be seen that the input proportion of each pollutant exceeded the proportion of water volume, indicating that the pollutants concentration transported by Fu River was higher than that of the other three rivers. A possible reason for this may be because there are more villages and farmland along the Fu River, especially in the section from east of Baoding City to Baiyangdian lake, resulting in more non-point source pollution dominated by nitrogen and phosphorus. The Bao River owned the best water quality, and the amount of each pollutant entering Baiyangdian lake was less than 10% of the total. Therefore, in general, the average pollutant content of Fu River was higher than that of the other three rivers, and should be given more attention for decreasing pollution.

Table 4. The amount of water and pollutants entering Baiyangdian lake in 2020.

River into the Baiyangdian Lake	Total Traffic (100 Million Cubic Meters)	Water Replenishment (100 Million Cubic Meters)	The Total Amount of Pollutants (t)		
			COD	Total Nitrogen	Total Phosphorus
Fu River	1.63	1.42	2813.3	844.8	27.6
Xiaoyi River	0.65	0.65	919.1	223.6	4.2
Baigouyin River	0.80	0.28	1063.2	150.5	3.7
Bao River	0.36	0.36	286.7	53.6	0.99
Total	3.44	2.71	5082.3	1272.5	36.5

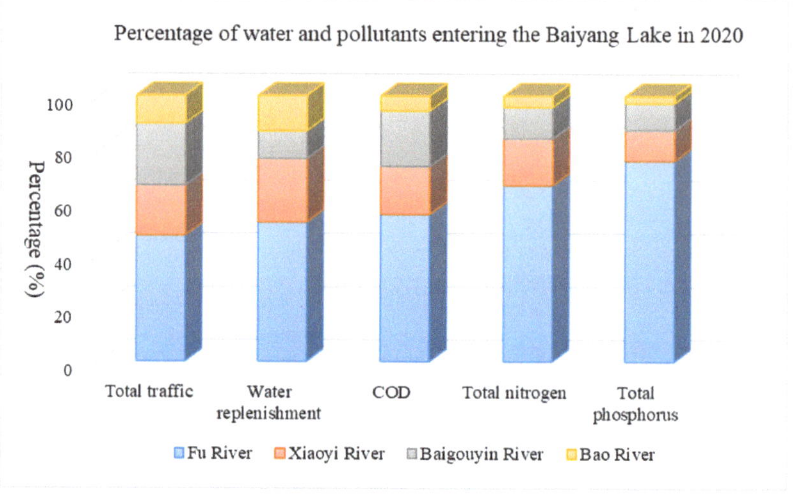

Figure 8. Percentage of water and pollutants entering Baiyangdian lake in 2020.

4. Discussion

In the 1970s, the water quality of Baiyangdian Lake began to deteriorate gradually [45]; since the 1990s, it has become a typical eutrophic lake [46]. The inflow of Baiyangdian Lake dropped sharply from 2000 to 2015, and the deterioration of the water quality intensified, reaching a peak in 2012 and 2013. During this period, chemical oxygen demand, ammonia nitrogen, and total phosphorus were the main pollutants [47]. After 2015, the degree of eutrophication in Baiyangdian Lake has changed from moderate to mild. Especially following the establishment of Xiong'an New Area in 2017, the nutrient load from point sources in Biayangdian lake basin has been significantly reduced [8].

The calculated pollution load entering the lake from the basin is shown in Table 5. The loads of total nitrogen, total phosphorus, and COD were 1885.35 t, 140.01 t, and 5832.35 t, respectively. Compared with similar studies, the results of calculation in this study are basically within the range of results of similar studies (Table 6). According to the spatial distribution map of COD load in the Baiyangdian watershed calculated by Li et al. (2021) through the ECM method, it was estimated that the COD pollution potential of the four sub-basins in this study was about 7000–10,000 t [4]. Those results were higher than ours, probably because those studies focused on the pollution production potential, rather than the pollution load flowing into the lake.

Table 5. The load of total pollutants entering Baiyangdian lake.

Load Sources	Total Nitrogen		Total Phosphorus		Total COD	
	Amount (t)	Percentage (%)	Amount (t)	Percentage (%)	Amount (t)	Percentage (%)
Point pollutants	271.01	14.37	21.49	15.35	1302.56	22.33
Agriculture and rural life	1073.82	56.96	102.72	73.37	3349.33	57.43
Runoff	107.91	5.72	3.47	2.48	564.94	9.69
Soil erosion	432.61	22.95	12.33	8.81	615.52	10.55
Total	1885.35		140.01		5832.35	
Flux of pollutants	1272.5	-	36.5	-	5082.3	-

Table 6. The comparison of total pollutants load entering Baiyangdian lake.

Total Nitrogen (t)	Total Phosphorus (t)	Year	Reference
2123	175	1992	Cui, 1995 [45]
1966	-	2007	Zhang et al., 2012 [48]
3463	405	2007	Zhu, 2011 [49]
1234	97	2008	Shi, 2012 [50]
1356	-	2005–2010	Zhao, 2013 [51]
2018	313	1995–2015	Zhang et al., 2021 [52]

The load ratio disclosed that the primary pollution sources of nitrogen, phosphorus, and organic pollutants were non-point sources, including agricultural and rural non-point sources, runoff, and soil erosion. The contribution rate of non-point sources all exceeded 70%. In Li et al. (2021), who calculated the results of pollution load in Baiyangdian lake basin in 2018, the proportion of total nitrogen load, total phosphorus load, and COD load of non-point sources were 77.11%, 86.93%, and 71.61%, respectively, which were close to the results of our study [4]. Zhao (2013) calculated the total nitrogen load of pollution sources in the Baiyangdian lake basin and found that the proportion of point source pollution loads decreased gradually [51]. With the establishment of Xiong'an New Area in 2017 and the tightening of environmental management policies, the quality of sewage from point sources has improved, and the discharge has been significantly less. As a result, the proportion of point source pollution loads has continued to decline.

Among all non-point sources, agricultural and rural non-point sources had the highest contribution rate of pollution load, all exceeding 50% for total nitrogen, total phosphorus, and COD. The pollution load of agricultural production and rural life activities imports large amounts of nitrogen and phosphorus into the environment, while extensive management methods lead to low utilization of those nutrients, making agricultural activities, especially agricultural planting activities, a huge potential source of nitrogen and phosphorus pollution [6]. The contribution rate to the total phosphorus load reaches were as high as 73.37%. The proportion of total phosphorus load was significantly higher than that of other pollutants, which may be due to the stricter restrictions on total phosphorus in industrial sewage and domestic sewage discharge standards and the large amount of phosphorus used in agricultural activities. Cui et al. (2011) found that the contribution rate of agricultural non-point sources in Baiyangdian lake basin to ammonia nitrogen, chemical oxygen demand, and other pollutants exceeded 50% [35]. Li et al. (2021) showed that the proportion of total nitrogen load, total phosphorus load, and COD load of agricultural and rural non-point sources were 50.63%, 52.03%, and 60.49%, respectively [4].

The load contribution of runoff to nitrogen, phosphorus, and organic pollutants was comparatively small, while the contribution of soil erosion to the total phosphorus was high, with a load ratio of 22.95%. The contribution rates of point source pollution to organic and phosphorus pollution were higher than that caused by runoff and soil erosion, reaching 22.33% and 15.35%, respectively. Normally, areas with low vegetation cover and farmland with only crops are prone to soil erosion occurring due to rainfall, which can cause dissolved pollutants and soil particles to contribute to agricultural non-point source pollution [29,53]. Thus, the total nitrogen and total phosphorus entering the Baiyangdian lake is mainly derived from agricultural non-point sources and corresponding soil erosion.

Non-point sources in agriculture and rural areas and point sources in enterprise sewage discharge are the main sources of COD in water bodies of Baiyangdian lake, accounting for 57.43% and 22.33% of the total COD, respectively. Among the non-point sources, livestock and poultry breeding in agricultural activities causes 70.16% of the COD pollution load, which is about 40% of the total COD load, as shown in Section 3.1 and Table 5. Thus, it is necessary to focus on the management of pollution discharge in breeding activities to reduce the total COD load entering Baiyangdian lake. In addition, industrial production activities produce sewage containing high concentrations of COD; thus, point source pollution is a main source for COD. However, the scale of pollution produced by industrial production is smaller than that from non-point agricultural sources (rural domestic sewage, aquaculture sewage, rainfall erosion, etc.), and the COD load was smaller than that of agricultural non-point sources.

Compared with the flux results of pollutants calculated by the LOADEST model with the inland four rivers in 2020, the load of total nitrogen, total phosphorus, and total COD from point source pollution and non-point source pollution (agriculture and rural areas pollution, runoff pollution, and soil erosion pollution) were significantly higher, at about 1.5 times, 3.8 times, and 1.15 times higher than the flux results, respectively. The main reason for the difference between the loads and the flux was the loss of pollutants in the process of migration and transportation, as the pollutants were gradually separated from the water body through decomposition, adsorption, sedimentation, and other self-purification processes. For example, ammonia nitrogen can be converted and absorbed through nitrification by microorganisms. Compared with total nitrogen, dissolved phosphorus is more likely to be adsorbed, precipitated, and separated from the water body, leading to a higher loss rate of total phosphorus in the process of long-distance transport [54]. Therefore, only about one third of the total phosphorus eventually entered Baiyangdian lake.

The non-point source pollution load in regions with significant spatial heterogeneity can be clarified from a global perspective using lumped non-point pollution models. It should be noted that although lumped models avoid describing the overcomplicated mechanisms that affect the implementation of the model, only the main influencing factors are used in the generalization process, which inevitably leads to deviations between the model results and reality. Further experiments need to be carried out in typical areas of Baiyangdian lake basin to optimize the pollution generation coefficient and river inflow coefficient of the ECM and to verify the applicability of RUSLE and PLOAD.

5. Conclusions

This study used the ECM model, RUSLE model, and LOADEST model to describe a systematic work for the identification of pollution released into Baiyangdian lake. It was calculated that a total nitrogen load of 1614.34 tons, total phosphorus load of 118.52 tons, and total COD load of 4529.79 tons were generated from non-point sources in the whole Baiyangdian lake basin in 2020. The primary pollution source was proven to be agricultural and rural pollution, with a load contribution rate exceeding 50%, of which the contribution rate to the total phosphorus load reached 73.37%. Among the sources of agricultural and rural pollution, the contribution rate of livestock and poultry breeding to COD exceeded 70%, while the contribution rate of agricultural planting to total phosphorus reached 73.12%. Among all the sub-basins, Dingzhou City, Tang County and Yi County were recognized as the most polluted areas, with the highest non-point source load in agriculture and rural areas. It was affirmed that Fu River delivered the most pollutants into the Baiyangdian lake, which should be paid attention to.

In conclusion, non-point source pollution was the main source of nitrogen, phosphorus, and COD. Such pollution sources are diverse and difficult to supervise and control. According to our study, agricultural activities, especially livestock and poultry breeding, had a relatively strong impact on water quality in Baiyangdian lake. Developing the ecological livestock breeding industry and realizing the resource recovery of livestock breeding manure would contribute to a harmonious environment. In order to ensure the environmental quality of Baiyangdian lake, it is necessary to strengthen the control of agricultural planting pollution discharge, standardize the pollution discharge of livestock and poultry breeding, and manage soil erosion and runoff with the aim of reducing the impact of pollution sources on eutrophication in Baiyangdian lake.

Supplementary Materials: The following supporting information can be downloaded at: https://www.mdpi.com/article/10.3390/w14213386/s1, Table S1. Data collection information; Table S2 Parameters sources used in RUSLE model; Figure S1. The soil type of Baiyangdian lake Basin; Figure S2. Non-point source pollution load of rural life in all districts and counties of the Baiyang Lake (a: Total nitrogen load; b: Total phosphorus load; c: COD load); Figure S3. Non-point source pollution load of livestock and poultry in various districts and counties of Baiyang Lake (a: Total nitrogen load; b: Total phosphorus load; c: COD load); Figure S4. Non-point source pollution load of agricultural planting in various districts and counties of Baiyang Lake (a: Total nitrogen load; b: Total phosphorus load; c: COD load).

Author Contributions: G.W.: Conceptualization, Data curation and analysis, Methodology, Writing—original draft preparation, Writing—review and editing. C.L.: Methodology, Writing—original draft preparation, Writing—review and editing. C.G.: Data analysis, Writing—review and editing. Y.Y.: Data investigation. Z.Y.: Data collection. Z.Z.: Data collection. C.T.: Conceptualization, Formal Calibration, Supervision, Writing—review. All authors have read and agreed to the published version of the manuscript.

Funding: This research was supported by the National Key R&D Program of China (No. 2018YFE0106500; 2021YFD1700500) and the National Natural Science Foundation of China (No.42071053).

Conflicts of Interest: The authors declare no conflict of interest.

References

1. Yang, Y.; Gao, B.; Hao, H.; Zhou, H.; Lu, J. Nitrogen and phosphorus in sediments in China: A national-scale assessment and review. *Sci. Total Environ.* **2017**, *576*, 840–849. [CrossRef] [PubMed]
2. Guan, Q.; Feng, L.; Hou, X.; Schurgers, G.; Zheng, Y.; Tang, J. Eutrophication changes in fifty large lakes on the Yangtze Plain of China derived from MERIS and OLCI observations. *Remote Sens. Environ.* **2020**, *246*, 111890. [CrossRef]
3. Derolez, V.; Bec, B.; Munaron, D.; Fiandrino, A.; Pete, R.; Simier, M.; Souchu, P.; Laugier, T.; Aliaume, C.; Malet, N.; et al. Recovery trajectories following the reduction of urban nutrient inputs along the eutrophication gradient in French Mediterranean lagoons. *Ocean Coast. Manag.* **2019**, *171*, 1–10. [CrossRef]
4. Li, Y.; Chen, H.; Sun, W. Load estimation and source apportionment of nitrogen, phosphorus and COD in the basin of Lake Baiyang. *China Environ. Sci.* **2021**, *41*, 366–376.
5. Zhu, Y.; Wu, F.; He, Z.; Guo, J.; Qu, X.; Xie, F.; Giesy, J.P.; Liao, H.; Guo, F. Characterization of organic phosphorus in lake sediments by sequential fractionation and enzymatic hydrolysis. *Environ. Sci. Technol.* **2013**, *47*, 7679–7687. [CrossRef]
6. Zhang, X.; Zhao, J.; Ding, L.; Li, Y.; Liu, H.; Zhao, Y.; Fu, G. Eutrophication evolution trajectory influenced by human activities and climate in the shallow Lake Gehu, China. *Ecol. Indic.* **2022**, *138*, 108821. [CrossRef]
7. Bao, L.; Kang, H.Y.; Hu, X.G.; Zhang, N.M.; Zhao, X.T.; Xu, H.J. The load estimation of nitrogen and phosphorus from non- point sources in the Songhuaba reservoir watershed. *Environ. Sci. Surv.* **2015**, *34*, 1–6.
8. Liu, S.C.; Wang, H.H.; Tian, K. Ecological environmental changes and influencing analysis in Baiyangdian Lake. *J. Agro Environ. Sci.* **2020**, *39*, 1060–1069.
9. Zhu, A.; Liu, P.; Gong, Y.; Li, M.; Su, J.; Liu, G. Residual levels and risk assessment of tetrabromobisphenol A in Baiyang Lake and Fuhe river, China. *Ecotoxicol. Environ. Saf.* **2020**, *200*, 110770. [CrossRef] [PubMed]
10. Li, W.; Shi, Y.; Gao, L.; Liu, J.; Cai, Y. Occurrence of antibiotics in water, sediments, aquatic plants, and animals from Baiyangdian Lake in North China. *Chemosphere* **2012**, *89*, 1307–1315. [CrossRef] [PubMed]
11. Yang, L.; Wang, T.; Zhou, Y.; Shi, B.; Bi, R.; Meng, J. Contamination, source and potential risks of pharmaceuticals and personal products (PPCPs) in Baiyangdian Basin, an intensive human intervention area, China. *Sci. Total Environ.* **2020**, *760*, 144080. [CrossRef] [PubMed]
12. Li, Y.; Ma, J.; Yang, Z.; Lou, I. Influence of non-point source pollution on water quality of Wetland Baiyangdian, China. *Desalination Water Treat.* **2011**, *32*, 291–296. [CrossRef]
13. Yang, J.; Strokal, M.; Kroeze, C.; Ma, L.; Bai, Z.; Teurlincx, S.; Janssen, A.B.G. What is the pollution limit? Comparing nutrient loads with thresholds to improve water quality in Lake Baiyangdian. *Sci. Total Environ.* **2022**, *807*, 150710. [CrossRef]

14. Sun, T.; Chen, J.; Wang, H.; Shi, Z.T. Study on non-point source pollution loads in villages along the Fuhe River, Baiyangdian watershed. *Res. Environ. Sci.* **2012**, *25*, 568–572.
15. Shen, Z.; Zhong, Y.; Huang, Q.; Chen, L. Identifying non-point source priority management areas in watersheds with multiple functional zones. *Water Res.* **2015**, *68*, 563–571. [CrossRef] [PubMed]
16. Chen, L.; Xu, J.; Wang, G.; Liu, H.; Zhai, L.; Li, S.; Sun, C.; Shen, Z. Influence of rainfall data scarcity on non-point source pollution prediction: Implications for physically based models. *J. Hydrol.* **2018**, *562*, 1–16. [CrossRef]
17. Zhou, Z.; Li, J. The correlation analysis on the landscape pattern index and hydrological processes in the Yanhe watershed, China. *J. Hydrol.* **2015**, *524*, 417–426. [CrossRef]
18. Chen, Y.; Xu, C.Y.; Chen, X.; Xu, Y.; Yin, Y.; Gao, L.; Liu, M. Uncertainty in simulation of land-use change impacts on catchment runoff with multi-timescales based on the comparison of the HSPF and SWAT models. *J. Hydrol.* **2019**, *573*, 486–500. [CrossRef]
19. Uribe, N.; Corzo, G.; Quintero, M.; van Griensven, A.; Solomatine, D. Impact of conservation tillage on nitrogen and phosphorus runoff losses in a potato crop system in Fuquene watershed, Colombia. *Agric. Water Manag.* **2018**, *209*, 62–72. [CrossRef]
20. Zema, D.A.; Lucas-Borja, M.E.; Carrà, B.G.; Denisi, P.; Rodrigues, V.A.; Ranzini, M.; Soriano Arcova, F.C.; de Cicco, V.; Zimbone, S.M. Simulating the hydrological response of a small tropical forest watershed (Mata Atlantica, Brazil) by the AnnAGNPS model. *Sci. Total Environ.* **2018**, *636*, 737–750. [CrossRef]
21. Li, P.; Lu, J.; Hou, W.; Pan, Y.; Wang, Y.; Khan, M.R.; Ren, T.; Cong, R.; Li, X. Reducing nitrogen losses through ammonia volatilization and surface runoff to improve apparent nitrogen recovery of double cropping of late rice using controlled release urea. *Environ. Sci. Pollut. Res.* **2017**, *24*, 11722–11733. [CrossRef]
22. Ouyang, W.; Gao, X.; Wei, P.; Gao, B.; Lin, C.; Hao, F. A review of diffuse pollution modeling and associated implications for watershed management in China. *J. Soils Sediments* **2017**, *17*, 1527–1536. [CrossRef]
23. Adu, T.J.; Kumarasamy, M.V. Assessing Non-Point Source Pollution Models: A Review. *Pol. J. Environ. Stud.* **2018**, *27*, 1913–1922. [CrossRef]
24. Rong, Q.; Cai, Y.; Chen, B.; Yue, W.; Yin, X.; Tan, Q. An enhanced export coefficient based optimization model for supporting agricultural nonpoint source pollution mitigation under uncertainty. *Sci. Total Environ.* **2017**, *580*, 1351–1362. [CrossRef] [PubMed]
25. Shen, Z.; Hong, Q.; Chu, Z.; Gong, Y. A framework for priority non-point source area identification and load estimation integrated with APPI and PLOAD model in Fujiang Watershed, China. *Agric. Water Manag.* **2011**, *98*, 977–989. [CrossRef]
26. Johnes, P.J. Evaluation and management of the impact of land use change on the nitrogen and phosphorus load delivered to surface waters: The export coefficient modelling approach. *J. Hydrol.* **1996**, *183*, 323–349. [CrossRef]
27. Ongley, E.D.; Zhang, X.; Yu, T. Current status of agricultural and rural non-point source pollution assessment in China. *Environ. Pollut.* **2010**, *158*, 1159–1168. [CrossRef] [PubMed]
28. Wischmeier, W.H.; Smith, D.D. *Predicting Rainfall Erosion Losses: A Guide to Conservation Planning. Agriculture Handbook*; United States Department of Agriculture, United States Government Printing Office: Washington, DC, USA, 1978.
29. Chen, H.; Teng, Y.; Wang, J. Load estimation and source apportionment of nonpoint source nitrogen and phosphorus based on integrated application of SLURP model, ECM, and RUSLE: A case study in the Jinjiang River, China. *Environ. Monit. Assess.* **2013**, *185*, 2009–2021. [CrossRef] [PubMed]
30. Zhang, W.; Fu, J. Rainfall Erosivity Estimation Under Different Rainfall Amount. *Resour. Sci.* **2003**, *1*, 35–41.
31. USEPA. *Compendium of Tools for Watershed Assessment and TMDL Development*; U.S. Environmental Protection Agency: Washington, DC, USA, 1997.
32. USEPA. *PLOAD Version 3.0: An ArcView GIS Tool to Calculate Nonpoint Sources of Pollution in Watershed and Stormwater Projects*; U.S. Environmental Protection Agency: Washington, DC, USA, 2001.
33. Kang, M.; Yoo, C. Application of the SCS-CN Method to the hancheon basin on the volcanic Jeju Island, Korea. *Water* **2020**, *12*, 3350. [CrossRef]
34. Li, J.C. *Simulation of Non-Point Source Pollution Load in Jialing River Basin*; Chongqing University: Chongqing, China, 2007.
35. Cui, H.M. Analysis on effect of agricultural non-point source pollution on the water environment in Baiyangdian basin. *Mod. Agric. Sci. Technol.* **2011**, *7*, 298–300.
36. Park, Y.S.; Engel, B.A. Analysis for regression model behavior by sampling strategy for annual pollutant load estimation. *J. Environ. Qual.* **2015**, *44*, 1843–1851. [CrossRef] [PubMed]
37. Duan, W.; Takara, K.; He, B.; Luo, P.; Nover, D.; Yamashiki, Y. Spatial and temporal trends in estimates of nutrients and suspended sediment loads in the Ishikari River, Japan, 1985 to 2010. *Sci. Total Environ.* **2013**, *461–462*, 499–508. [CrossRef] [PubMed]
38. Cohn, T.A.; Delong, L.; Gilroy, E.J. Estimating constituent loads. *Water Resour. Res.* **1989**, *25*, 937–942. [CrossRef]
39. Cohn, T.A.; Gilroy, E.J.; Baier, W.G. *Estimating Fluvial Transport of Trace Constituents Using a Regression Model with Data Subject to Censoring*; Joint Statistical Meeting: Boston, MA, USA, 1992; pp. 142–151.
40. Powell, J.L. Least absolute deviations estimation for the censored regression model. *J. Econom.* **1984**, *25*, 303–325. [CrossRef]
41. Chen, T.; Xia, J.; Zou, L.; Yan, Q. Spatiotemporal variations of NDVI of different vegetation types in the Baiyangdian Basin under the background of climate change. *Resour. Sci.* **2021**, *43*, 1248–1259. [CrossRef]
42. Wu, X.H.; Wang, S.J.; Peng, J.Q. *Hebei Economic Yearbook China*; China Statistics Press: Beijing, China, 2018.
43. Ouyang, W.; Wu, Y.; Hao, Z.; Zhang, Q.; Bu, Q.; Gao, X. Combined impacts of land use and soil property changes on soil erosion in a mollisol area under long-term agricultural development. *Sci. Total Environ.* **2018**, *613–614*, 798–809. [CrossRef] [PubMed]

44. Meng, B.; Zhi, X.; Wang, C.; Yin, S.; Zhou, Y.; Zhao, Q. Remote sensing monitoring of dry river beds in Baiyangdian Lake Basin. *Sci. Technol. Eng.* **2022**, *22*, 3827–3833.
45. Cui, X. Investigation of Eutrophication Pollution Sources in Baiyangdian Lake. *Environ. Ecol.* **1995**, *16*, 17–18.
46. Liu, X.; Shi, B.; Meng, J.; Zhou, Y.Q.; Ke, X.; Wang, T.Y. Spatio-temporal variation characteristics of water eutrophication and sediment pollution in Baiyangdian Lake. *Environ. Sci.* **2019**, *41*, 2127–2136.
47. Zhao, X.; Zhang, X. Study on policy tool selection in water pollution control in Xiong'an New Area—based on the policy text of Baiyangdian Basin (1984–2018). *J. Party Sch. CPC Ningbo Munic. Comm.* **2019**, *41*, 103–111.
48. Zhang, H.; Xi, B.; Gao, R.; Jiang, L.; Wang, J.; Xu, Q. Water environmental capacity accounting and upstream capacity allocation of Baiyangdian Lake. *J. Environ. Eng. Technol.* **2012**, *2*, 313–318.
49. Zhu, M. *Study on Agricultural NPS Loads of Haihe Basin and Assessment on its Environmental Impact*; Chinese Academy of Agricultural Sciences: Beijing, China, 2011.
50. Shi, J. *Characteristics of Agricultural Non-Point Source Pollution and Farmland Nutrient Management in the Plain Area of Baiyangdian Watershed*; Beijing Normal University: Beijing, China, 2012.
51. Zhao, Z. Simulation Study of TN Load in Baiyangdian Basin Based on the SWAT Model. *Water Sci. Eng. Technol.* **2013**, *3*, 33–37.
52. Zhang, X.; Yi, Y.; Liu, H.; Yang, Z. Modelling nutrient loadings to the Baiyangdian Lake under the background of urbanization. *Acta Ecol. Sin.* **2021**, *41*, 7521–7529.
53. Wen, G.C.; Zhao, M.J.; Xie, H.B.; Zhang, Y.; Zhang, J. Analysis of land vegetation cover evolution and driving forces in the western part of the Ili River Valley. *Arid. Zone Res.* **2021**, *38*, 843–854.
54. Trolle, D. *The Influence of Sediment Nutrient Dynamics on the Response of Lake Ecosystems to Restoration and Climate Change*; University of Waikato: Hamilton, New Zealand, 2009.

water

MDPI

Article

Hydrochemical Characteristics, Water Quality, and Evolution of Groundwater in Northeast China

Tao Zhang [1,2,†], Pei Wang [3,†], Jin He [1], Dandan Liu [1], Min Wang [4], Mingguo Wang [1,*] and Shibin Xia [2,*]

[1] Center for Hydrogeology and Environmental Geology, China Geological Survey, Baoding 071051, China; zhangtao1@mail.cgs.gov.cn (T.Z.); hejin@mail.cgs.gov.cn (J.H.); liudandan@mail.cgs.gov.cn (D.L.)
[2] School of Resources and Environmental Engineering, Wuhan University of Technology, Wuhan 430070, China
[3] State Key Laboratory of Freshwater Ecology and Biotechnology, Institute of Hydrobiology, Chinese Academy of Sciences, Wuhan 430072, China; pwang@ihb.ac.cn
[4] Zhoukou Vocational and Technical College, Zhengzhou 466000, China; wangmin2060@163.com
* Correspondence: wangmingguo@mail.cgs.gov.cn (M.W.); xiashibin@126.com (S.X.); Tel.: +86-153-928-80268 (S.X.)
† These authors contributed equally to this work.

Abstract: Groundwater is vital to local human life and agricultural irrigation, and the quality of the water is critical to human health. As a result, it is critical to investigate the hydrochemical evolution and water quality of groundwater in the Sanjiang Plain. There were 259 samples obtained. Furthermore, hydrogeochemical simulation was performed to highlight groundwater's hydrochemical features, evolution process, and water quality. The analytical results show that the groundwater in the study area is somewhat alkaline with a mean TDS of 285.94 mgL^{-1} and the primary contributing ions being Ca^{2+} and HCO_3^-. The closer the concentration of TDS and NO_3^- is to the city, the higher the concentration, indicating that the chemical composition of the water body has been affected by certain human activities. The Piper diagram, Gibbs diagram, and correlation analysis results demonstrate that the chemical type of groundwater is mostly HCO_3-Ca and the hydrochemistry is primarily regulated by weathering and carbonate and silicate dissolution. According to the entropy-weighted water quality index, the groundwater quality in this location is pretty acceptable. This study could help strengthen groundwater quality monitoring based on local conditions, identify the source of nitrate, provide data support for the safe use of local water resources, and serve as a reference for global water chemical evolution and water quality evaluation in cold regions.

Keywords: groundwater; hydrochemistry; water quality; WQI; Sanjiang Plain

Citation: Zhang, T.; Wang, P.; He, J.; Liu, D.; Wang, M.; Wang, M.; Xia, S. Hydrochemical Characteristics, Water Quality, and Evolution of Groundwater in Northeast China. *Water* **2023**, *15*, 2669. https://doi.org/10.3390/w15142669

Academic Editors: Christos S. Akratos, Guilin Han and Xiaolong Liu

Received: 13 June 2023
Revised: 4 July 2023
Accepted: 5 July 2023
Published: 24 July 2023

1. Introduction

Water is essential for the survival and growth of both animals and plants, and it also serves as the foundation for industrial and agricultural development. Water safety is critical to human existence [1–4]. Groundwater is frequently employed in industrial and agricultural production because of its steady water supply and generally acceptable water quality [4–8]. Groundwater has traditionally been used as a primary source of water, particularly in locations where rivers and lakes are underdeveloped [9–11]. Understanding the variance trend in regional groundwater quality and the variables that influence it is critical for fully using and rationally maintaining groundwater resources, avoiding and managing groundwater contamination, protecting people's health, and fostering economic growth. As a result, the researchers performed extensive study on groundwater contamination [12,13], focusing on three-nitrogen pollution and heavy metal pollution [14–16]. Pollutant movement, transformation, source, and health risk assessment were researched using statistical analysis, ion ratio, isotope analysis, and entropy weight analysis, yielding numerous findings [17–19].

Nitrate contamination is one of the most serious environmental geological issues that has to be addressed immediately [18,20]. Groundwater nitrate-nitrogen pollution tends to be severe when urbanization speeds up and human activities intensify. NO_3-N is frequently used to assess the toxicity and eutrophication of water [21,22], and its high concentration will result in degradation of water quality, ecological damage, and a hazard to human health. NO_3-N undergoes a sequence of modifications that result in the formation of carcinogenic and teratogenic chemicals. Long-term use of nitrate-rich water raises the chance of people developing disorders such as methemoglobin. Furthermore, high nitrate levels have a negative impact on the biological environment, causing eutrophication and algal bloom. As a result, removing nitrate from groundwater has become a research priority.

Sanjiang Plain has an abundance of black land, which is ideal for cultivating food crops. Groundwater is frequently utilized to irrigate crops in this region [23,24]. The exploitation of groundwater is expanding as urbanization and industrialization progress, and human activities have an increasing influence on groundwater quality. There are many dry and paddy fields in the region, and groundwater is mostly utilized to irrigate crops. The hydrochemical composition of groundwater is continually changing as a result of overexploitation of groundwater, fertilization, irrigation, and other agricultural operations, and the deterioration of water quality will impair the safety of local agricultural and household water [25–27].

So far, the primary groundwater study paths in Sanjiang Plain have been a groundwater pollution risk model [28], shallow groundwater quality [29], and water resource allocation [30], with assessment research on the combination of groundwater pollution being very rare. The regional variable features and sources of major ions, as well as the overall water quality and control factors in the research area, need to be investigated further. As a result, this work focuses on the source and chemical evolution of groundwater in the Sanjiang Plain, and it investigates groundwater hydrochemistry using hydrochemical theory, multivariate statistical techniques, and a water quality index. In order to explain the mechanism of groundwater hydrochemistry evolution, the following three contents are largely investigated: (1) spatial distribution characteristics of groundwater chemistry, (2) groundwater hydrochemical evolution, and (3) groundwater quality. It is expected to provide a hydrochemical foundation for better revealing the variation trend of groundwater quality and environmental factors, as well as realizing the Sanjiang Plain groundwater management objectives of food security, ecological security, and water supply security. This work will serve as a resource for assessing water chemical evolution in frigid places across the world.

2. Materials and Methods

2.1. Study Area

The Sanjiang Plain is situated in the Heilongjiang Province, China (Figure 1). The Sanjiang Plain's mountains rose and the plain sank as a result of tectonic and neotectonic activity, and quaternary strata 100–300 m thick were deposited in the lower plain region. The principal river that flows through the study region is the Naoli River, which is a tributary of the Wusuli River. It rises in the Nadanhada Mountains of the Wanda Mountains, between Qitaihe and Mishan in the Heilongjiang Province, and runs through Qitaihe, Baoqing, Fujin, and Raohe before joining the Wusuli River in Dong'an Town. The overall length is 596 km, the main river channel is 20–100 m wide, the curvature coefficient is 1.4–3.0, and the gradient of the river channel is 1/2000–1/8000. There are nine rivers that flow into the upper reaches, eleven rivers that flow into the middle and lower sections, and a total of twenty tributaries in the basin. The river channels are meandering and serpentine, with crisscrossing river branches. Both sides were previously highly forested, with plentiful plankton growth and great fishing productivity conditions. This area's climate belongs to the temperate zone's humid and semi-humid continental monsoon climatic zone, which is marked by rapid spring warming, frequent wind and little rain, and the occurrence of spring drought. Summer is warm and humid, with frequent rainstorms and flash floods. The sky is high and the air is cool in fall; the temperature drops swiftly, and frost and cold waves arrive early. Winter is cold and dry, with ice and snow, although it is mainly bright.

The annual average temperature is 3.5 °C, with July being the warmest month and the monthly average temperature being 21.9 °C; January is the coldest month, with a monthly average temperature of −18.1 °C, an annual sunshine duration of approximately 2500 h, a frost-free period of approximately 140 days, and a maximum depth of 2.2 m. The yearly precipitation average is 537.2 mm. The yearly evaporation rate is 670.8 mm.

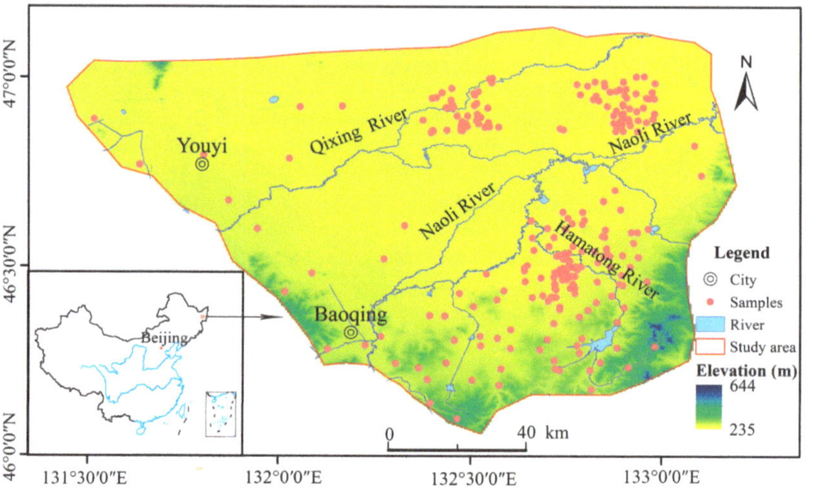

Figure 1. Spatial distribution of groundwater samples.

The regional aquifer system is divided into three compartments based on topography, landform, hydrology, and aquifer geological structure: the Quaternary pore aquifer system, the Paleogene–Neogene pore fissure aquifer system, and the Quaternary bedrock fissure aquifer system. The aquifer formations from bottom to top are the Lower Pleistocene Suibin Formation, the Middle Pleistocene Nongjiang Formation, the Upper Pleistocene Xiangyangchuan Formation, and the Biela Formation, according to the lithology, origin, and distribution of the Quaternary pore aquifer system. The sediments of different ages and sources in the alluvium and strata of the Honghe River are overlying structures between aquifers. There is no obvious water barrier in the middle, and the weak permeable layer is discontinuous, which is a uniform and large-thickness aquifer. The quaternary pore water may be separated into three strata based on groundwater circulation depth and runoff intensity: shallow, medium, and deep. Extracted from the Heilongjiang River, groundwater in the region flows generally from the southwest to the northeast. The major source of groundwater recharge is atmospheric precipitation, followed by lateral runoff replenishment of rivers, recharge of river water during flood times, and infiltration of swamp water. The main excretion routes are runoff, evaporation, and artificial mining.

Sanjiang Plain features flat topography, abundant land resources, little pollution, and significant agricultural development potential. This area has a relative scarcity of water resources, which are mostly used to irrigate crops by pumping groundwater. Since the end of the twentieth century, agricultural activities have increased, particularly the planting of well-irrigated rice. Groundwater alone cannot cover the water requirements of agricultural irrigation due to its scarcity. As a result, a huge number of tube wells have been developed and groundwater extraction has expanded fast, resulting in various ecological concerns such as wetland degradation, a major reduction in grassland area compared to previous periods, the loss of pump wells, and a reduction in rice output. The issue of groundwater quality is also fairly prevalent in this area. Trace indicators such as NH_4^+, NO_3^-, NO_2^-, COD_{Mn}, Fe^{2+}, and Mn^{2+} exceed the limit at a very high rate, which is the primary source of groundwater quality degradation.

2.2. Sample Collection and Analysis

A total of 259 groundwater samples were taken (Figure 1). The water sample was filtered and placed in two plastic bottles. Nitric acid was added to one bottle of sample to make the pH of water sample 2 for the cation test, and another bottle was added to make the pH of water sample 2 for the other index test. The cation was measured using a flame atomic absorption spectrometer (ContrAA300, Thuringia, Germany) [31]. Cl^-, NO_3^-, and SO_4^{2-} were quantified using ion chromatographs (883i, Herisau, Swiss) [32]. Titration with hydrochloric acid was used to determine HCO_3^- and CO_3^{2-} [29]. Titration with disodium (EDTA) was used to calculate total hardness (TH) [32]. Total dissolved solids (TDSs) were measured using a weighing technique [32]. pH was measured in the field by the glass electrode method [32]. The relative standard deviation of all test indices was less than 5%.

2.3. Entropy-Weighted Water Quality Index (EWQI)

The entropy technique is an objective weighing approach that determines an index's dispersion degree by computing the entropy value [33–35]. The entropy technique was used to calculate the weights of hydrochemical indicators such as Na^+, Cl^-, SO_4^{2-}, NO_3^-, TDS, TH, and pH, and groundwater quality was assessed using the groundwater quality standard: excellent (WQI < 25), acceptable (25 < WQI < 50), medium (50 < WQI < 100), poor (100 < WQI < 150), or severely bad (WQI > 150) [36]. Before computing the overall score using these indications, because the influence on groundwater chemistry varies depending on the content of the test indicators, various data were translated to the same measurement standard to remove the inaccuracy caused by variable content. The entropy weight method's quality index evaluation value (EWQI) is the product of w_i and C_j is the concentration of j in groundwater, and S_j is the standard value of j in groundwater, and was calculated as Equation (1) [37]:

$$\mathrm{EWQI} = \sum_{j=1}^{m} w_i \times \frac{C_j}{S_j} \quad (1)$$

3. Results and Discussion

3.1. Hydrochemical Composition and Hydrochemical Types

3.1.1. Chemical Composition Characteristics of Groundwater

The primary cations in groundwater are Ca^{2+} and Na^+. The concentration of groundwater ions shows the relationship $Ca^{2+} > Na^+ > Mg^{2+} > K^+$, with Ca^{2+} and Na^+ content ranges of 0.34–142.28 $mg \cdot L^{-1}$ and 2.75–65.22 $mg \cdot L^{-1}$, respectively, with mean values of 45.76 $mg \cdot L^{-1}$ and 23.23 $mg \cdot L^{-1}$ (Table 1).

Table 1. Statistics of major parameters in groundwater.

Chemical Parameter	Mean	Standard Deviation	Min	Max	NSPRC (NSPRC, 2017)	Coefficient Variation
$K^+/mg \cdot L^{-1}$	3.01	4.83	0.08	61.17	—	1.61
$Na^+/mg \cdot L^{-1}$	23.23	12.53	2.75	65.22	200.00	0.54
$Ca^{2+}/mg \cdot L^{-1}$	45.76	25.59	0.34	142.28	—	0.56
$Mg^{2+}/mg \cdot L^{-1}$	16.13	10.74	2.52	67.47	—	0.67
$Cl^-/mg \cdot L^{-1}$	12.45	15.31	1.77	109.02	250.00	1.23
$SO_4^{2-}/mg \cdot L^{-1}$	9.31	32.04	0.00	450.00	250.00	0.54
$HCO_3^-/mg \cdot L^{-1}$	255.28	138.65	0.00	682.63	—	3.16
$NO_3^-/mg \cdot L^{-1}$	10.21	24.57	0.00	205.00	88.60	2.41
$TDS/mg \cdot L^{-1}$	285.94	118.59	85.24	839.64	1000.00	0.41
$TH/mg \cdot L^{-1}$	178.33	93.62	20.80	474.95	450.00	0.52
pH	7.15	0.42	5.46	9.04	6.5–8.5	0.06

The anion is mostly HCO_3^-, and groundwater concentrations show a relationship of $HCO_3^- > Cl^- > SO_4^{2-} > NO_3^-$. The concentration ranges for HCO_3^- and NO_3^- were 0–682.63 mg·L^{-1} and 0–205.0 mg·L^{-1}, respectively, with mean values of 255.27 mg·L^{-1} and 10.21 mg·L^{-1}. In general, the greater the content, the closer to the metropolitan area. This is linked to the aggregation of urban populations, demonstrating that NO_3^- is heavily influenced by human activity.

The pH value, as an essential parameter, can represent information about hydrogeochemical equilibrium. The pH ranges from 5.46 to 9.04, with a mean value of 7.15, showing that the regional diversity of pH values is limited and that the groundwater samples are generally slightly alkaline. TDS and TH concentration variation ranges were 85.24–839.64 mg·L^{-1} and 20.80–474.95 mg·L^{-1}, respectively, with mean values of 285.94 mg·L^{-1} and 178.33 mg·L^{-1}, and the coefficients of variation were 41.47% and 52.50%, showing that the regional disparities in TDS and TH concentrations were considerable. TDS was found to be higher in the north–south direction and lower in the east–west direction. The greater the TDS of densely inhabited populations, the closer they were to towns, indicating that the high TDS of certain water samples was induced by human activity (Figure 2).

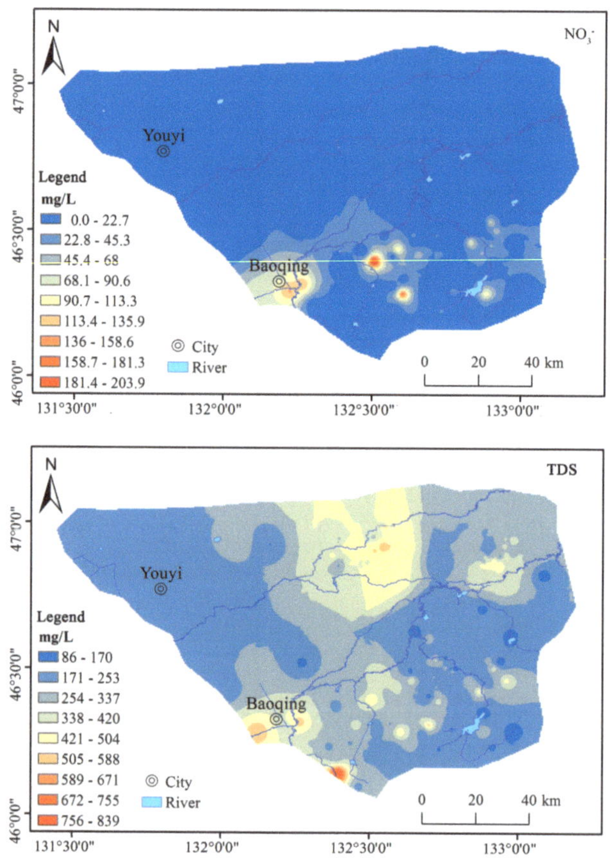

Figure 2. Distribution characteristics of NO_3^- and TDS.

3.1.2. Hydrochemical Type

The hydrochemistry and hydrochemical types were evaluated using a piper triplot, and the majority of the water samples fell towards Ca^{2+}, showing that Ca^{2+} is prevalent in groundwater, followed by Mg^{2+}. All of the water sample points are near the HCO_3^- end of the anion diagram, indicating that HCO_3^- ions are common and that the chemical types are mostly HCO_3-Ca·Mg (Figure 3a). Ca^{2+} and HCO_3^-, on the other hand, are primarily derived from the dissolution of carbonate or silicate, indicating that this area is primarily governed by weathering and rock dissolution.

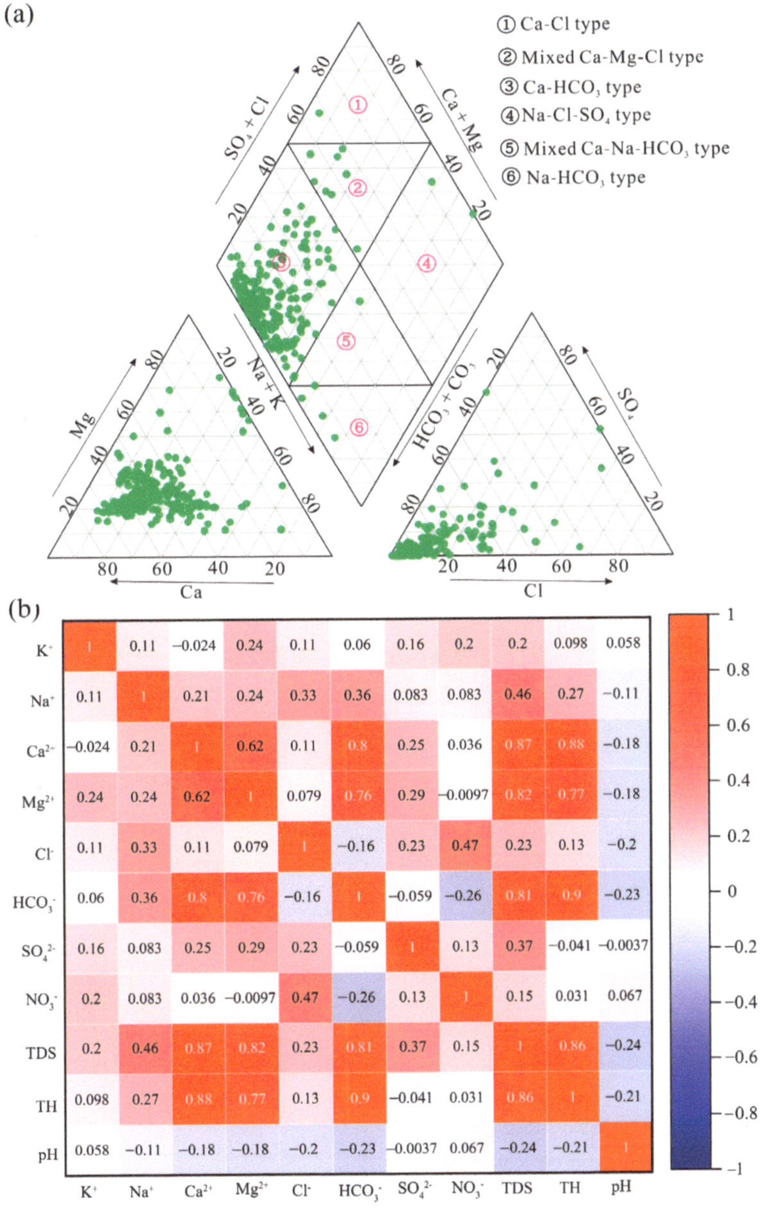

Figure 3. The piper plot of groundwater (**a**) and correlation between chemical parameters in groundwater of Sanjiang Plain (**b**).

3.1.3. Relationship between Chemical Indexes

Correlation analysis is frequently used to investigate the origin of ions in the evolution of hydrochemistry [24]. There is a strong link between the components of the same source. TDS has a high association with Na^+, Ca^{2+}, Mg^{2+}, SO_4^{2-}, and HCO_3^- (Figure 3b), indicating that Na^+, Ca^{2+}, Mg^{2+}, SO_4^{2-}, and HCO_3^- are the primary sources of groundwater TDS, among which TDS has a substantial correlation with Ca^{2+} ($R^2 = 0.866$, $p < 0.01$) and HCO_3^- ($R^2 = 0.813$, $p < 0.01$). The correlation coefficients between SO_4^{2-} and Ca^{2+}, Mg^{2+} are 0.251 and 0.288, respectively, showing that they have a similar material source, namely weathering dissolution and sulfuric acid dissolution of calcite, dolomite, and other carbonate minerals.

3.2. *Control Factors of Groundwater Hydrochemistry*

3.2.1. Natural Factors

The Gibbs model is commonly used to determine the primary water regulating parameters [38–42]. It consists of three governing endmembers: precipitation, rock weathering, and evaporative [43–45]. The atmospheric precipitation control area has a lower total dissolved solid concentration and a greater ratio of $Na^+/(Na^+ + Ca^{2+})$ and $Cl^-/(Cl^- + HCO_3^-)$, both of which are normally in the 0.5–1 range, and is dispersed in the bottom right corner of the Gibbs plot. The rock weathering area is in the middle left, the TDS value ranges from 70 to 300 $mg \cdot L^{-1}$, and the ratio is less than 0.5. The upper right of the diagram shows the evaporative crystallization control region, which has a greater TDS and a higher ratio (0.5–1). The TDS and ratios of the groundwater samples in this location are lower, and all samples fall within the rock weathering control area, indicating that the major ions in groundwater are caused by rock weathering (Figure 4a,b).

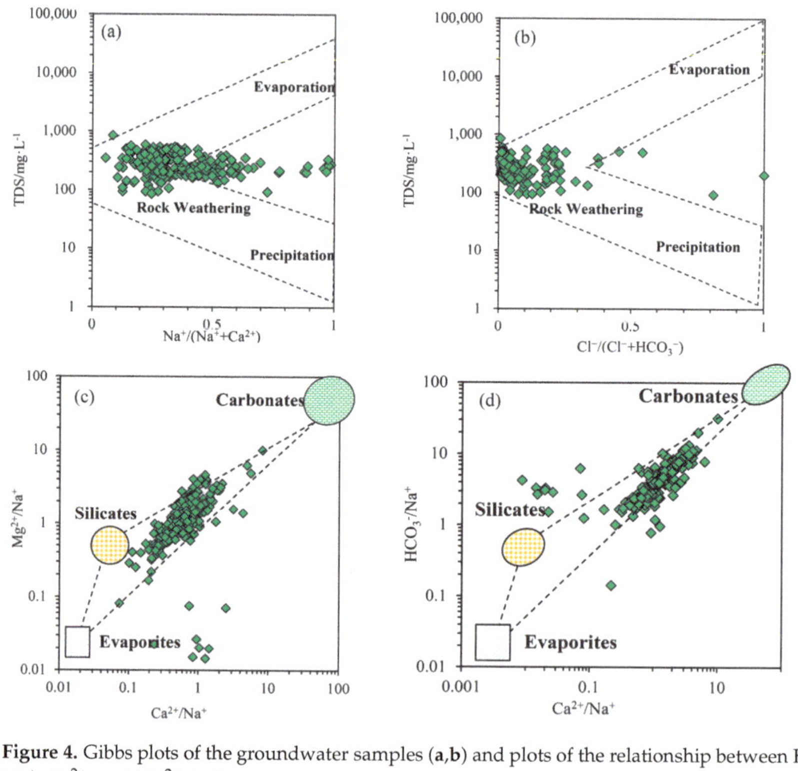

Figure 4. Gibbs plots of the groundwater samples (**a**,**b**) and plots of the relationship between HCO_3^-, Na^+, Ca^{2+}, and Mg^{2+} (**c**,**d**).

Various rocks will create different ions as they weather. The weathering of carbonate rock, silicate rock, and evaporite produces Ca^{2+} and Mg^{2+} in abundance. The main elements Na^{+} and K^{+} are formed through the weathering and disintegration of silicate rock and evaporite. Weathering and dissolution of carbonate and silicate rocks produce HCO_3^{-} in significant amounts. As a result of the breakdown of evaporated, SO_4^{2-} and Cl^{-} significant compounds are formed [46,47]. A mixing map generally reveals the source of ions created by chemical weathering in the watershed. Because the Ca^{2+}/Na^{+}, Mg^{2+}/Na^{+}, HCO_3^{-}/Na^{+} relationships are unaffected by flow rate, dilution, or evaporation, they can reveal the hydrochemical origin and dissolution of minerals from which the major ions originate. The samples are located in the silicate rock and carbonate areas, demonstrating that groundwater is influenced by water–rock interaction (Figure 4c,d).

Cation alternating adsorption occurs when particles adsorb some cations in water and transform some of their initially adsorbed cations into water components under particular circumstances [48]. ($Na^{+} - Cl^{-}$) and [($2SO_4^{2-} + HCO_3^{-}$) − $2(Ca^{2+} + Mg^{2+})$] (mmol·L^{-1}) concentrations can be used to determine if $Na^{+} - Ca^{2+}$ exchange occurs. The ratio ($Na^{+} - Cl^{-}$)/[($2SO_4^{2-} + HCO_3^{-}$) − $2(Ca^{2+} + Mg^{2+})$] should be close to 1:1 if the hydrogeochemical process is mostly cation exchange [49]. Water sample molar concentration ratios ($Na^{+} - Cl^{-}$)/[($2SO_4^{2-} + HCO_3^{-}$) − $2(Ca^{2+} + Mg^{2+})$] are typically near the 1:1 line, indicating that cation exchange is important in the subsurface hydrogeochemical process (Figure 5a).

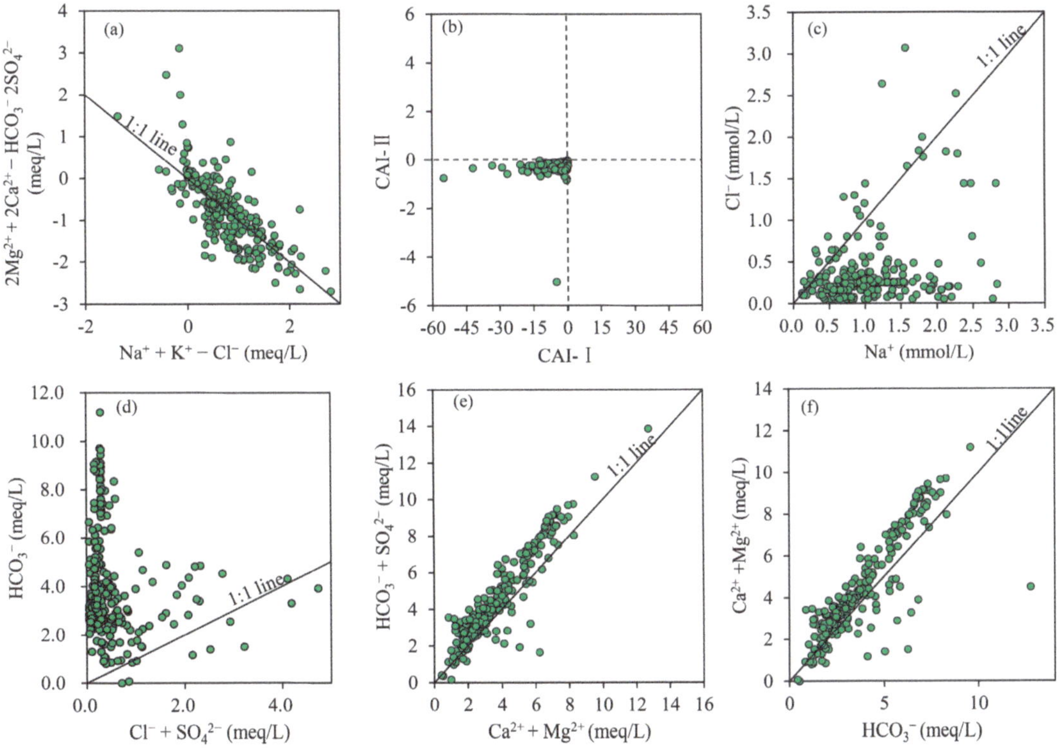

Figure 5. Plots of (**a**) ($Na^{+} - Cl^{-}$) vs. [($2SO_4^{2-} + HCO_3^{-}$) − $2(Ca^{2+} + Mg^{2+})$], (**b**) chlor-alkali indices CAI-Ivs. CAI-II, (**c**) Cl^{-} vs. Na^{+}, (**d**) HCO_3^{-} vs. $Cl^{-} + SO_4^{2-}$, (**e**) $HCO_3^{-} + SO_4^{2-}$ vs. $Ca^{2+} + Mg^{2+}$, (**f**) $Ca^{2+} + Mg^{2+}$ vs. HCO_3^{-}.

The chlor-alkali index (CAI-I and CAI-II) is a water ion exchange test. If CAI-I and CAI-II are negative, then Ca^{2+} and Mg^{2+} exchange with Na^+ and K^+ in groundwater; if CAI-I and CAI-II are positive, then Na^+ and K^+ exchange with Ca^{2+} and Mg^{2+} in groundwater [39,50]. The majority of chlor-alkali indices in groundwater are less than zero, indicating that calcium and magnesium are mostly exchanged for sodium and potassium in groundwater in the research region. It aids in the enrichment of sodium and potassium in groundwater (Figure 5b).

The major source of Na^+ is the dissolution of halite and silicate, and the dissolution of halite results in the same molar concentration of Na^+ and Cl^- [51]. However, the ratio of Na^+ to Cl^- in the water sample is far from the 1:1 line and near that of Na^+ (Figure 4c), showing that the Cl^- content is definitely lower than that of Na^+ and that there should be other anions to balance the excess sodium, indicating that sodium should have other sources. The molar concentration ratio of Na^+/Cl^- ranges from 0.47 to 55.67, with an average of 5.56. This demonstrates that Na^+ is not only derived from the dissolving of rock salt, whereas sodium is mostly derived from the weathering breakdown of silicate rock.

The majority of locations are above and deviate significantly from the $Cl^- + SO_4^{2-}$ and HCO_3^- 1:1 lines (Figure 5d), indicating that the carbonate source is significantly greater than the evaporate source. The average $(Ca^{2+} + Mg^{2+})/(Na^+ + K^+)$ ratio of groundwater samples in Sanjiang Plain is 3.93, indicating that weathering and carbonate and silicate dissolution are dominant in this region. The correlation coefficient between sodium and calcium is 0.2, suggesting that Na^+ and Ca^{2+} are derived from the same source, implying that some of them are derived via silicate breakdown.

In order to identify whether the hydrochemistry in this region is governed by carbonate rock or gypsum dissolution, $(Ca^{2+} + Mg^{2+})$ and $(HCO_3^- + SO_4^{2-})$ are usually used to examine the hydrochemical process at the watershed scale [52]. If the equivalent concentration ratio of $(HCO_3^- + SO_4^{2-})/(Ca^{2+} + Mg^{2+})$ in the water sample is 1:1, carbonate rock or gypsum dissolution controls the hydrochemistry [53,54]. The ratio of $(HCO_3^- + SO_4^{2-})/(Ca^{2+} + Mg^{2+})$ is greater than 1:1. $(HCO_3^- + SO_4^{2-})$ exceeds $(Ca^{2+} + Mg^{2+})$ (Figure 5e), indicating that other cations should be present to balance the amount of surplus anions and that there are other sources of HCO_3^- in groundwater, some of which originate from silicate dissolution. The correlation coefficients between HCO_3^- and Ca^{2+}, Mg^{2+} are 0.806 ($p < 0.01$) and 0.757 ($p < 0.01$), respectively, indicating that they share a common material source, primarily carbonate dissolution. When combined with the Sanjiang Plain's geological history, it suggests that weathering and dissolution of carbonate rocks contribute to Ca^{2+} and HCO_3^-.

The majority of the points in the figure are in the upper part of the 1:1 ratio (Figure 5f), and the amount of Mg^{2+} and Ca^{2+} is greater than that of HCO_3^-, with a ratio of around 1.19, indicating that carbonate dissolution cannot explain the composition of Mg^{2+} and Ca^{2+} in groundwater entirely. Cl^- and SO_4^{2-} may balance out additional Mg^{2+} and Ca^{2+}. Ca^{2+} and SO_4^{2-} have a significant correlation value of 0.251 ($p < 0.01$), indicating that a small amount of halite dissolution is occurring to balance Mg^{2+} and Ca^{2+}.

3.2.2. Human Activity Input

Human actions have a significant impact on the evolution of water chemistry [32,55]. It alters the chemical composition of water by transporting wastewater, waste, and waste gas. NO_3^-, Cl^-, and SO_4^{2-} levels will be higher in areas with high human activity. Of course, weathering can produce Cl^- and SO_4^{2-}. When water is impacted by human activity, the Cl^-/Na^+ and NO_3^-/Na^+ ratios are often high. The ratios are rather high (Figure 6), demonstrating that human production and life have an effect on groundwater in this area, which is connected to urban home sewage discharge and large-scale agricultural output.

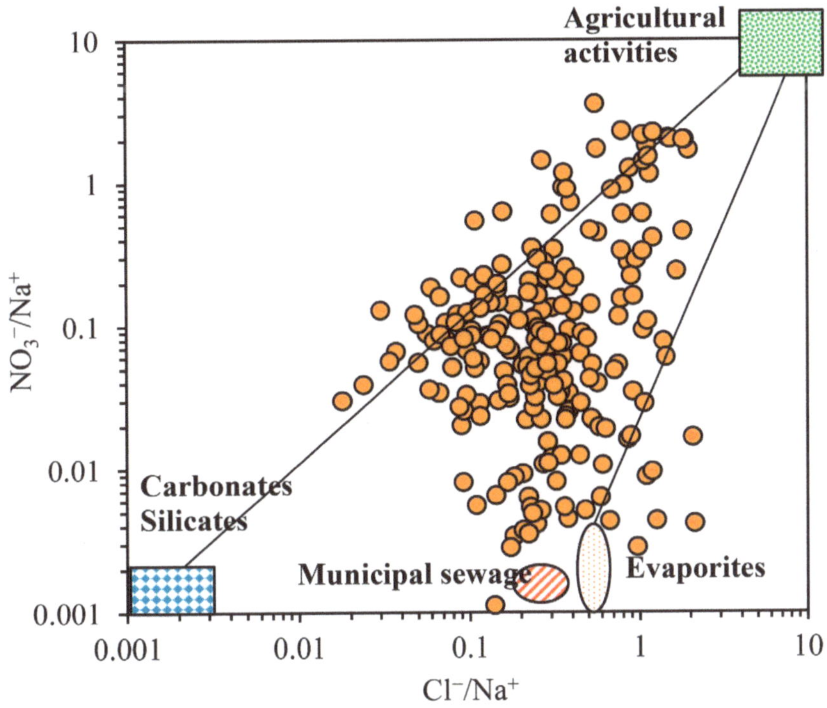

Figure 6. The ratio of Cl^-/Na^+ vs. NO_3^-/Na^+.

3.3. Groundwater Quality

According to the findings of the water quality assessments, the majority of the samples are of high quality (Figure 7a). The EWQI value varied from 2.27 to 91.69, with a mean value of 9.82 obtained from 259 samples. There are 243 high-quality water sources in the research region, accounting for approximately 93.8% of the total. Another nine groundwater samples were categorized as having good water quality, accounting for approximately 3.5% of the total. These two types of water are useful for a variety of applications. Seven samples are rated as medium-grade water, accounting for around 2.7% of the total, which is within the recommended drinking range. There are no samples of water that are of bad or extremely poor quality. The higher the content of NO_3^- in water, the higher the corresponding WQI, indicating that NO_3^- is the main influencing factor of WQI (Figure 7a). The higher the concentration of NO_3^- in water, the higher the corresponding WQI (Figure 7a), indicating that NO_3^- is the primary influencing factor of WQI. The samples with low water quality were primarily dispersed in cities with dense populations, indicating that water quality was directly tied to human activities (Figure 7b).

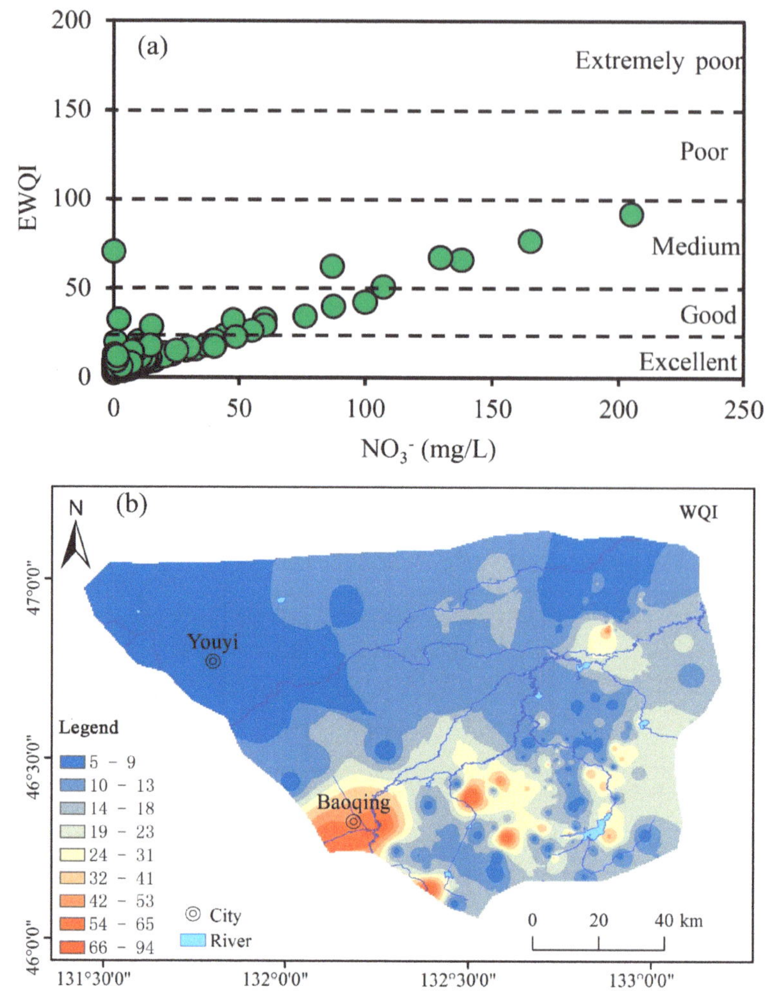

Figure 7. The relationship between WQI and nitrate (**a**) and distribution characteristics of WQI (**b**).

4. Conclusions

Based on the regional scale, this study analyzed the hydrochemical characteristics and hydrochemical evolution of groundwater in Sanjiang Plain, Heilongjiang Province, China and grasped the hydrochemical evolution law of this area. The chemical type of groundwater is Ca-HCO_3, with a cation concentration of $Ca^{2+} > Na^+ > Mg^{2+} > K^+$ and an anion concentration of $HCO_3^- > Cl^- > SO_4^{2-} > NO_3^-$. The chemical composition of water is mainly affected by the dissolution of carbonate rock and silicate rock, and the alternative adsorption of cations and the impact of human activities also play a role. The groundwater quality index results show that the EWQI value is lower than the World Health Organization's drinking water standards, making it suitable for drinking. This study provides a hydrochemical basis for better revealing the interaction mechanism between groundwater and environment and achieving the goal of groundwater management for food security, ecological security, and water supply security in the Sanjiang Plain, and it can serve as a reference for groundwater chemical evolution and water quality evaluation in the global cold area.

Author Contributions: T.Z.: Writing—review and editing; P.W.: Writing—original draft; D.L. and M.W. (Min Wang): Visualization; J.H.: Funding acquisition; M.W. (Mingguo Wang): Funding acquisition: S.X.: Supervision. All authors have read and agreed to the published version of the manuscript.

Funding: This research was funded by China Geological Survey, grant number [DD20230456; DD20230424], and also funded by the National Key R&D Program of China, grant number [2021YFA0715901]. It was also funded by the Medical Education Research Project of Henan Province (Wjlx2020372). And The APC was funded by China Geological Survey.

Data Availability Statement: Not applicable.

Conflicts of Interest: The authors declare that they have no competing interest.

References

1. Döll, P.; Schmied, H.M.; Schuh, C.; Portmann, F.T.; Eicker, A. Global-scale assessment of groundwater depletion and related groundwater abstractions: Combining hydrological modeling with information from well observations and GRACE satellites. *Water Resour. Res.* **2014**, *50*, 5698–5720. [CrossRef]
2. Islam, N.; Irshad, K. Artificial ecosystem optimization with Deep Learning Enabled Water Quality Prediction and Classification model. *Chemosphere* **2022**, *309*, 136615. [CrossRef] [PubMed]
3. Zhang, T.; Cai, W.; Li, Y.; Geng, T.; Zhang, Z.; Lv, Y.; Zhao, M.; Liu, J. Ion chemistry of groundwater and the possible controls within Lhasa River Basin, SW Tibetan Plateau. *Arab. J. Geosci.* **2018**, *11*, 510. [CrossRef]
4. Zhang, H.; Cheng, S.Q.; Li, H.F.; Fu, K.; Xu, Y. Groundwater pollution source identification and apportionment using PMF and PCA-APCA-MLR receptor models in a typical mixed land-use area in Southwestern China. *Sci. Total Environ.* **2020**, *741*, 140383. [CrossRef] [PubMed]
5. Changmai, M.; Pasawan, M.; Purkait, M.K. A hybrid method for the removal of fluoride from drinking water: Parametric study and cost estimation. *Sep. Purif. Technol.* **2018**, *206*, 140–148. [CrossRef]
6. Zhong, C.; Wang, H.; Yang, Q. Hydrochemical interpretation of groundwater in Yinchuan basin using self-organizing maps and hierarchical clustering. *Chemosphere* **2022**, *309*, 136787. [CrossRef]
7. Huang, G.; Song, J.; Han, D.; Liu, R.; Liu, C.; Hou, Q. Assessing natural background levels of geogenic contaminants in groundwater of an urbanized delta through removal of groundwaters impacted by anthropogenic inputs: New insights into driving factors. *Sci. Total Environ.* **2023**, *857*, 159527. [CrossRef]
8. Zhang, Y.; Chen, Z.; Huang, G.; Yang, M. Origins of groundwater nitrate in a typical alluvial-pluvial plain of North China plain: New insights from groundwater age-dating and isotopic fingerprinting. *Environ. Pollut.* **2023**, *316*, 120592. [CrossRef]
9. Du, Y.; Deng, Y.; Ma, T.; Lu, Z.; Shen, S.; Gan, Y.; Wang, Y. Hydrogeochemical evidences for targeting sources of safe groundwater supply in arsenic-affected multi-level aquifer systems. *Sci. Total Environ.* **2018**, *645*, 1159–1171. [CrossRef]
10. Smith, R.G.; Majumdar, S. Groundwater storage loss associated with land subsidence in western United States mapped using machine learning. *Water Resour. Res.* **2020**, *56*, e2019WR026621. [CrossRef]
11. Harkness, J.S.; Jurgens, B.C. Effects of imported recharge on fluoride trends in groundwater used for public supply in California. *Sci. Total Environ.* **2022**, *830*, 154782. [CrossRef]
12. Gill, B.; Webb, J.; Stott, K.; Cheng, X.; Wilkinson, R.; Cossens, B. Economic, social and resource management factors influencing groundwater trade: Evidence from Victoria, Australia. *J. Hydrol.* **2017**, *550*, 253–267. [CrossRef]
13. Chen, R.H.; Tenga, Y.G.; Chen, H.Y.; Hu, B.; Yue, W.F. Groundwater pollution and risk assessment based on source apportionment in a typical cold agricultural region in Northeastern China. *Sci. Total Environ.* **2019**, *696*, 133972. [CrossRef] [PubMed]
14. Yang, Q.; Li, Z.; Ma, H.; Wang, L.; Martín, J.D. Identification of the hydrogeochemical processes and assessment of groundwater quality using classic integrated geochemical methods in the Southeastern part of Ordos basin, China. *Environ. Pollut.* **2016**, *218*, 879–888. [CrossRef] [PubMed]
15. Emenike, P.C.; Tenebe, I.T.; Neris, J.B.; Omole, D.O.; Afolayan, O.; Okeke, C.U.; Emenike, I.K. An integrated assessment of land-use change impact, seasonal variation of pollution indices and human health risk of selected toxic elements in sediments of River Atuwara, Nigeria. *Environ. Pollut.* **2020**, *265*, 114795. [CrossRef]
16. Goswami, R.; Neog, N.; Bhagat, C.; Hdeib, R.; Mahlknecht, J.; Kumar, M. Arsenic in the groundwater of the Upper Brahmaputra floodplain: Variability, health risks and potential impacts. *Chemosphere* **2022**, *306*, 135621. [CrossRef]
17. Devaraj, N.; Panda, B.; Chidambaram, S.; Prasanna, M.V.; Singh, D.K.; Ramanathan, A.L.R.; Sahoo, S.K. Spatio-temporal variations of uranium in groundwater: Implication to the environment and human health. *Sci. Total Environ.* **2021**, *775*, 145787.
18. Li, X.; Huang, X.; Zhang, Y. Spatio-temporal analysis of groundwater chemistry, quality and potential human health risks in the Pinggu basin of North China Plain: Evidence from high-resolution monitoring dataset of 2015–2017. *Sci. Total Environ.* **2021**, *800*, 149568. [CrossRef]
19. Elumalai, V.; Rajmohan, N.; Sithole, B.; Li, P.; Uthandi, S.; Tol, J. Geochemical evolution and the processes controlling groundwater chemistry using ionic ratios, geochemical modelling and chemometric analysis in uMhlathuze catchment, KwaZulu-Natal, South Africa. *Chemosphere* **2022**, *312*, 137179. [CrossRef]

20. Lin, L.; Zhang, Y.; Qian, X.; Wang, Y. Hydrochemical Characteristics and Human Health Risk Assessment of Surface Water in the Danjiang River Source Basin of the Middle Route of China's South-to-North Water Transfer Project. *Water* **2023**, *15*, 2203. [CrossRef]
21. Wu, J.J.; Bian, J.M.; Wan, H.L.; Ma, Y.X.; Sun, X.Q. Health risk assessment of groundwater nitrogen pollution in Songnen Plain. *Ecotoxicol. Environ. Saf.* **2021**, *207*, 111245. [CrossRef] [PubMed]
22. Zhao, C.; Zhang, X.G.; Fang, X.; Zhang, N.; Xu, X.Q.; Li, L.H.; Liu, Y.; Su, X.; Xia, Y. Characterization of drinking groundwater quality in rural areas of Inner Mongolia and assessment of human health risks. *Ecotoxicol. Environ. Saf.* **2022**, *234*, 113360. [CrossRef] [PubMed]
23. Jia, Z.; Bian, J.M.; Wang, Y. Impacts of urban land use on the spatial distribution of groundwater pollution, Harbin City, Northeast China. *J. Contam. Hydrol.* **2018**, *215*, 29–38. [CrossRef] [PubMed]
24. Zhao, G.; Li, W.; Li, F.; Zhang, F.; Liu, G.C. Hydrochemistry of waters in snowpacks, lakes and streams of Mt. Dagu, eastern of Tibet Plateau. *Sci. Total Environ.* **2018**, *610–611*, 641–650. [CrossRef]
25. Li, Q.; Zhao, D.; Yin, J.; Zhou, X.; Li, Y.; Chi, P.; Han, Y.; Ansari, U.; Cheng, Y. Sediment Instability Caused by Gas Production from Hydrate-bearing Sediment in Northern South China Sea by Horizontal Wellbore: Evolution and Mechanism. *Nat. Resour. Res.* **2023**, *32*, 1595–1620. [CrossRef]
26. Wang, F.; Liu, X.; Jiang, B.; Zhuo, H.; Chen, W.; Chen, Y.; Li, X. Low-loading Pt nanoparticles combined with the atomically dispersed FeN4 sites supported by FeSA-N-C for improved activity and stability towards oxygen reduction reaction/hydrogen evolution reaction in acid and alkaline media. *J. Colloid Interface Sci.* **2023**, *635*, 514–523. [CrossRef]
27. Li, Q.; Zhang, C.; Yang, Y.; Ansari, U.; Han, Y.; Li, X.; Cheng, Y. Preliminary experimental investigation on long-term fracture conductivity for evaluating the feasibility and efficiency of fracturing operation in offshore hydrate-bearing sediments. *Ocean Eng.* **2023**, *281*, 114949. [CrossRef]
28. Nixdorf, E.; Sun, Y.Y.; Lin, M.; Kolditz, O. Development and application of a novel method for regional assessment of groundwater contamination risk in the Songhua River Basin. *Sci. Total Environ.* **2017**, *605–606*, 598–609. [CrossRef]
29. Zhang, B.; Song, X.F.; Zhang, Y.H.; Han, D.M.; Tang, C.Y.; Yang, L.H.; Wang, Z.L. The renewability and quality of shallow groundwater in Sanjiang and Songnen Plain, Northeast China. *J. Integr. Agric.* **2017**, *16*, 229–238. [CrossRef]
30. Liu, Y.L.; Jin, M.G.; Wang, J.J. Insights into groundwater salinization from hydrogeochemical and isotopic evidence in an arid inland basin. *Hydrol. Process.* **2018**, *32*, 3108–3127. [CrossRef]
31. *GB/T 5750*; Standards for Drinking Water Quality. Ministry of Health of the People's Republic of China: Beijing, China, 2006.
32. *DZ/T 0064*; MAGQ, Methods for Analysis of Groundwater Quality. Ministry of Natural Resources of the People's Republic of China: Beijing, China, 2021.
33. Paca, J.M.; Santos, F.M.; Pires, J.C.M.; Leitão, A.A.; Boaventura, R.A.R. Quality assessment of water intended for human consumption from Kwanza, Dande and Bengo rivers (Angola). *Environ. Pollut.* **2019**, *254*, 113037. [CrossRef] [PubMed]
34. Adimalla, N. Application of the entropy weighted water quality index (EWQI) and the pollution index of groundwater (PIG) to assess groundwater quality for drinking purposes: A case study in a rural area of Telangana State, India. *Arch. Environ. Contam. Toxicol.* **2021**, *80*, 31–40. [CrossRef] [PubMed]
35. Liu, J.T.; Gao, Z.J.; Zhang, Y.Q.; Sun, Z.B.; Sun, T.Z.; Fan, H.B.; Wu, B.; Li, M.B.; Qian, L.L. Hydrochemical evaluation of groundwater quality and human health risk assessment of nitrate in the largest peninsula of China based on high-density sampling: A case study of Weifang. *J. Clean. Prod.* **2021**, *322*, 129164. [CrossRef]
36. *GB/T 14848-2017*; Standards for Groundwater Quality. National Standard of People's Republic of China: Beijing, China, 2017.
37. Chai, N.; Yi, X.; Xiao, J.; Liu, T.; Liu, Y.; Deng, L.; Jin, Z. Spatiotemporal variations, sources, water quality and health risk assessment of trace elements in the Fen River. *Sci. Total Environ.* **2021**, *757*, 143882. [CrossRef]
38. Gibbs, R.J. Mechanisms controlling world water chemistry. *Science* **1970**, *170*, 1088–1090. [CrossRef] [PubMed]
39. Zhang, Y.; Dai, Y.; Wang, Y.; Huang, X.; Xiao, Y.; Pei, Q. Hydrochemistry, quality and potential health risk appraisal of nitrate enriched groundwater in the Nanchong area, southwestern China. *Sci. Total Environ.* **2021**, *784*, 147186. [CrossRef]
40. Ke, X.; Li, Y.; Wang, W.; Niu, F.; Gao, Z. Hydrogeochemical characteristics and processes of thermokarst lake and groundwater during the melting of the active layer in a permafrost region of the Qinghai-Tibet Plateau, China. *Sci. Total Environ.* **2022**, *851*, 158183. [CrossRef]
41. Wang, Y.; Li, P. Appraisal of shallow groundwater quality with human health risk assessment in different seasons in rural areas of the Guanzhong Plain (China). *Environ. Res.* **2022**, *207*, 112210. [CrossRef]
42. Xiao, J.; Lv, G.R.; Chai, N.P.; Hu, J.; Jin, Z.D. Hydrochemistry and source apportionment of boron, sulfate, and nitrate in the Fen River, a typical loess covered area in the eastern Chinese Loess Plateau. *Environ. Res.* **2022**, *206*, 112570. [CrossRef]
43. Gao, Y.Y.; Qian, H.; Zhou, Y.H.; Chen, J.; Wang, H.K.; Ren, W.H.; Qu, W.G. Cumulative health risk assessment of multiple chemicals in groundwater based on deterministic and Monte Carlo models in a large semiarid basin. *J. Clean. Prod.* **2022**, *352*, 131567. [CrossRef]
44. Sunkari, E.D.; Seidu, J.; Ewusi, A. Hydrogeochemical evolution and assessment of groundwater quality in the Togo and Dahomeyan aquifers, Greater Accra Region, Ghana. *Environ. Res.* **2022**, *208*, 112679. [CrossRef] [PubMed]
45. Wang, M.; Wang, M.; Yang, L.; Yang, T.; Li, J.; Chen, Y. Distribution Characteristics and Genesis of Iron and Mangnese Ions in Groundwater of Eastern Sanjiang Plain, China. *Water* **2023**, *15*, 2068. [CrossRef]

46. Gaillardet, J.; Dupre, B.; Louvat, P.; Allegre, C.J. Global silicate weathering and CO_2 consumption rates deduced from the chemistry of large rivers. *Chem. Geol.* **1999**, *159*, 3–30. [CrossRef]
47. Shen, B.B.; Wu, J.L.; Zhan, S.E.; Jin, M.; Saparov, A.S.; Abuduwaili, J. Spatial variations and controls on the hydrochemistry of surface waters across the Ili-Balkhash Basin, arid Central Asia. *J. Hydrol.* **2021**, *600*, 126565. [CrossRef]
48. Zhang, X.; Miao, J.; Hu, B.X.; Liu, H.; Zhang, H.; Ma, Z. Hydrogeochemical characterization and groundwater quality assessment in intruded coastal brine aquifers (Laizhou Bay, China). *Environ. Sci. Pollut. Res.* **2017**, *24*, 21073–21090. [CrossRef] [PubMed]
49. Argamasilla, M.; Barbera, J.A.; Andreo, B. Factors controlling groundwater salinization and hydrogeochemical processes in coastal aquifers from southern Spain. *Sci. Total Environ.* **2017**, *580*, 50–68. [CrossRef]
50. Ha, Q.K.; Ngoc, T.D.T.; Vo, P.L.; Nguyen, H.Q.; Dang, D.H. Groundwater in Southern Vietnam: Understanding geochemical processes to better preserve the critical water resource. *Sci. Total Environ.* **2022**, *807*, 151345. [CrossRef]
51. Borrok, D.M.; Lenz, R.M.; Jennings, J.E.; Gentry, M.L.; Steensma, J.; Vinson, D.S. Applied Geochemistry the origins of high concentrations of iron, sodium, bicarbonate, and arsenic in the Lower Mississippi River Alluvial Aquifer. *Appl. Geochem.* **2018**, *98*, 383–392. [CrossRef]
52. Wang, H.L.; Hussaini, K.M.; Zhang, H.; Rinklebe, J. Hydrogeochemical and health risk evaluation of arsenic in shallow and deep aquifers along the different floodplains of Punjab, Pakistan. *J. Hazard. Mater.* **2021**, *402*, 124074.
53. Brindha, K.; Paul, R.; Walter, J.; Tan, M.L.; Singh, M.K. Trace metals contamination in groundwater and implications on human health: Comprehensive assessment using hydrogeochemical and geostatistical methods. *Environ. Geochem. Health* **2020**, *42*, 3819–3839. [CrossRef]
54. Wang, Z.Y.; Su, Q.; Wang, S.; Gao, Z.J.; Liu, J.T. Spatial distribution and health risk assessment of dissolved heavy metals in groundwater of eastern China coastal zone. *Environ. Pollut.* **2021**, *290*, 118016. [CrossRef] [PubMed]
55. Adimalla, N.; Qian, H. Groundwater chemistry, distribution and potential health risk appraisal of nitrate enriched groundwater: A case study from the semi-urban region of South India. *Ecotoxicol. Environ. Saf.* **2021**, *207*, 111277. [CrossRef] [PubMed]

water

MDPI

Article

Evolution Characteristics of Water Quality in Plain Reservoirs and Its Relationship with the Economic Development Response: A Case Study of Daheiting Reservoir in Northern China

Budong Li [1,2,3], Kaiqi Chen [4], Xiaobo Liu [1,2,3,*], Chang Liu [1,2,3] and Shiyan Wang [1,2,3]

1 State Key Laboratory of Simulation and Regulation of Water Cycle in River Basin, China Institute of Water Resources and Hydropower Research, Beijing 100038, China; libd_smile@126.com (B.L.); liuch@iwhr.com (C.L.); wangsy@iwhr.com (S.W.)
2 Department of Water Ecology and Environment, China Institute of Water Resources and Hydropower Research, Beijing 100038, China
3 Key Laboratory of Water Safety for Beijing-Tianjin-Hebei Region of Ministry of Water Resources, Beijing 100038, China
4 Appraisal Center for Environment and Engineering, Ministry of Environmental Protection, Beijing 100012, China; kqchen010@163.com
* Correspondence: xbliu@iwhr.com

Abstract: In order to explore the evolution characteristics of TP and NH_3-N in Daheiting reservoir since its construction, and their response to economic development, the monitoring data of water quality from 1992 to 2018 and statistical data of socio-economic development in Qianxi County were analyzed to examine the interannual evolution of TP and NH_3-N and their correlation with upstream water quality, various economic indicators, and the scale of cage fish culture. The results show that, influenced by economic development, the evolution process of TP and NH_3-N in Daheiting reservoir can be divided into three stages. In Stage I, the economic development of Qianxi County was slow, and the water quality of upstream water and the reservoir was good, with TP and NH_3-N concentrations remaining relatively stable. In Stage II, Qianxi County entered a period of rapid economic development, and the TP and NH_3-N in upstream water and Daheiting reservoir both increased significantly, with TP exceeding the standard limit. In Stage III, the intensity of external pollution control increased, and all cages were removed from the reservoir. Both TP and NH_3-N showed a downward trend, but TP still exceeded the standard limit. Pearson correlation analysis and RDA analysis revealed that the levels of TP and NH_3-N in Daheiting reservoir were mainly affected by the water quality of upstream water and the development of primary industry (including cage fish culture).

Keywords: cage fish culture; Daheiting reservoir; economic development; NH_3-N; TP

Citation: Li, B.; Chen, K.; Liu, X.; Liu, C.; Wang, S. Evolution Characteristics of Water Quality in Plain Reservoirs and Its Relationship with the Economic Development Response: A Case Study of Daheiting Reservoir in Northern China. *Water* **2023**, *15*, 3229. https://doi.org/10.3390/w15183229

Academic Editors: Guilin Han and Xiaolong Liu

Received: 22 July 2023
Revised: 1 September 2023
Accepted: 4 September 2023
Published: 11 September 2023

1. Introduction

As global socio-economic development progresses, the demand for water resources is gradually increasing, and the quantity and quality of water resources have become important factors constraining social development. China's per capita freshwater resources are only one-fourth of the world's average level, and water use in northern China is severe. The construction of reservoirs has not only alleviated the problem of drinking water, but also brought certain economic benefits. In recent decades, China has accelerated the pace of dam construction [1]. According to data released by the Chinese Ministry of Water Resources, more than 98,000 reservoirs of various types have been built nationwide, with a total storage capacity of 898.3 billion m^3. However, eutrophication is a widespread problem in China and around the world [2,3]. Many research scholars believe that the eutrophication of lakes and reservoirs is mainly caused by the excessive input of exogenous nutrients, such as nitrogen and phosphorus, which is closely related to the economic development model at the time [4]. Taking China as an example, in the early stage of reform and opening up, people had weak environmental awareness, and the country's environmental protection policies were insufficient. Economic development was achieved at the expense of the environment, following the development concept of "pollute first and then treat".

In 1991, American economists Grossman and Krueger [5] first empirically studied the relationship between environmental quality and per capita income, pointing out that the relationship between pollution and per capita income is "pollution rises with per capita GDP at low income levels, declines with GDP growth at high income levels". In 1993, Panayotou [6] first called the relationship between environmental quality and per capita income the Environmental Kuznets Curve (EKC). And it has been verified in many research areas. Paudel (2005) [7], revisiting the environmental Kuznets curve (EKC) hypothesis for water pollution, showed evidence of the inverted U-shaped EKC relationships' existence in America and Europe, while Lee (2010) [8] and Thompson (2014) [9] did the same. However, some recent studies have found that the relationship between the environment and economic growth is not necessarily in an inverted "U" shape due to environmental policy adjustments and other factors, such as "N" type and other modes [10–12]. Currently, there are three types of basic EKC models: the quadratic function (1), cubic function (2) and logarithmic function (3), as follows [13]:

$$y = a + b_1x + b_2x^2 \quad (1)$$

$$y = a + b_1x + b_2x^2 + b_3x^3 \quad (2)$$

$$y = a + b_1 \ln x + b_2(\ln x)^2 + b_3(\ln x)^3 \quad (3)$$

Tianjin and Tangshan cities are located in the North China region of China, with urban populations of 13.73 million and 7.70 million, respectively, in 2021. In order to alleviate the water supply problems of the two cities, the "Luanhe River Diversion Project" was proposed in 1973, with the newly built Panjiakou Reservoir and Daheiting reservoir cascade reservoir serving as water sources [14]. Among them, construction of the Panjiakou Reservoir began in 1975, and it began impounding water in 1979. In April 1980, the project began to produce benefits, supplying agricultural water to the Tangshan area. In July 1983, it was ready to supply water to Tianjin, and officially began supplying water to Tianjin in September of that year. The Daheiting reservoir began construction in October 1973, and was basically completed and put into use in 1983. The Panjiakou Reservoir can regulate 1.95 billion m^3 water annually, and supply 1.0 billion m^3 water to Tianjin and Tangshan each year. Over the past 30 years, it has supplied a cumulative total of approximately 31.6 billion m^3, effectively guaranteeing the water supply for industry, agriculture, and urban life in both cities. From the 1980s to before the 21st century, the water quality of the two reservoirs was good and met Class III of the Chinese Environmental Quality Standards for Surface Water (EQS 2002) [15]. Starting from the year 2000, the decade of gold development, the concentration of pollutants in the reservoir has rapidly increased and eutrophication has become increasingly severe, posing a significant threat to the water supply [16,17]. From June 2016 to November 2017, the water supply to Tianjin was stopped due to water quality issues. Therefore, the water quality standards and eutrophication problem of Daheiting reservoir urgently need to be improved.

Currently, the research on Daheiting reservoir has mainly focused on the monitoring and evaluation of water quality and eutrophication status, and has rarely analyzed the water-quality evolution process and influencing factors since the reservoir was built. Only by clarifying the historical causes of the deterioration of reservoir water quality can an accurate and effective treatment plan be proposed. This study is based on the TP and NH_3-N monitoring data of Daheiting reservoir from 1992 to 2018, and sorts out the economic development trend and industrial structure changes of Qianxi County during this period, as well as the scale of cage fish culture and the historical changes of upstream water quality, and attempts to analyze the historical evolution characteristics and causes of water quality in Daheiting reservoir, in order to provide a theoretical basis for the improvement of water quality in the reservoir.

2. Materials and Methods

2.1. Study Area

The Daheiting reservoir is located in Qianxi County, Tangshan City, Hebei Province (Figure 1). The construction of the Daheiting reservoir started in October 1973, was basically completed and put into operation in 1983, and was completed in 1986. Daheiting reservoir is an annual regulating reservoir, with a total surface area of 25 km^2 and storage capacity of 337 million m^3. The crest elevation of the main dam is about 138.8 m (Yellow Sea elevation), the normal water storage level is 133.00 m, and the dead water level is 121.50 m. The maximum water depth in front of the dam can reach about 26 m. About 30 km upstream of it is the Panjiakou Reservoir, which is connected to the Panjiakou Reservoir through the lower pool, collectively referred to as the Pan Da Reservoir. The Daheiting reservoir controls a watershed area of 3.51×10^4 km^2, of which the watershed area between Panjiakou and Daheiting reservoir is 1400 km^2. It is one of the important sources of drinking water in the Luanhe River system, and also one of the main sources of water in Tianjin.

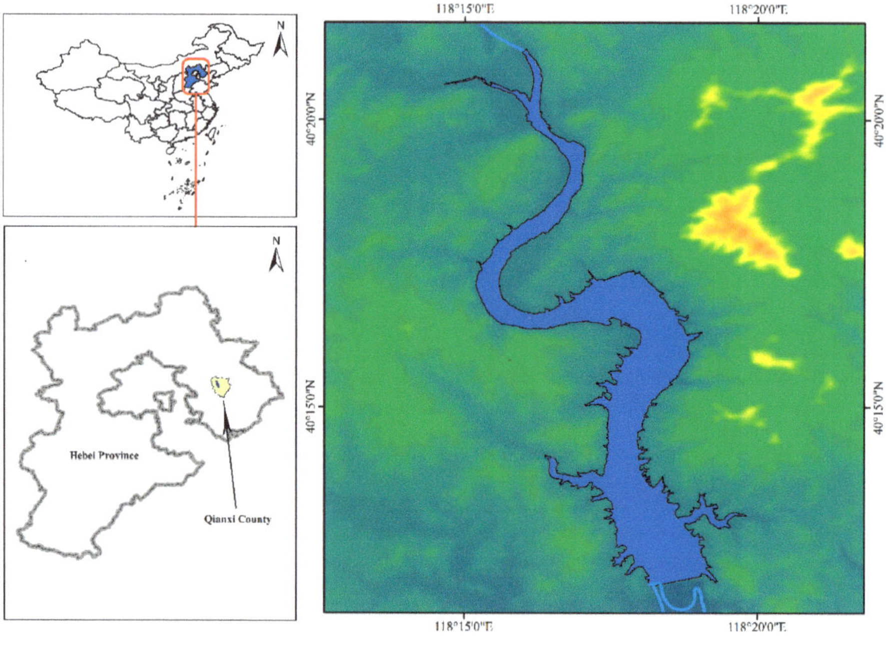

Figure 1. Map of Daheiting reservoir.

2.2. Data Sources and Analysis

The water quality monitoring data of Pan Da Reservoir comes from the routine monitoring data of the Luanhe River Diversion Project Management Bureau, and the social and economic development data of Qianxi County are obtained through the Tangshan City Statistical Yearbook, including per capita GDP, GDP of the primary, secondary and tertiary industries, cage fish culture area and cage fish culture production (CFCP), etc. Pearson correlation analysis was performed in SPSS 20.0, $p < 0.05$ was used to determine significance, and origin 2018b was used for regression fitting and drawing. Redundancy analysis (RDA) was performed using CANOCO 5 to explain the relationship between water quality indicators and influencing factors. The map of the study area was drawn using ArcGIS 10.4.

3. Results and Analysis

3.1. Analysis of Water Quality Change

3.1.1. Analysis of Annual Changes in Water Quality

The interannual variation of TP in Daheiting reservoir showed a trend of increasing first and then decreasing (Figure 2a). Specifically, it can be divided into three stages. The first stage is from 1992 to 2006. The concentration of TP remained basically unchanged, and the overall level remained at about 0.04 mg/L, which met the water quality requirements of EQS Class III; in the second stage, from 2007 to 2016, the overall trend increased exponentially, from 0.054 mg/L in 2007 to 0.382 mg/L in 2016. It is generally believed that a TP concentration greater than 0.1 mg/L is a state of severe eutrophication [17–20]. It can be seen that the reservoir was in a state of severe eutrophication from 2010 to 2016. The third stage is from 2017 to 2018. The concentration of TP dropped rapidly. In 2018, it was 0.069 mg/L, which was Class IV of the EQS, and the TP still exceeded the standard (Figure 2c).

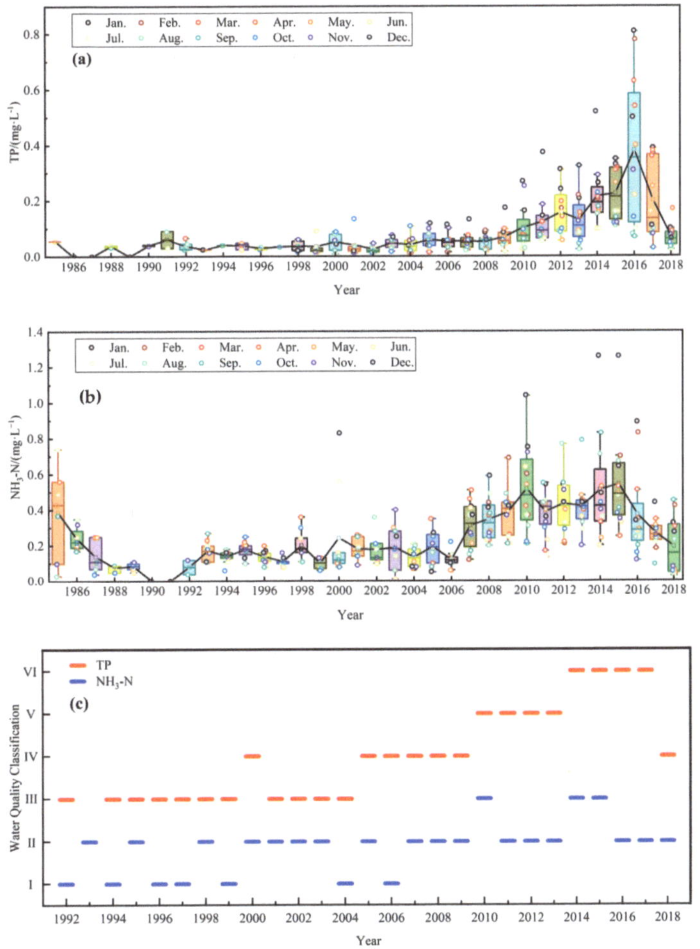

Figure 2. Interannual changes of TP (**a**) and NH_3-N (**b**) and water quality categories (**c**) in Daheiting reservoir from 1992 to 2018.

The interannual change trend of NH_3-N in Daheiting reservoir is consistent with that of TP, which is also divided into three stages (Figure 2b). The first stage is from 1992 to 2006. The annual average value of this stage fluctuates, with 0.08 mg/L in 1992 and 0.12–0.24 mg/L in other years. It belongs to Class II or Class I water quality. The second stage is from 2007 to 2015. The NH_3-N showed a significant upward trend, from 0.31 mg/L in 2007 to 0.54 mg/L in 2015, and the ammonia nitrogen concentration reached the highest value since the establishment of the reservoir. Among them, in 2010, 2014 and in 2015, it was Class III water quality. The third stage is from 2016 to 2018, and the NH_3-N shows a linear downward trend, with an annual average of 0.19 mg/L in 2018. It can be seen that although the NH_3-N concentration in the reservoir increased significantly in the second stage, it did not exceed the standard, and it met the Class III water quality requirements (Figure 2c).

3.1.2. Analysis of Monthly Changes in Water Quality

Both TP (Figure 3a) and NH_3-N (Figure 3b) in Daheiting reservoir showed obvious seasonal changes. Overall, TP is high in winter with a large interannual variation, and lowest in summer with small interannual variation. Among them, the maximum value of TP appeared in January, which was 0.231 mg/L, showing a state of severe eutrophication, which was inferior to Class V water quality, while it was the lowest in July to September, about 0.05 mg/L, meeting the requirements of Class III water quality. However, NH_3-N is generally high in winter and low in other seasons, and the interannual variation range is slightly higher in winter than in other seasons. The maximum annual average concentration appeared in December, which was 0.42 mg/L, and the lowest occurred in October, which was 0.20 mg/L.

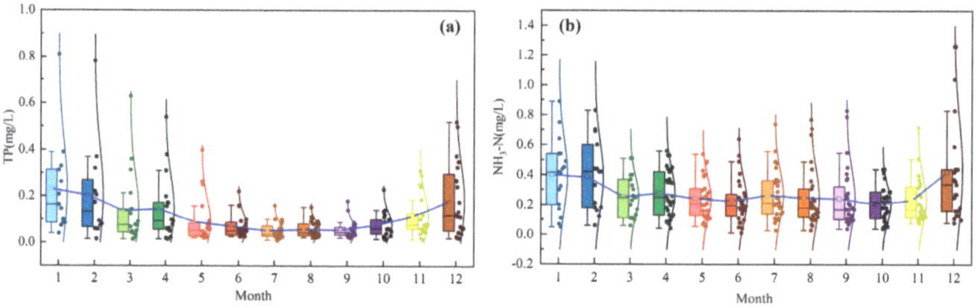

Figure 3. Monthly changes of TP (**a**) and NH_3-N (**b**) in Daheiting reservoir from 1992 to 2018.

3.2. Analysis of Social and Economic Development Trends

3.2.1. Trend Analysis of GDP

The output value of the three industries in Qianxi County and per capita GDP are shown in Figure 4a, which maintained a rapid growth rate from 1992 to 2013, and its GDP increased from CNY 7.05 $\times$ 10^4 ten thousand to CNY 422.28 $\times$ 10^4 ten thousand, an increase of about 60 times. From 2013 to 2018, the economic development slowed down, showing the characteristics of fluctuations. The per capita GDP also showed an obvious upward trend, especially during the period from 2003 to 2013: the per capita GDP increased from CNY 15,900 to CNY 107,000. From 2014 to 2018, economic development slowed down, and fluctuated around CNY 100,000.

In China, industries are divided into three categories. The primary industry includes agriculture, forestry, animal husbandry and fishery, while the secondary industry refers to the manufacturing, mining and construction sectors. The tertiary industry refers to industries other than the primary and secondary industries, including commerce, finance, trust, service industries and other industries that provide production and consumption services [21]. The industrial structure of Qianxi County has changed significantly, which is mainly reflected in the decline of the proportion of primary industry before 2004, from 30.9% to 7.23% (Figure 4b); The proportion of the secondary industry gradually increased, from 41.59% to 63.84%, and has always accounted for more than 50% of the total output value since 2002; the proportion of the tertiary industry did not change. After 2004, the proportion of the industrial structure in Qianxi County remained basically stable, the secondary industry had a slight downward trend, and the tertiary industry showed just the opposite.

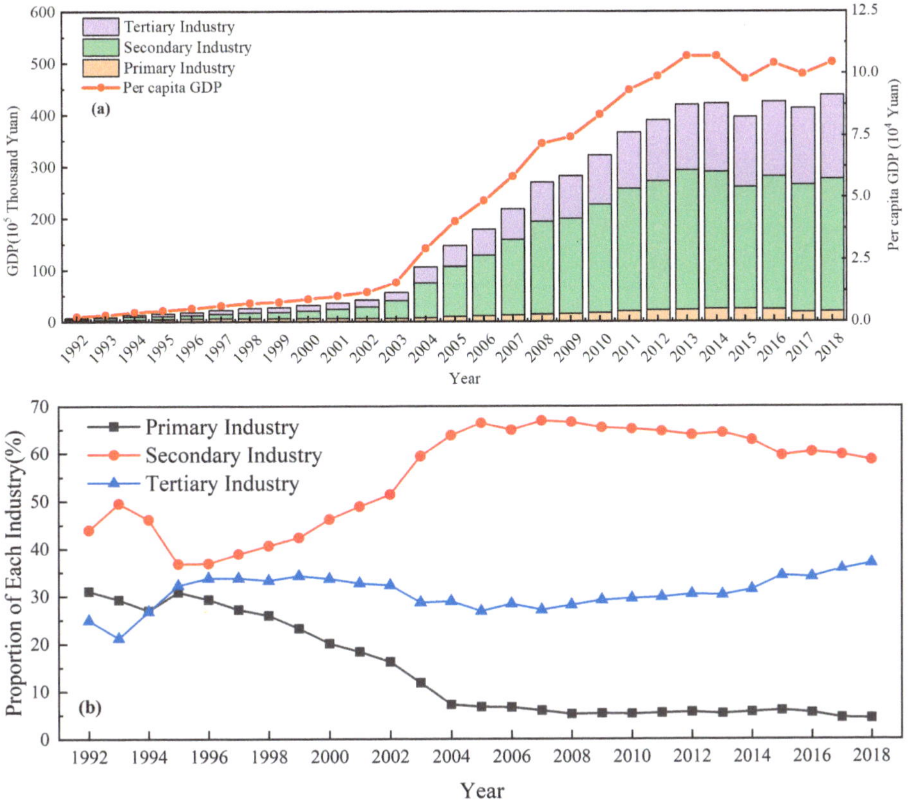

Figure 4. Economic development trend (**a**) and industrial structure (**b**) changes in Qianxi County from 1992 to 2018.

3.2.2. Analysis of Cage Fish Culture Development Trend in Qianxi County

After the completion of the Daheiting reservoir, in order to drive the economic development of the surrounding areas, in 1990, the development of the "cage fish culture" was encouraged. In this study, the development of cage fish culture in Qianxi County was divided into three stages according to the changes in fish species and farming area (Figure 5). Stage I was from 1992 to 2006. In order to improve the income level of local residents, the government encouraged the development of cage fish culture [22]. At this stage, filter-feeding silver carp and bighead carp, which feed on phytoplankton and zooplankton, were mainly cultured, and no artificial feeding was required. Their farming areas were 25.77×10^6 m^2 and 27.74×10^6 m^2 in 1992 and 1993, respectively. In 1994, the total area of farming of silver carp and bighead carp were developed to 41.73×10^6 m^2 and remained until 2006. The production of aquatic products increased from 1200 tons in 1991 to 26,500 tons in 2006. The second stage was from 2007 to 2015. Affected by the large-scale fish-kill incidents in 2007, the aquaculture area in Qianxi County was greatly reduced to about 10.75×10^6 m^2. At the same time, in order to maintain a high cage-fish culture production, Grass carp were cultured in the reservoir, and fish food was artificially added for feeding. The annual production increased from 21,631 tons to 44,900 tons. The third stage was from 2016 to 2018. A series of water quality-deterioration problems, such as large-scale fish-kill incidents and algal blooms, occurred continuously in the Daheiting reservoir. In order to prevent further deterioration of the water quality, cages were removed from November 2016 to May 2017, and the aquaculture area in Qianxi County dropped again to 4.23×10^6 m^2. At the same time, the production of Grass carp dropped sharply to 12,610 tons and 12,720 tons.

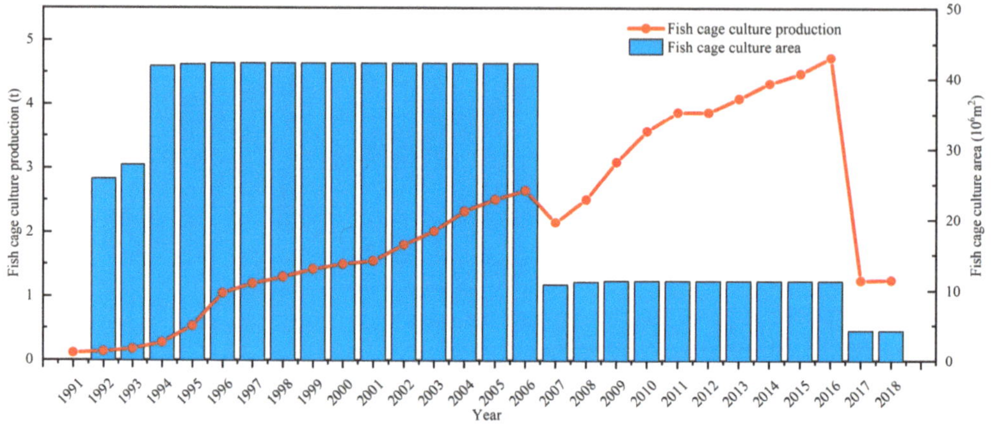

Figure 5. Changes in the scale of cage fish culture industry in Qianxi County from 1992 to 2018.

3.3. *Effects of Upstream Water on Water Quality Change of Daheiting Reservoir*

The Daheiting reservoir mainly receives water from the Sahe and Panjiakou Reservoirs. Among them, the Sahe River has a much smaller flow rate, with an average annual runoff of 280 million m^3, and its TP (Figure 6a) and NH_3-N (Figure 6b) are significantly lower than those of the Panjiakou Reservoir [23]. Panjiakou Reservoir is located about 30 km upstream of Daheiting reservoir, with a total reservoir capacity of 2.93 billion m^3, nine times the capacity of Daheiting reservoir [24]. It is about 1.33 billion m^3 [25], meaning it can fill the Daheiting reservoir four times. Therefore, Daheiting reservoir mainly receives water from Panjiakou Reservoir, and the water quality of Panjiakou Reservoir will directly affect the water quality of Daheiting reservoir.

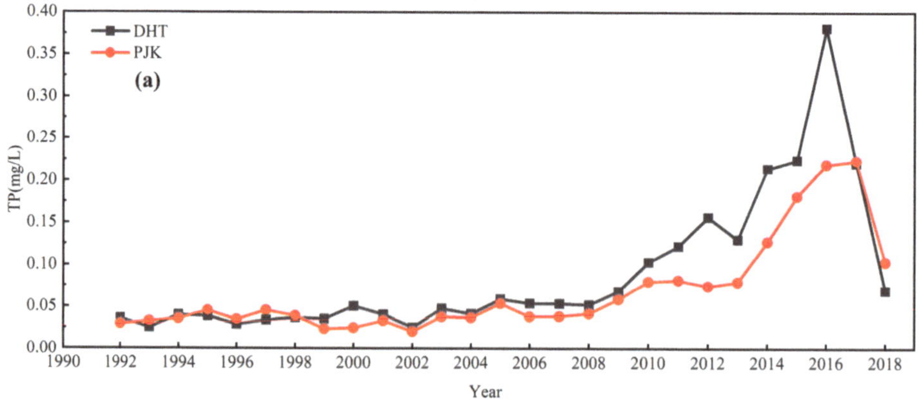

Figure 6. *Cont.*

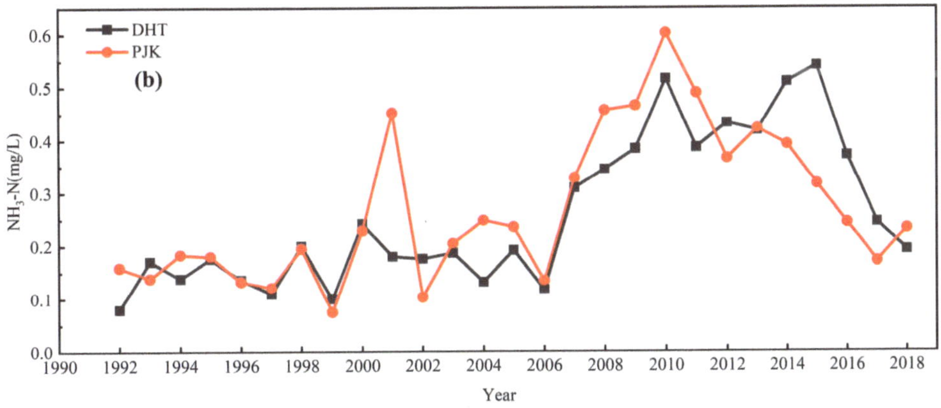

Figure 6. Interannual changes of TP (**a**) and NH_3-N (**b**) in Pan Da Reservoir from 1992 to 2018.

Domagalski [23] found that the main source of nutrients in Panjiakou Reservoir and Daheiting reservoir is the Luan River, and the water from the Luan River flows directly into the Panjiakou Reservoir. Therefore, using the monitoring data of TP and NH_3-N in front of the dam of Panjiakou Reservoir as the upstream water quality, the impact of upstream water on the water quality change of Daheiting reservoir was analyzed. The results showed that the annual average change trend of TP and NH_3-N in the upstream water was basically consistent with that of Daheiting reservoir, and there was a significant positive correlation between TP (Figure 7a) and NH_3-N (Figure 7b) and the upstream water quality (R^2 = 0.84, $p < 0.01$; R^2 = 0.60, $p < 0.01$). It is one of the main sources of TP and NH_3-N in Daheiting reservoir.

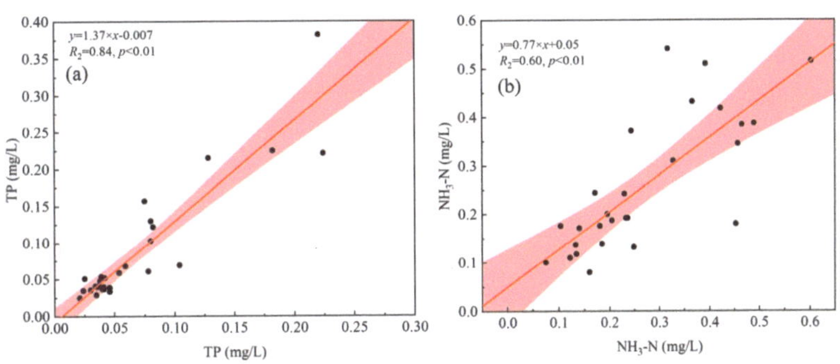

Figure 7. Correlation analysis of TP (**a**) and NH_3-N (**b**) between Daheiting reservoir and upstream water quality.

3.4. Effects of Cage-Fish Culture on Water Quality Changes in Daheiting Reservoir

Cage-fish culture in Daheiting reservoir is divided into three stages. In order to more accurately analyze the impact of cage-fish culture at each stage on the water quality of Daheiting reservoir, correlation analysis and regression analysis were carried out in stages using the basic data of the aquatic product yield in Qianxi County and the TP and NH_3-N of Daheiting reservoir from 1992 to 2018(Table 1, Figure 8). The results showed that there was a significant correlation ($p < 0.05$) between TP and cage-fish culture production in stage one (1992–2006), but the fitting effect was not satisfactory (R^2 = 0.36); there was no correlation between NH_3-N and cage-fish culture production. At this stage, the reservoir mainly cultured silver carp and bighead carp. Bighead carp and silver carp have been proven to effectively inhibit the reproduction of Microcystis [26], thereby indirectly absorbing nitrogen and phosphorus nutrients. For example, Benjamin et al. found that phytoplankton accounted for about 63.5% of silver carp food [27]. Huang et al. [28] took Panjiakou Reservoir as the research object and found that bighead carp can reduce the concentration of TP in water by preying on phytoplankton. Therefore, although the cage culture area was larger at this stage, it did not have a significant impact on TP and NH_3-N. The second stage was from 2007 to 2015. Compared with the first stage, the aquaculture area in this stage was greatly reduced, while the cage-fish culture production increased significantly. Both TP and NH_3-N had a very significant positive correlation with cage-fish culture production ($p < 0.01$). Because the feed contains a large amount of pollutants such as nitrogen, phosphorus, and organic matter, after being put into the water body, the organic nitrogen gradually mineralizes to form NH_3-N, which causes the concentration of NH_3-N and TP in the water to increase [29,30]. Therefore, "cage fish culture" at this stage was one of the important factors that caused the obvious increase in reservoir TP and NH_3-N. In the third stage (2016~2018), since the cages were completely removed, there was no correlation between TP and NH_3-N and cage-fish culture production.

Table 1. Fitting results of water quality indicators and cage-fish culture production in Daheiting reservoir.

Water Quality Factor	Stage	Curve Fitting	R^2	p
TP	Stage I	$y = 7.28 \times 10^{-3} \times x + 0.03$	0.36	<0.05
	Stage II	$y = 0.07 \times x - 0.13$	0.81	<0.01
	Stage III	-	-	-
NH_3-N	Stage I	-	-	-
	Stage II	$y = 0.08 \times x + 0.14$	0.67	<0.01
	Stage III	-	-	-

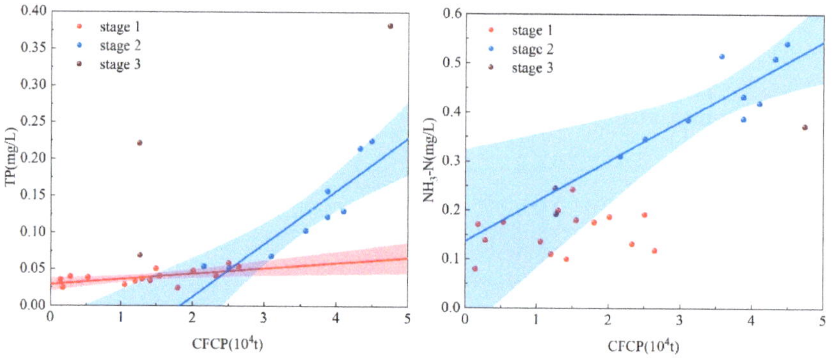

Figure 8. Correlation analysis between water quality index and cage-fish culture production of Daheiting reservoir.

4. Discussions

We selected Qianxi County's per capita GDP from 1991 to 2018 as the independent variable, and TP and NH_3-N as the dependent variable, and used the mathematical model of EKC theory to conduct regression analysis. The results are as follows.

TP and per capita GDP show a "U"-shaped change pattern (Figure 9), and are in the right half of the "U" shape. This shows that with the development of the economy, the TP of the reservoir maintained an upward trend. The possible reasons for this phenomenon are the following: (1) There are many and complex factors affecting the TP change of Daheiting reservoir. Although there is a positive correlation between the two, it is still uncertain whether economic development is the main influencing factor. (2) The maximum water depth in front of the dam of Daheiting reservoir can reach 26 m, and there is a stable thermal stratification phenomenon from May to September. Phosphorus content in the mud is relatively high, gradually shifting from "sink" to "source". During thermal stratification, the lack of oxygen at the bottom will cause a large amount of phosphorus to be released from the sediment to the overlying water body. After the thermal stratification disappears, TP quickly diffuses to the surface water body, resulting in an increase in the concentration of TP in the water body [15]. Therefore, there may be some hysteresis in the improvement of water quality, and this hypothesis can be further verified according to the monthly change law of TP in Daheiting reservoir.

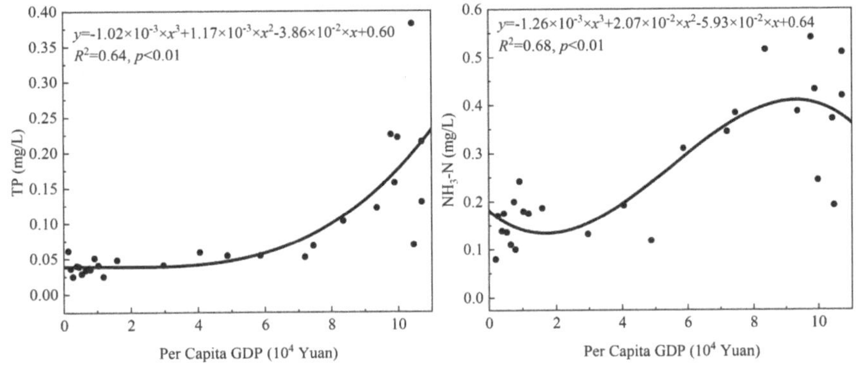

Figure 9. EKC fitting curve of water quality index and per capita GDP of Daheiting reservoir.

NH_3-N and per capita GDP are in an inverted "N" shape (Figure 9), and the inflection points appear when the per capita GDP is CNY 16,900 and CNY 92,600, corresponding to 2003 and 2011, respectively. Before 2003, the economic development of Qianxi County was slow, the industrial structure was in the adjustment stage, and the industry was still in the rising stage of development. At this time, economic development did not significantly increase the concentration of NH_3-N. After 2003, the proportion of each industry structure was basically stable, and the secondary industry accounted for more than 60%. In 2007, the influence of artificial feed for cage fish culture was superimposed, which further intensified the input of exogenous NH_3-N, and the concentration of NH_3-N in the reservoir increased rapidly. From 2011 to 2015, China began to attach importance to and strengthen water-environment protection, and the State Council successively issued water-environment protection policies, such as "Opinions on Strengthening Key Environmental Protection Work" and "Notice on National Environmental Protection "Twelfth Five-Year Plan"" planning [14], especially after the release of the Action Plan for the Rule of Law in Water Pollution in 2015; this was the first time that a basic national policy of environmental protection had been implemented as a specific and strict government responsibility. The NH_3-N in Daheiting reservoir fluctuated up and down. After the implementation of the net cage cleanup operation in Daheiting reservoir in 2016, NH_3-N began to decrease year by year.

Through the above analysis, it can be found that the water quality change of Daheiting reservoir is closely related to the upstream water, economic development and cage-fish culture. However, the priority relationship between TP and NH_3-N and the response of the above influencing factors is still unknown. For this reason, this study conducted further analysis through RDA (Redundancy analysis), and the results showed that the concentration of TP and NH_3-N in the upstream Panjiakou Reservoir and the primary industry (including cage fish culture industry) are the factors that affect the TP and NH_3-N concentration of Daheiting reservoir (Figure 10). Among them, upstream water is the primary factor affecting TP, and the primary industry (cage-fish farming) has a greater impact on NH_3-N. It can be shown that the exogenous input of nutrients was the main factor of TP and NH_3-N concentration during the second stage of the reservoir.

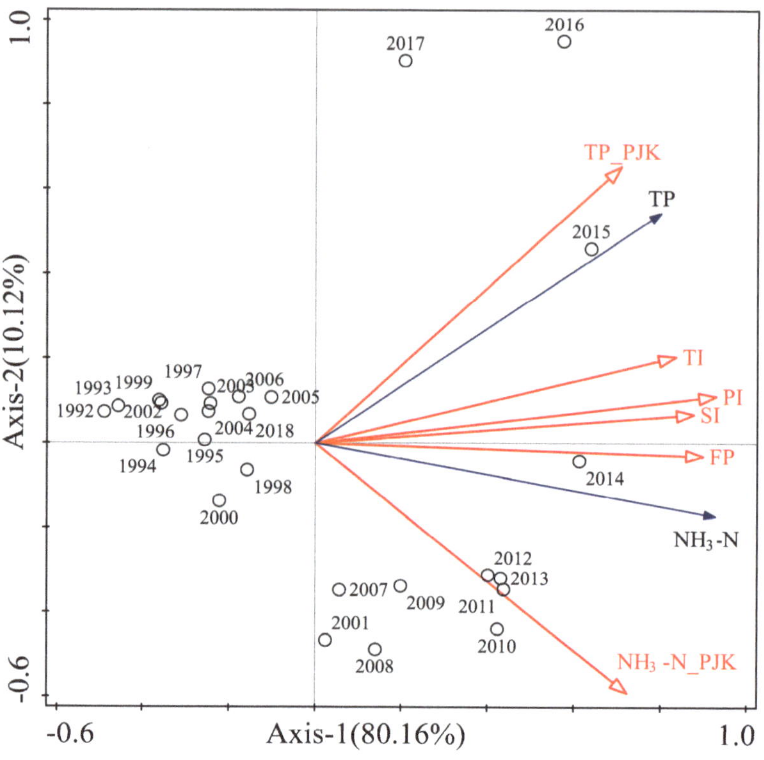

Figure 10. RDA analysis between water quality index and various influencing factors in Daheiting reservoir.

5. Conclusions

The interannual changes of TP and NH_3-N in Daheiting reservoir can be divided into three stages: 1992–2006, 2007–2016 and 2017–2018. These periods demonstrated the characteristics of basically unchanged, a significant increase and a significant decrease in turn. Among them, both TP and NH_3-N in the first stage met the water quality target requirements, and the change trend was basically not affected by economic development. In the second stage, there was a significant positive correlation. With the improvement in the economic level of China, the concentration of TP and NH_3-N increased rapidly; in particular the content of TP was significantly exceeded, and the water body was in a state of severe eutrophication. In the third stage, the speed of economic development slowed down, environmental protection efforts were strengthened, and TP and NH_3-N showed significant improvement again.

The TP and NH_3-N of Daheiting reservoir were significantly correlated with the upstream water, cage-fish culture and the output value of various industries. Among them, the quality of upstream water quality and the primary industry (including cage-fish culture) were the primary factors affecting the changes in TP and NH_3-N concentration, which was mainly reflected in the second and third stages, when the concentration of upstream water quality indicators increased and the net cages increased. The concentration of TP and NH_3-N in the reservoir increased sharply due to the feeding of fish culture, and the TP and NH_3-N decreased significantly after the cages were removed, but there was a certain lag in TP due to the influence of endogenous pollution.

Author Contributions: Writing—original draft preparation, B.L.; writing—review and editing, X.L. and B.L.; methodology, K.C.; visualization, C.L. and B.L.; data curation, S.W. All authors have read and agreed to the published version of the manuscript.

Funding: This research was funded by the National Basic Research Program of China (2021YFC3200903), independent research project of State Key Laboratory of Simulation and Regulation of Water Cycle in River Basin (NO. SKL2022ZD02).

Data Availability Statement: The data have been explained in the paper.

Acknowledgments: We thank Shaoming Wang and Enling Zhao from the Luanhe River diversion project management bureau for providing water quality data.

Conflicts of Interest: The authors declare no conflict of interest.

References

1. Li, Z.; Ma, J.; Guo, J.; Paerl, H.W.; Brookes, J.D.; Xiao, Y.; Fang, F.; Ouyang, W.; Lunhui, L. Water quality trends in the Three Gorges Reservoir region before and after impoundment (1992–2016). *Ecohydrol. Hydrobiol.* **2019**, *19*, 317–327. [CrossRef]
2. Shourian, M.; Moridi, A.; Kaveh, M. Modeling of eutrophication and strategies for improvement of water quality in reservoirs. *Water Sci. Technol.* **2016**, *74*, 1376–1385. [CrossRef]
3. Yang, L.; Peng, S.; Zhao, X.; Li, X. Development of a two-dimensional eutrophication model in an urban lake (China) and the application of uncertainty analysis. *Ecol. Model.* **2017**, *345*, 63–74.
4. Longyang, Q. Assessing the effects of climate change on water quality of plateau deep-water lake-A study case of Hongfeng Lake. *Sci. Total Environ.* **2019**, *647*, 1518–1530. [CrossRef] [PubMed]
5. Grossman, G.M.; Krueger, A.B. *Environmental Impacts of a North American Free Trade Agreement*; National Bureau of Economic Research: Cambridge, MA, USA, 1991.
6. Panayotou, T. *Empirical Tests and Policy Analysis of Environmental Degradation at Different Stages of Economic Development*; International Labour Office: Geneva, Switzerland, 1993.
7. Paudel, K.P.; Zapata, H.; Susanto, D. An empirical test of environmental Kuznets curve for water pollution. *Environ. Resour. Econ.* **2005**, *31*, 325–348. [CrossRef]
8. Lee, C.C.; Chiu, Y.B.; Sun, C.H. The environmental Kuznets curve hypothesis for water pollution: Do regions matter? *Energy Policy* **2010**, *38*, 12–23. [CrossRef]
9. Thompson, A. Environmental Kuznets curve for water pollution: The case of border countries. *Mod. Econ.* **2014**, *5*, 66–69. [CrossRef]
10. Tenaw, D.; Beyene, A.D. Environmental sustainability and economic development in sub-Saharan Africa: A modified EKC hypothesis. *Renew. Sustain. Energy Rev.* **2021**, *143*, 110897. [CrossRef]
11. Al Sayed, A.; Sek, S.K. Environmental Kuznets curve: Evidences from developed and developing economies. *Appl. Math. Sci.* **2013**, *7*, 1081–1092. [CrossRef]
12. Özokcu, S.; Özdemir, Ö. Economic growth, energy, and environmental Kuznets curve. *Renew. Sustain. Energy Rev.* **2017**, *72*, 639–647. [CrossRef]
13. Ongan, S.; Isik, C.; Ozdemir, D. Economic growth and environmental degradation: Evidence from the US case environmental Kuznets curve hypothesis with application of decomposition. *J. Environ. Econ. Policy* **2021**, *10*, 14–21. [CrossRef]
14. Wei, M.; Huang, S.; Li, L.; Zhang, T.; Akram, W.; Khatoon, Z.; Renaud, F.G. Evolution of water quality and biota in the Panjiakou Reservoir, China as a consequence of social and economic development: Implications for synergies and trade-offs between Sustainable Development Goals. *Sustain. Sci.* **2021**, *17*, 1385–1404. [CrossRef]
15. Liu, C.; Wang, S.; Liu, X.; Zhou, H.; Li, B.; Du, Y.; Wang, L. Characteristics of water quality response to hypolimnetic anoxia in Daheiting Reservoir. *Water Sci. Technol.* **2022**, *85*, 2065–2075. [CrossRef]
16. Yang, W.; Yao, J.; He, Y.; Huang, Y.; Liu, H.; Zhi, Y.; Qian, S.; Yan, X.; Jian, S.; Li, W. Nitrogen removal enhanced by benthic bioturbation coupled with biofilm formation: A new strategy to alleviate freshwater eutrophication. *J. Environ. Manag.* **2021**, *292*, 112814. [CrossRef] [PubMed]
17. Smith, V.H.; Tilman, G.D.; Nekola, J.C. Eutrophication: Impacts of excess nutrient inputs on freshwater, marine, and terrestrial ecosystems. *Environ. Pollut.* **1999**, *100*, 179–196. [CrossRef]
18. Nürnberg, G.K. Trophic state of clear and colored, soft-and hardwater lakes with special consideration of nutrients, anoxia, phytoplankton and fish. *Lake Reserv. Manag.* **1996**, *12*, 432–447. [CrossRef]
19. Dodds, W.K.; Jones, J.R.; Welch, E.B. Suggested classification of stream trophic state: Distributions of temperate stream types by chlorophyll, total nitrogen, and phosphorus. *Water Res.* **1998**, *32*, 1455–1462. [CrossRef]
20. Håkanson, L. A Review on Effect-Dose-Sensitivity Models for Aquatic Ecosystems. *Int. Rev. Gesamten Hydrobiol. Hydrogr.* **1994**, *79*, 621–667. [CrossRef]
21. Jing, C.; Su, B.D.; Zhai, J.Q.; Wang, Y.J.; Lin, Q.G.; Gao, M.N.; Jiang, S.; Chen, Z.Y.; Jiang, T. Gridded value-added of primary, secondary and tertiary industries in China under Shard Socioeconomic Pathways. *Sci. Data* **2022**, *9*, 309. [CrossRef]

22. Kang, G.; Yin, J.; Cui, N.; Ding, H.; Wang, S.; Wang, Y.; Qi, Z. The long-term and retention impacts of the intervention policy for cage aquaculture on the reservoir water qualities in Northern China. *Water* **2020**, *12*, 3325. [CrossRef]
23. Domagalski, J.; Lin, C.; Luo, Y.; Kang, J.; Wang, S.; Brown, L.R.; Munn, M.D. Eutrophication study at the Panjiakou-Daheiting Reservoir system, northern Hebei Province, People's Republic of China: Chlorophyll-a model and sources of phosphorus and nitrogen. *Agric. Water Manag.* **2007**, *94*, 43–53. [CrossRef]
24. Liu, C.; Liu, X.; Zhou, H.; Li, B.; Wang, S.; Wang, L. Study on the threshold condition to suppress anoxic zone by large flow operation process. *J. Hydraul. Eng.* **2021**, *52*, 1217–1228. [CrossRef]
25. Liu, C.; Liu, X.; Zhou, H.; Wang, S.; Li, B. Temporal and spatial evolution characteristics and driving factors of reservoir anoxic zone. *J. Hydraul. Eng.* **2019**, *50*, 1479–1490. [CrossRef]
26. Geletu, T.T. Lake eutrophication: Control of phytoplankton overgrowth and invasive aquatic weeds. *Lakes Reserv. Res. Manag.* **2023**, *28*, e12425. [CrossRef]
27. Tumolo, B.B.; Flinn, M.B. Diet of invasive silver carp (Hypophthalmichthys molitrix) in a mainstem reservoir ecosystem. *J. Ky. Acad. Sci.* **2019**, *79*, 3–11.
28. Huang, S.; Wu, M.; Zang, C.; Du, S.; Domagalski, J.; Gajewska, M.; Gao, F.; Lin, C.; Guo, Y.; Liu, B. Dynamics of algae growth and nutrients in experimental enclosures culturing bighead carp and common carp: Phosphorus dynamics. *Int. J. Sediment Res.* **2016**, *31*, 173–180. [CrossRef]
29. Hargreaves, J.A. Nitrogen biogeochemistry of aquaculture ponds. *Aquaculture* **1998**, *166*, 181–212. [CrossRef]
30. Wu, M.; Huang, S.; Zang, C.; DU, S. Enclosure experimental research with fish culturing in the Panjiakou ReservoirII.Effects of fish food and fish species on phosphorus flux. *J. Hydraul. Eng.* **2013**, *44*, 1204–1209. [CrossRef]

water

MDPI

Article

Sustaining the Pearl River: A Critical Review of Changes in Fluvial Geomorphological Processes and the Driving Forces in the Pearl River Basin

Haidong Ou [1], Shirong Cai [1], Wei Fan [1], Junliang Qiu [2], Xiaolin Mu [1], Tao Zhou [1], Xiankun Yang [1,3,*] and Lorenzo Picco [2]

[1] School of Geography and Remote Sensing, Guangzhou University, Guangzhou 510006, China; haidong_ou@e.gzhu.edu.cn (H.O.)
[2] Department of Land, Environment, Agriculture and Forestry (TESAF), University of Padova, Agripolis, Viale dell'Università 16, 35020 Legnaro, Italy
[3] Rural Non-Point Source Pollution Comprehensive Management Technology Center of Guangdong Province, Guangzhou University, Guangzhou 510006, China
* Correspondence: yangxk@gzhu.edu.cn

Abstract: The Pearl River is one of China's large rivers, the second-largest river and the fourth-longest river in China. Its unique geography, landform, and climate conditions create unique fluvial geomorphological processes. Affected by human activities and climate change, the fluvial geomorphological processes in the Pearl River Basin have undergone significant changes in recent decades, seriously affecting the river's sustainable development. This paper critically reviews changes in fluvial geomorphological processes and analyzes influencing factors in the Pearl River Basin with a focus on possibilities for policy overhaul and strategic adjustments.

Keywords: Pearl River Basin; fluvial geomorphological processes; anthropogenic impacts; climate change; land cover change

check for updates

Citation: Ou, H.; Cai, S.; Fan, W.; Qiu, J.; Mu, X.; Zhou, T.; Yang, X.; Picco, L. Sustaining the Pearl River: A Critical Review of Changes in Fluvial Geomorphological Processes and the Driving Forces in the Pearl River Basin. *Water* **2024**, *16*, 1001. https://doi.org/10.3390/w16071001

Academic Editor: Salvatore Ivo Giano

Received: 1 March 2024
Revised: 26 March 2024
Accepted: 27 March 2024
Published: 29 March 2024

1. Introduction

Due to the increasingly dramatic impacts of global climate change and human activities, the ecological balance characterized by the material and energy cycles via fluvial geomorphological processes has received widespread attention in recent decades [1–3]. Rivers are a primary component in fluvial geomorphological cycles; they serve as important sources of water resources and critical functions for terrestrial ecosystems [4,5]. Understanding the new status in fluvial geomorphological processes is valuable in mitigating natural hazards such as floods and droughts, as well as assessing the anthropogenic impact on natural ecosystems [6]. Asia is one of the most densely populated regions on the Earth. The influence of human activities on fluvial geomorphological processes is particularly pronounced in Asia, with China's large rivers being the most representative [7].

The Pearl River is one of the large rivers in China, composed of the West (Xijiang) River, North (Beijing) River, East (Dongjiang) River, and the Pearl River Delta (PRD). It spans six provinces in China, namely Yunnan, Guizhou, Guangxi, Guangdong, Hunan, and Jiangxi provinces, as well as the special administrative regions of Hong Kong and Macao and an extremely small northeastern part in Vietnam [5,8]. The total area of the Pearl River Basin (PRB) is 45.37×10^4 km^2, with 44.21×10^4 km^2 in China [9]. The Pearl River ranks as the second-largest river and the fourth-longest river in China [10]. The main stem of the Pearl River is the West River, originating from the eastern foothills of the Wumeng Mountains in the Qujing Prefecture of Yunnan province. Flowing from west to east, it traverses Yunnan, Guizhou, Guangxi, and Guangdong provinces, with a total length of 2075 km and a drainage area of 35.31×10^4 km^2 (Figure 1). Major tributaries include the Beipan River, Liu River, Yu River, Gui River, and He River. Another major tributary, the

North River, originates from the Xiaomao Mountain in Xinfeng County, Jiangxi Province. It flows through the provinces of Hunan, Jiangxi, and Guangdong, with a total length of 468 km and a drainage area of 4.67×10^4 km^2. Major tributaries include the Wushui River, Lian River, and Sui River. The West River and North River converge at Sanshui District in Foshan (Figure 1b), Guangdong Province. The third major tributary, the East River, originates from the Yajiboshan Mountain in Xunwu County, Jiangxi Province, and flows southward into Guangdong Province. It has a total length of 520 km and a drainage area of 2.70×10^4 km^2. Major tributaries include the Xinfeng River and Xizhi River. The PRD, situated in the downstream area, exhibits a dense network of waterways and intricate channels, covering a drainage area of 2.68×10^4 km^2. With nearly 100 small creeks, it forms a typical subtropical river delta [11,12].

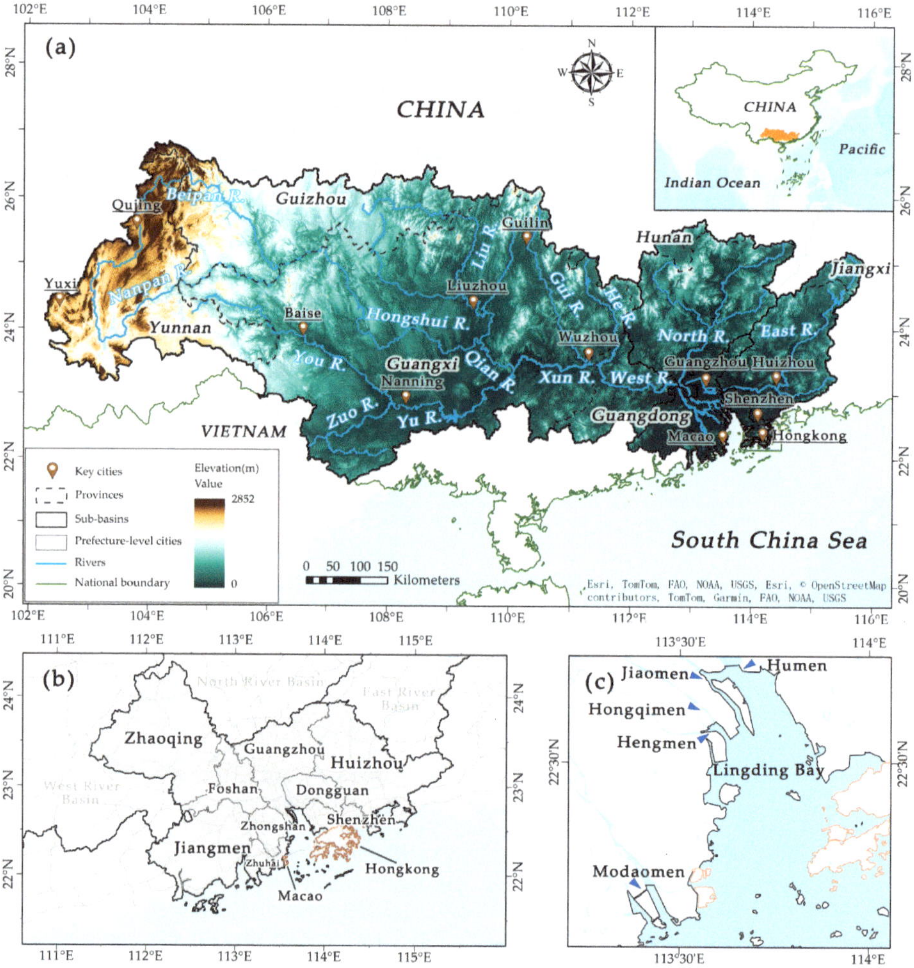

Figure 1. (**a**) Geographical configuration of the Pearl River Basin, (**b**) nine cities and two special administrative regions in the Pearl River Delta, and (**c**) main outlets in the Pearl River Estuary. The PRD, serving as an individual study unit, encompasses 11 cities, as depicted in (**b**). It constitutes the study area for numerous research endeavors mentioned in this paper. Its delineation is primarily driven by economic considerations rather than geographical considerations, as will be discussed in Section 3.1.1. In this study, the term "PRD" refers to the region depicted in (**b**).

The increase in extreme precipitation events can lead to changes in fluvial geomorphological processes, which are particularly pronounced in tropical and subtropical regions. The PRB is situated in a subtropical zone with a mild and monsoon climate. The average annual temperature ranges from 14 to 22 °C, with average annual precipitation from 1200 to 2200 mm [13]. From 1954 to 2022, the annual average water discharge in the PRB reached over 290 billion m^3, with an annual average sediment load of over 66 million tons [14]. The temporal distribution of water discharge and sediment load is uneven, with high observations in the wet season, typically from April to September, contributing approximately 80% of the annual water discharge [15,16] and 91% to 95% of the sediment load. Because of the uneven distribution, the PRB often experiences frequent water and drought disasters [17,18]. Issues such as soil erosion and karst rocky desertification are prominent, leading to a decline in water quality and degradation of the basin ecosystem. Severe soil erosion, covering an area of 79,928 km^2 [19], is mainly in the middle and upper reaches. The substantial soil erosion and karst rocky desertification have created various issues in ecological security, food security, flood control, and human settlement safety in the basin. Consequently, governance, development, and protection remain quite challenging tasks [20,21].

In addition, there is an extreme imbalance in economic development within the basin. Provinces in the upstream regions, such as Yunnan, Guizhou, and Guangxi, experienced relatively slow economic growth. In contrast, the downstream PRD region saw a fast economic development and stands as one of China's three major megaregions benefiting from the national policy of Reform and Opening Up. It is also recognized as a world-class center for advanced manufacturing and modern service industries. However, rapid economic development has also brought various challenges, including the deterioration of the river environment and human settlement environment [22].

The PRB, with a total population of 118 million, is one of China's densely populated areas, which needs substantial natural resources and thus may exert a highly complex influence on the river environment. Activities related to basin governance and development, such as the construction of reservoirs, river channelization, embankment construction, and sand mining, can also alter the basin's runoff and hydrological processes. Moreover, the basin is experiencing an increased frequency of floods, droughts, and other natural hazards due to climate change, posing higher challenges in flood control and drought prevention and thus hindering the pursuit of rapid and sustainable socio-economic development [23].

In the past few decades, significant changes in fluvial geomorphological processes have occurred in the PRB. Understanding these changes is crucial for promoting the sustainable development of the Pearl River. Currently, the new status of hydrogeomorphic processes in the PRB under the dramatic impacts of global climate change and human activities is still not clear. It is necessary to provide a comprehensive review of such processes and identify the issues to propose possible policy adjustments in river management. This information is pivotal in promoting flood protection and hazard mitigation, ensuring water security, protecting water resources, restoring ecological balance, and achieving integrated basin management in the PRB.

2. Changes over Recent Decades

The wet season in the PRB spans from April to September, while the dry season extends from October to March of the following year. The seasonal distribution of precipitation significantly influences the water discharge and water levels of rivers throughout the entire basin. Based on long-term monitoring data from hydrological stations, previous studies have found that over the past few decades, there has been a reduction in the difference in water discharge between the wet and dry seasons in the PRB. Specifically, there has been a slight decrease in water discharge during the wet season, concurrent with a significant increase during dry seasons [24–27]. However, no statistically significant changes in interannual water discharge have been identified [28–31], except for a declining trend observed in the northern part of the West River Basin (WRB) in the Liu River and the southwestern Nanpan River [32].

The Pearl River is characterized by a relatively low sediment load among China's seven large rivers. Sediment concentration tends to be higher in the west than in the east and the northwest than in the southeast. Tributaries with high sediment concentration, such as the Nanpan River, Beipan River, Hongshui River, and Yu River, are all located in the WRB. Unlike the relatively consistent stability in water discharge, sediment load displays notable variations over different periods. From the 1950s to the mid-1980s, visible increases in sediment load were observed in the estuarine zones of the West River and North River, notably between the late 1970s and the mid-1980s [23,24]. However, since the mid-1980s, there has been a distinct decline in sediment concentration [16,33–35] and a faster reduction from the 1990s to the early 21st century [31,34,36]. The East River Basin (ERB) has also experienced an overall decrease in sediment load since the 1960s [29,37]. Several studies conducted in the Pearl River Estuary (PRE) have reported a consistent decline in suspended sediment concentrations (SSC) in recent decades [38–44]. This decline significantly increases estuarine erosion and alters the coastal geomorphology of the PRE. The variations in both water discharge and sediment load across the entire basin are influenced by multiple factors, broadly categorized as anthropogenic and natural ones. This section aims to provide a detailed analysis of these impacts.

2.1. Human-Induced Changes

2.1.1. Land-Use Change

Previous studies on land-use/cover change (LUCC) across the PRB have identified a common trend over recent decades. With the development of urbanization, a large amount of cropland has been reclaimed for build-ups. The total area of grassland and forestland initially decreased, followed by an increase after 2000, but overall, both have reported a net decrease. Unused land remained decreasing. Previous studies also revealed [45] that the decrease in grassland mainly happened in the northwestern part, but the decrease in cropland was mainly observed in the central part as well as in the coastal and delta areas. The increase in build-ups was concentrated in the PRD (Figure 2). To be specific:

In the WRB, cropland and forestland in Guangxi Province (Figure 1) accounted for about 88% of the province's total area [46], which decreased substantially during 2005–2015. Build-up land increased rapidly, which was the fastest change in land use during this period, and it was mainly transferred from cropland. The total area of waterbody, grassland, and barren land also increased. In the past 20–30 years, studies on LUCC in Yunnan Province (mainly distributed in the Nanpan River basin) and Guizhou Province [47–49] found that cropland and forestland accounted for the largest proportion in land uses, but their areas tended to decrease. On the contrary, the build-up area has increased rapidly, with a significantly higher growth rate from 2010 to 2020. Other land uses have undergone relatively minor changes.

In the ERB, forestland is still the most important land-use type, accounting for a constant share of over 58% from the 1970s to the beginning of the 21st century [50]. As changes in the WRB, there has been a continuous decline in cropland in recent years. Build-up land area has steadily increased [51,52] and experienced the fastest growth among all land-use types, which primarily came from forestland, cropland, and unused land [51].

Likewise, in the North River Basin (NRB), forestland is also the predominant land-use type, encompassing over 63% of the basin's total area, while cropland makes up approximately 20% [53]. From 2000 to 2015, both cropland and forestland witnessed a decline, with a total reduction of 3.5% in forestland and approximately 4.45% in cropland. Changes in grassland and water bodies were relatively inconspicuous. Similarly, built-up land has expanded significantly during the same period. In 2000, built-up land accounted for approximately 3.96% of the North River Basin, while in 2015, the proportion reached 7.23%, marking an increase of 82.93%.

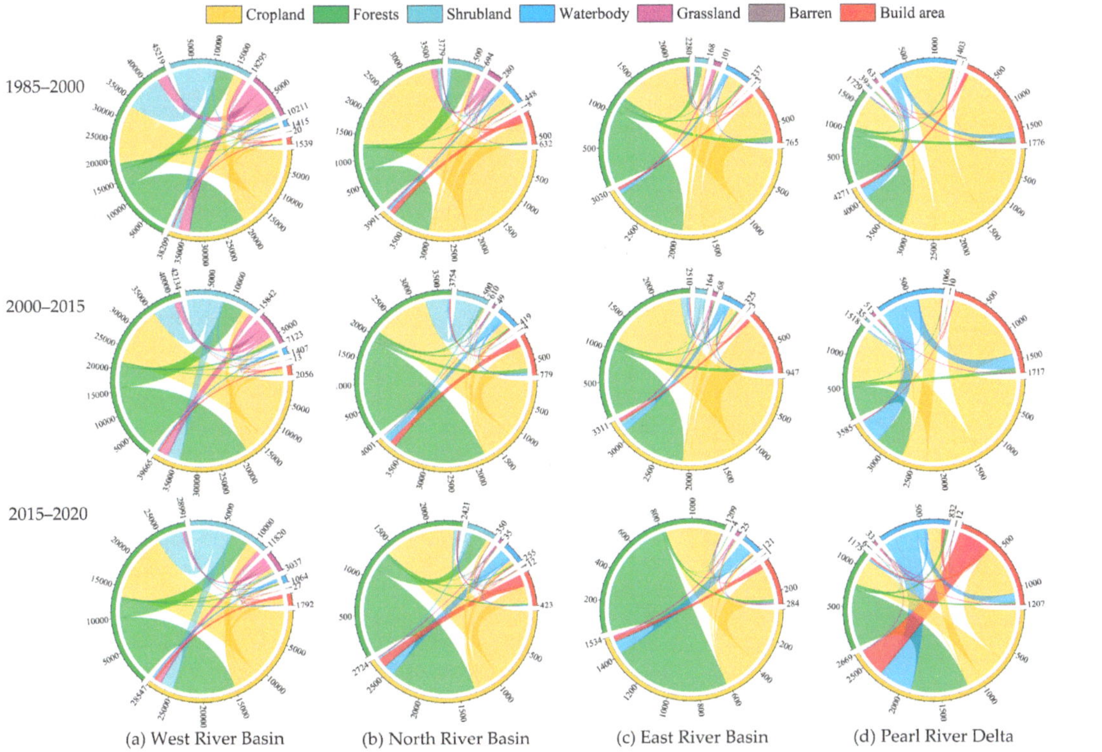

Figure 2. Land-use type transformation network in (**a**) WRB, (**b**) NRB, (**c**) ERB, and (**d**) PRD. (data from Ref. [54]).

Owing to the consistent economic importance, cities in the PRD exhibited a similar trend in land-use change. From the late 20th century to the first two decades of the 21st century, LUCC in PRD cities has shown a significant expansion in build-up land accompanied by rapid decreases in cropland and forestland [55–57]. The proportion of build-up land increased from 5.47% in 1990 to 14.56% in 2020 [57], with a loss of 31.74% of cropland from 1990 to 2006. Forestland remains the dominant land use in the PRD, constituting over 53% of the total area, but the majority of forestland is located in the outer PRD (Zhaoqing, Huizhou, and northern Guangzhou areas) instead of the central PRD [58]. The areas of waterbody and grassland fluctuated, with a decreasing trend similar to unused land. The changes in build-up land and cropland are statistically negatively correlated [59]. During the period from 2010 to 2015, the rate of expansion of build-up land slowed down, but during this period, there was also an increase in urbanization based on unused land.

Changes in land use can alter hydrological processes, consequently impacting soil erosion, especially in the areas with the reduction in forestland [60,61]. Over the past few decades, forestland has undergone an initial decrease followed by an increase in the PRB. Previous studies [62] indicate that during the period from 1990 to 2020, soil erosion in the PRB exhibited a decreasing trend, with an annual reduction rate of 13.44 ($\pm$1.53) t/(km^2·a). Soil erosion primarily occurs in the western tributaries of the West River, especially in the areas with slopes greater than 15°, low vegetation cover, and inadequate forest management.

2.1.2. River Channel Alterations

The PRB has numerous tributaries, and over the years, the river channels have undergone varying degrees of modifications. Since the mid-1980s, dramatic channel downcutting has occurred in the lower reaches of the West River and North River [11,16,63–66]. The average channel downcutting depths are 0.59 to 1.73 m and 0.34 to 4.43 m, respectively [11]. The West River and North River converge at Sanshui District, Guangdong Province. The downcutting in the North River is greater than in the West River, leading to a significant increase in the diversion ratio at the confluence of the three rivers at Sanshui District—this results in water supply initially from the North River as a provider to the West River shifting to water acquisition from the West River as a provider to the North River [65]. Since the 1990s, substantial downcutting also occurred in the lower reaches of the East River, with an average depth ranging from 1.77 to 6.0 m, surpassing that of the West River and North River [11]. Despite the unclear trend in water discharge changes in Pearl River's three major tributaries, the downcutting of river channels has significantly lowered water levels downstream in the three major tributaries [24,63,67–69]. Additionally, fluctuations in sediment load and water discharge contribute to sedimentation and erosion along different sections of each tributary, with erosion being a common phenomenon in the lower reaches. For example, the middle and lower reaches of the East River experienced siltation under previous natural conditions. The annual accumulated sediment transport volume reaches 280.66 million m^3 per year. However, from 2003 to 2009, the riverbed underwent significant downcutting with an average erosion depth of 1.9 m [70]. In the middle reaches of the North River, the area of sandbars increased 19.75 km^2 from 1975 to 1984. However, after 1988, the area decreased rapidly, and some sandbars at the lower end were eroded, resulting in a widening of the river channel [71,72]. Changes in siltation and erosion have led to channel shifts and water flow alterations, primarily occurring in sections adjacent to urban areas, such as the lower reaches of the North River and the Huizhou section of the East River (Figure 3).

Figure 3. Satellite images of parts of the East River and North River obtained from Google Earth.

The PRD is characterized by a highly branched, tree-like river network with frequent bifurcations and mergers [12]. The temporal variation in water discharge and sediment load within the delta is primarily influenced by the redistribution of upstream water and sediment [64], exacerbating changes in the PRD river network. Since the mid-1980s, there has been an overall uneven downcutting in the river channels of the PRD [73], with cross-section narrowing and deepening [74–76]. The downcutting became more evident from the 1990s to the early 21st century, leading to a dramatic decrease in water levels [11]. Due to human activities such as land reclamation and the construction of embankments and sluice gates, the coastline of the PRE has continued to extend seaward [77]. Between 1976 and 2006, the average annual extension was approximately 20 m [78]. This extension is manifested near the estuary as the narrowing and elongation of river channels, with typical areas near Modaomen and the northern part of Lingding Bay (Figure 1c) [79–81].

2.1.3. Reservoir and Dike Construction

Since the 1950s, over 9000 reservoirs have been constructed in the PRB, collectively holding approximately 21% of the basin's annual runoff [82]. Among these, 72 are large reservoirs (storage capacity $\geq$ 100 million m^3), with 36 located in the WRB. The two largest reservoirs in the WRB, Tianshengqiao Reservoir (completed in 1997) and Longtan Reservoir (completed in 2006), have capacities of 10.8 billion m^3 and 27.3 billion m^3, respectively. In the NRB, the total storage capacity of large reservoirs has exceeded 5 billion m^3, with Nanshui Reservoir (constructed in 1971) and Feilaixia Reservoir (completed in 1999) having capacities of 1.243 billion m^3 and 1.95 billion m^3, respectively. Additionally, the ERB possesses three major regulating reservoirs—Xinfengjiang, Fengshuba, and Baipenzhu—constructed in 1962, 1973, and 1985, respectively, with storage capacities of 13.896 billion m^3, 1.94 billion m^3, and 1.22 billion m^3, respectively [37,83] (Figure 4).

Most of the dikes in the middle and upper reaches of the PRB were constructed between the mid-1950s and the 1970s, primarily consisting of small-scale dikes. In the absence of effective flood storage reservoirs, dike construction served as the principal defense infrastructure against frequent floods. Key dike constructions included the Zhanyi, Qujing, and Donghedi projects along the Nanpan River in the upper reaches, while the middle reaches featured the Yongjiang dike project along the Yu River and the Sidan dike project along the Xun River. Downstream in the PRD, major dike construction includes the Beijiang, Zhongshun, and Jiangxinlian dike projects (Figure 4). Given the intricate river network and proximity to the South China Sea, large-scale construction projects on dikes and sluices were crucial in mitigating floods. Between 1949 and 1976, more than 10,000 low-rise riverine and coastal dikes in the PRD were connected, forming 494 sections [84]. This enables water drainage downstream within the dikes, facilitating separate flood prevention and drainage irrigation. It concentrates the flow regulation, carrying sediment to the estuary while also bolstering defenses against coastal saltwater erosion.

The construction of dikes and reservoirs on rivers has direct implications, leading to nonlinear alterations in river runoff and sediment load dynamics [43]. These alterations often display a long time lag before the impacts are evident. Reservoir construction causes a decrease in maximum annual water discharge while significantly increasing the minimum water discharge. Interannual changes in water discharge are predominantly influenced by precipitation, without apparent impacts from reservoir construction [16,40,63,64]. However, alterations in sediment load in the mainstreams primarily result from large reservoirs or dams, while tributaries' sediment changes mainly from extreme climatic events and soil conservation measures [85]. Post-1990s, the effects of reservoir construction and soil conservation practices have surpassed the impacts of deforestation, significantly reducing sediment load in lower reaches [29]. For instance, after the closure of the Yantan Dam on the Hongshui River in 1992, the annual sediment load at the downstream Qianjiang station decreased by 27.6 Mt/a (58.7%) from 1992 to 2002 compared to the previous period of 1989–1991 [40]. Since the closure of the Longtan and Baise reservoirs in 2006, sediment load in the Pearl River has decreased by 70% compared to the 1950s to the 1980s [30]. Overall, subsequent to the completion of numerous large reservoirs on the mainstem and major tributaries, sediment load in the 2000s and 2010s decreased to less than half and one third of that in the 1950s, respectively [31]. This may exacerbate channel downcutting and erosion in downstream floodplains.

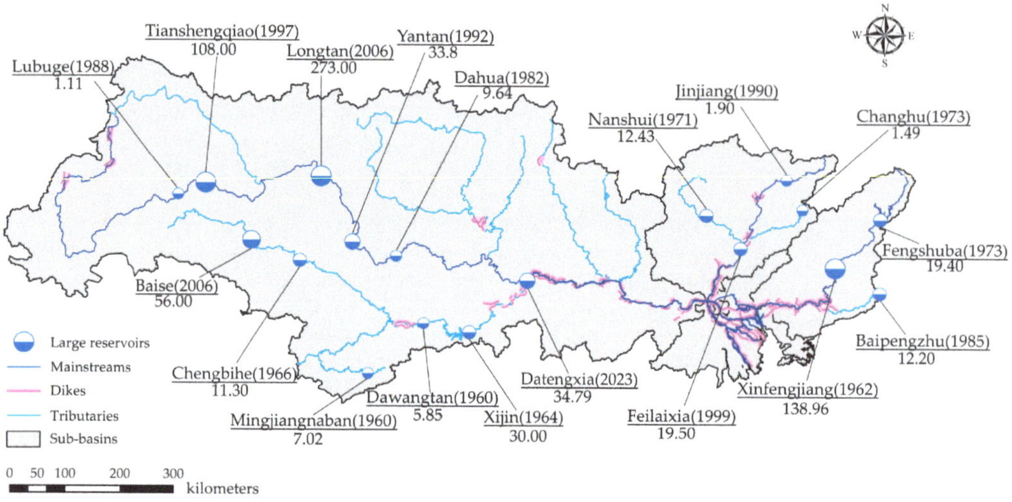

Figure 4. Major large reservoirs in the PRB (modified from Refs. [31,33,86]). Dam locations are denoted by circles with the dam name, year of closure, and storage capacity of the corresponding reservoir (unit in 100 million m^3) under each dam name.

2.2. Natural Climate Variations

Temperature and precipitation are two crucial climatic factors regulating fluvial geomorphological processes. Variations in temperature can affect processes such as water body evaporation, vegetation transpiration, and soil evaporation, therefore influencing runoff generation and resultant fluvial geomorphological processes. The intensity, frequency, and spatial-temporal distribution of precipitation can directly regulate river water discharge.

2.2.1. Temperature

Since 1954, the annual average temperature in the PRB has shown a significant upward trend [87–92], generally increasing from west to east PRB [90,91]. The seasonal average temperature in the PRB has experienced varying degrees of increase [89,90]. As seen in Figure 5, the warming rate in winter is higher than in summer [91], and the temperature increase during the monsoon season is lower than during the non-monsoon season [92]. Additionally, most months have shown an upward trend in temperatures [87,88,92]. In terms of spatial distribution, the monthly average temperature has increased at a faster rate in coastal areas [93], and there has been an increase in May, July, August, and September in almost all regions [94]. Overall, the average temperature in most areas of the PRB is increasing, with the most significant warming in the PRD and ERB [87,88,90].

The PRB generally shows an increase in extremely high-temperature events and a decrease in extremely low-temperature events [95]. The monthly extreme minimum temperature TNn, the monthly minimum temperature maximum TNx, the monthly extreme maximum temperature TXx, and the monthly maximum temperature minimum TXn have all shown an upward trend [95–99]. Spatially, there are notable spatial differences in the trend of extremely high temperature [98], with an increasing trend from west to east, particularly in the eastern part [100]. The number of warm days increases and cold days decreases, which is significant and consistent in most areas in the PRB, especially in the Yu

River, Zuo River, You River basins, and the PRD [95]. Furthermore, the daily maximum temperature shows a significant upward trend, with a significant shift occurring around 2002 [100]. Generally, extreme temperature has a high correlation with average temperature, and their spatiotemporal trends are relatively similar [98].

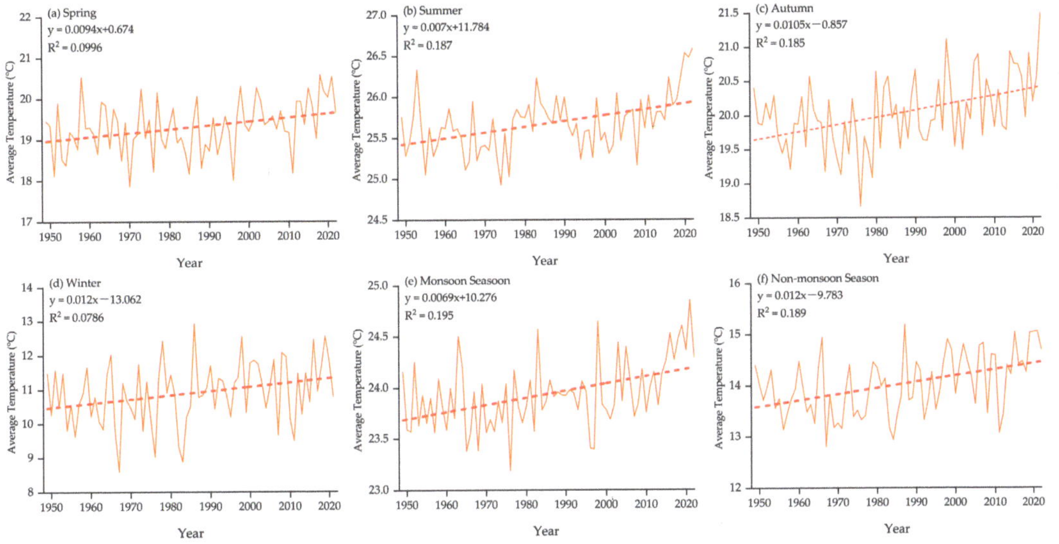

Figure 5. Temporal variations in average temperature in (**a**) Spring, (**b**) Summer, (**c**) Autumn, (**d**) Winter, (**e**) Monsoon Season, and (**f**) Non-Monsoon Season from 1949 to 2022 (Data from Ref. [101]).

2.2.2. Precipitation

Unlike the variation in temperature, precipitation in the PRB is characterized by small interannual variability and uneven spatiotemporal distribution, which affects the effective use of water resources.

The annual precipitation in the PRB shows non-significant changes [92,102–106]. In terms of spatial distribution, the eastern PRB has higher annual average precipitation but fewer rainy days, while the western areas have lower precipitation and more rainy days. The number of rainy days is decreasing in almost the entire basin [67,102,107]. There is a non-significant temporal trend in annual precipitation, and the number of rainy days is decreasing, leading to an increase in precipitation intensity in the eastern PRB [105]. In seasonal precipitation (Figure 6), there is a slight upward trend in summer and winter [102,108], a steady downward trend in spring and autumn [109], and a significant decrease in the number of rainy days in autumn [105]. Additionally, it was found that there is an insignificant decrease in precipitation during the monsoon season, while the non-monsoon season shows an insignificant increasing trend [108].

For extreme precipitation, the extreme precipitation index gradually decreases from east to west in the PRB [110]. The threshold of extreme precipitation is higher in the east and lower in the west, and there was a significant change in the entire basin in the late 1960s and 1980s [111] (Figure 7). The upper reaches show a significant increasing trend in extreme precipitation in both volume and frequency [67,112] and are more prone to extreme events compared to the lower reaches [104]. The number of days of extreme events in the entire basin shows a non-significant decreasing trend, while the duration of consecutive non-rainy days exhibits a significant increasing trend [106].

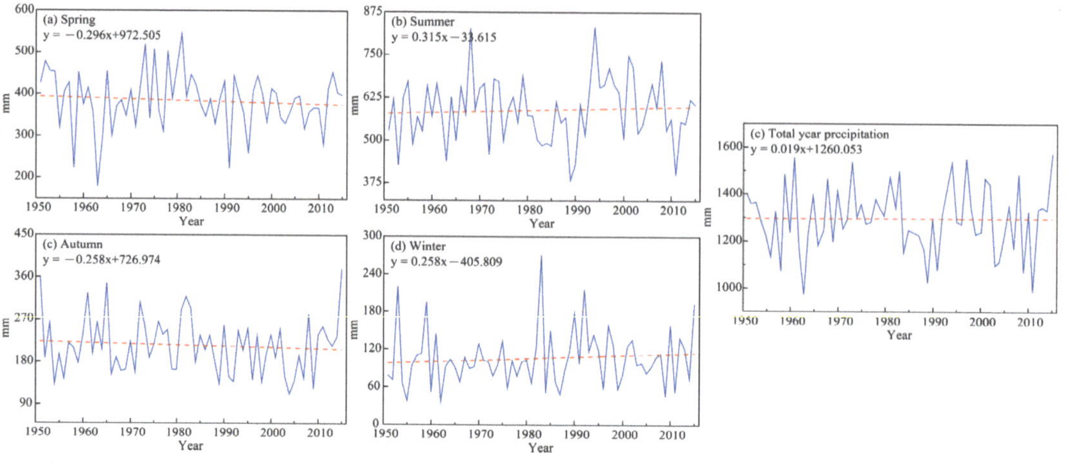

Figure 6. Temporal variations in precipitation in (**a**) Spring, (**b**) Summer, (**c**) Autumn, (**d**) Winter, and (**e**) Total year (modified from Ref. [109]).

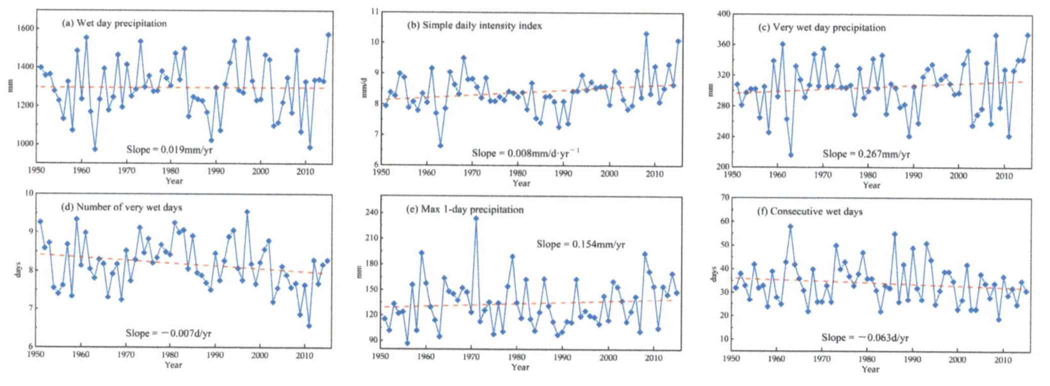

Figure 7. Interannual variation of extreme precipitation magnitude in the PRB during 1951–2015 (modified from Ref. [109]).

3. Analysis of Influencing Factors

3.1. Impacts of Anthropogenic Factors on Basin Changes

3.1.1. Socio-Economic Development

The changes in the ecological environment of the river basin are closely linked to the socio-economic development within the PRB. Since the establishment of the People's Republic of China, the economic policies have profoundly influenced the ecological environment of the PRB, which can be roughly divided into the following stages:

Early years of the establishment of the People's Republic of China (1949–1978): During this period, the PRB was in a state of destruction after World War II. The economy of the PRB was mainly focused on agriculture and the handicraft industry. The urbanization process was thus slow. To quickly change the country's poor situation, the government proposed the "Great Leap Forward" policy at the end of the 1950s to promote industrial development. Large-scale deforestation took place for steel production, leading to substantial deforestation [113]. Water environment problems also emerged due to large-scale cropland improvement, land reclamation, and reservoir construction [114]. However, due to the relatively low technology in hydraulic engineering, only small tributary reservoirs were built, with a relatively small impact on the variation of river runoff and sediment [115].

From 1978 to 2008, China began implementing its Reform and Opening-Up policy, which provided new opportunities for economic development in the PRB: The government increased investment in rural water conservancy infrastructure and expanded the irrigation area for cropland. In the 1980s, the Chinese government implemented the Forest Contract Responsibility System [113]. This system transfers the ownership of forests from the government to individual farmers under contract. Some farmers began to harvest forests for immediate profits, particularly in the WRB. During this period, there was a significant decrease in forestland and a substantial increase in cropland in the WRB, which exacerbated the soil erosion issue [116]. In the late 20th century, the government began to implement measures such as returning cropland to forestland and water and soil conservation in the WRB. These measures had a significant effect on vegetation recovery and soil erosion regulation in the WRB from 2002 to 2014 [117].

After 1995, with the support of the central government, the Guangdong provincial government formally proposed the establishment of the Pearl River Delta Economic Zone. As industry and labor force continued to expand, the urbanization process in the PRD region advanced rapidly. From 2008 to the present, guided by the active direction of the Reform and Opening-Up policy, urbanization construction in the PRD region has been basically completed. The size of cities also nearly doubled from 1995 to 2015 [55]. Urbanization has led to adjustments and optimizations in land use, but it has also exerted certain pressures on the ecological environment of the PRB. For example, urban development can cause changes in regional surface water processes, therefore affecting the entire river system's ecosystem.

3.1.2. Water Resources Management Policies Development

The Pearl River is renowned for its numerous tributaries and dense waterways, with abundant water resources and strong potential for hydropower utilization, which have significant applications in agriculture, forestry, hydroelectric power generation, and transportation. However, due to uneven spatiotemporal distribution in river runoff, the lower reaches face dramatic flood protection pressures in the monsoon season, while the upper reaches, such as the Yunnan-Guizhou Plateau and the northwest mountains of Guangxi, often suffer from severe droughts during the dry season. Basin policymakers have tried various measures to cope with the uneven spatiotemporal distribution of water resources.

In the early stages of the establishment of the People's Republic of China, flood-control reservoirs were mostly constructed with high dams, underestimating the role of dikes in flood control. After the 1960s, cascade development plans with runoff-type power stations became popular. Due to technological glitches, the constructed reservoirs often failed to run with full storage capacity, except for a few hydropower stations that showed satisfactory benefits [118]. During this period, flood control and water infrastructure in the PRB remained weak, with only a few large cities having piped water supply facilities and small-scale self-built water conservancy projects being the main reliance for agricultural irrigation by farmers.

In the 1970s, the construction of embankments moved to another extreme. The constructed reservoirs were relatively small and could not meet the storage capacity required for flood control. In addition to insufficient storage capacity, the existing reservoirs also suffered from varying degrees of siltation, with some reservoirs even becoming fully silted and abandoned. This led to a significant loss of effective storage capacity and even dam collapses. For example, the Baidonghe Reservoir, built in 1958 in the Baise prefecture, had accumulated sediment of 11.81 million m^3 by 1982, equivalent to 32% of its effective storage capacity. Likewise, the Xiangshuiba Reservoir in Yunnan impounded sediment of 71.2 million m^3 within 20 years of its completion, resulting in a loss of 36% of its effective storage capacity [119].

In the 1980s, flood-control planning in the PRB proposed the principle of "balancing upstream and downstream, giving priority to flood water release with a combination of flood regulation" [120]. The role of existing dikes received comprehensive attention. Key dikes were strengthened, and several problematic and poorly constructed reservoirs underwent reinforcement. At the same time, it was proposed that key mainstem reservoirs in the upper and middle reaches be constructed to reduce peak flood flow, therefore increasing the flood-control capacity in the middle and lower reaches.

Since the early 1990s, more large-scale dams and dikes have been constructed in the PRB, with a sharp increase in total reservoir capacity on both mainstem and tributaries. Compared to the reservoirs constructed in the 1980s and earlier, the total reservoir capacity in the PRB in the 1990s almost doubled [31]. For instance, The Tianshengqiao Reservoir in the XRB was completed in 1997, with a total storage capacity of 10.8 billion m^3, greater than the combined capacity of all the larger reservoirs (storage capacity > 10 million m^3) previously built in the PRB.

It also should be highlighted that from the mid-1980s to the early 21st century, driven by profits from urban expansion, large-scale sand mining activities became increasingly frequent in the PRB. Previous studies suggested that from the 1950s to the 1980s, the increase in sediment discharge in the three major tributaries was mainly attributed to intensified deforestation and resultant soil erosion [28–33,64]. The small-scale reservoirs have been constructed on tributaries and had a relatively weak impact on sediment trapping. However, after the mid-1980s, the sediment trapping by large reservoirs significantly increased, coupled with large-scale sand mining activities, resulting in a drastic reduction in sediment load in the delta region [11,16,33,34,40,65] (Figure 8). In addition, large-scale river dredging also occurred since the 1990s to guarantee maritime shipping [63]. All these factors have caused significant riverbed incisions in the downstream channels. Although climate change also played an important role in river flow change, anthropogenic impacts contributed more than 80% of the changes [30,82].

Figure 8. (**a**) Water level decreased in December in the segment of the North River; (**b**) The Baipenzhu Reservoir; (**c**) Sand mining vessel in the PRD; (**d**) Feilaixia water conservancy hub. All the photos were taken by the authors of this article.

At the beginning of the 21st century, the rapid economic and social development, as well as the impact of climate factors, have caused a series of ecological issues and natural hazards in the river basin. As a result, the water resources development and utilization shifted towards an overall consideration of development, protection, and regulation, which was the major feature of water resources planning during this period.

Several factors have contributed to this shift. First, changes in runoff have dramatically affected the self-purification capacity of rivers. Since the release of the policy of Reform and Opening Up in 1978, economic development and population growth have led to an increase in domestic and industrial water consumption, as well as an increase in sewage discharge. The Nanpan and Beipan river basins in the upper reaches, serving as an industrial and energy base for Yunnan and Guizhou provinces, have large sewage discharge and lack effective governance measures. During dry seasons, the low runoff and poor self-purification capacity in these areas have exacerbated river pollution. In the middle reaches of the Pearl River, pollution in some small and medium-sized rivers in Guangxi is also prominent. In the 1990s, during dry seasons, the water in the Gui River in the middle and lower reaches turned black, leading to a complete halt in the water supply for local residents [121].

Second, the problem of annual imbalance in water discharge has become more prominent. Despite the construction of large-scale dams and embankments, the basin has experienced several regional flood events since the 1990s. In the PRD region, three consecutive years of low water levels from 2003 to 2005 resulted in saltwater intrusion into the estuary, severely affecting the water supply of 15 million local residents in Zhuhai, Zhongshan, and Macao [122]. This situation caused an urgent water dispatch from upstream reservoirs.

In addition, intensive human activities such as hydropower projects, navigation, and sewage treatment have had a significant impact on the river ecosystems [123]. For example, the West River's largest tributary, the Yu River, has been blocked by the Xijin hydropower dam, disrupting the migration routes of some fish species and damaging the spawning grounds of fish both upstream and downstream, resulting in reduced fish populations. Although some fish species that can live in shallow waters or pass through sluices have increased their yields due to expanded habitats, overall, the negative impacts of dam construction on fish survival have outweighed the corresponding benefits, especially for rare and endangered fish species that require long-distance migration [124].

In 2003, the Chinese government proposed the concept of "scientific development" as the national policy guideline. "Maintaining river health and building a sustainable Pearl River" became the guideline for water management in the new century. In October 2007, the Pearl River Water Resources Commission (PRWRC) organized the signature of the "Trans-province River Water Affairs Convention in the Pearl River Basin", which regulated trans-province water activities through a contractual framework [125]. Disorderly sand mining activities in the PRD region were completely prohibited in 2008. Overall, with the implementation of scientific, systematic, and sustainable development concepts, basin governance has emphasized the coordinated management among provinces, cities, and river sections. Outdated notions of water resource exploitation have been gradually abandoned. Various state-of-the-art perspectives, such as river ecology, upstream–downstream interactions, and adaptation to climate change, have been incorporated into river management.

After 2010, the concept of scientific and sustainable development has become a consensus across society and the entire basin, emphasizing the harmony between humans and nature as a top priority in river management. Within the legal framework, various administrative departments are performing their duties and carrying out administrative law enforcement activities in an orderly manner, gradually forming an administrative law enforcement system and a scientific water resources management system. At the end of June 2018, the "River-Director System" policy was established nationwide, with the heads of local governments serving as "River Directors" responsible for organizing and leading the management and protection of corresponding rivers and lakes [126]. This further clarified the regulatory and accountability issues in the management of rivers in the

PRB. Overall, the concept of scientific and sustainable development has provided valuable guidelines for river management in the PRB.

3.2. Impacts of Atmospheric Oscillations on Climate Variability

Under global warming and enhanced human activities, climate change in different regions is significantly influenced by amplified changes in global and regional-scale water cycles. Therefore, many studies [15,127–131] have attempted to investigate its large-scale patterns, which are characterized by the fluctuations in sea surface temperature. The El Niño-Southern Oscillation (ENSO) is a fluctuation between warm El Niño and cold La Niña events in the ocean. During El Niño, annual precipitation increased in southeastern China and decreased in southwestern China, while the trend reversed during La Niña events [130]. The impact of ENSO on precipitation varies significantly with seasons and months. In the lower reaches of the Pearl River, winter precipitation is relatively lower in summer and autumn during El Niño events than in La Niña events [132].

Previous studies investigated the influences of ENSO, North Atlantic Oscillation (NAO), Indian Ocean Dipole (IOD), and Pacific Decadal Oscillation (PDO) on climate change across the PRB. It has been found that ENSOs are positively correlated with autumn and winter precipitation and have a strong impact on the occurrence of drought events [133], especially during El Niño events [134]; NAO has a significant negative correlation with summer precipitation and a positive correlation with winter precipitation [135] and serves as the dominant climatic indicator affecting droughts in the WRB [131]. PDO has a significant positive correlation with annual precipitation in the Beipan and Hongshui tributary basins [136]. The average winter precipitation showed an upward trend when IOD was observed [137]. In ERB, NRB, and the southwestern WRB, the average annual runoff and maximum daily runoff were found to be very sensitive to IOD events [138]. Generally, ENSO and IOD are two large-scale climate indicators that have a strong impact on climate change in the PRB. Both of them play crucial roles in rainfall and drought events [139]. It is worth noting that these climate events do not happen independently but often interact simultaneously [137,140,141], which may contribute to the differences observed in various studies.

Due to global climate change, the spatial and temporal changes in precipitation are becoming more uncertain. Therefore, studying the impacts of climate change is of great significance for water resources management in the PRB [8,15]. Previous studies on atmospheric circulation [127,142,143] have shown that the Western Pacific Subtropical High and the Eurasian High have a significant impact on the regulation of the East Asian monsoon by ENSO. The variation of monsoon affects the changes in water vapor transport, leading to changes in rainfall patterns in coastal areas of China [144]. In southeastern China, water vapor primarily originates from the Indian Ocean and the tropical Pacific [145]. The southwest monsoon and southeast monsoon transport water vapor from the Indian Ocean and Pacific Ocean to the Chinese monsoon regions, respectively [146]. Sea surface temperature changes in the central tropical Pacific are crucial to East Asian summer monsoon precipitation [147]. For example, the northwest Pacific subtropical high and the East Asian summer monsoon were weak during El Niño events, resulting in insufficient water vapor transport to the PRB, triggering the occurrence of drought events [134]. However, when the Northwestern Pacific subtropical high moves westward or enhances, the East winds in the equatorial convergence zone will intensify, leading to a continuous weak East Asian southwest monsoon. As a result, water vapor transported by the southwest monsoon from the Indian Ocean is unable to reach North China, and a large amount of water vapor is retained in South China [148]. Moreover, the wind direction in the eastern PRB shifted to the east, facilitating the convergence and uplift of water vapor, increasing the amount and frequency of precipitation, and causing floods in the coastal areas of the PRB [139,149].

4. Major Issues Faced by the Pearl River Basin

4.1. Increased Extreme Precipitation Events

Since 1980, the frequency of extreme flood events in the PRB has significantly increased [150]. In comparison to the period from 1951 to 1980, the occurrence of extreme flood events in the NRB and the northern WRB showed an increasing trend from 1981 to 2010, while a decreasing trend was observed in the ERB. Larger-scale floods may occur in the northeastern part of the WRB and the NRB [83,151–153]. Zhang et al. reported a significant increase in peak flow and flood magnification in the WRB and the NRB [21]. Over the past century, there has been an increasing trend in flood risk in the middle and lower reaches of the PRB. Following the frequent occurrence of large floods in the mid-1990s, an unprecedented surge in dike construction projects took place across the region, altering the natural hydrological conditions. Additionally, due to uneven regional development, flood-control facilities (such as embankments) are more sparse in remote areas, potentially exposing farmers in these remote regions to greater flood threats [86].

4.2. Desertification and Soil Loss

The PRB is a region with the most concentrated distribution and the most severe degree of karst rocky desertification in China. The presence of karst rocky desertification and soil erosion directly impacts the safety of water resources and hydropower facilities, as well as the aquatic ecological security in the PRB, therefore influencing the overall sustainable development of the entire basin [154]. The karst geomorphology is widely distributed in the WRB, where frequent droughts and severe rocky desertification are influenced by climatic factors [155]. Anthropogenic activities, such as indiscriminate logging, steep slope cultivation, excessive grazing, as well as unreasonable agricultural production, mining, and road construction, all contribute to the exacerbation of rocky desertification [33]. Since the 1990s, the WRB has undergone multiple rounds of measures, including reforestation and grassland restoration, in response to such issues as karst rocky desertification and urban expansion [154]. The forestland initially decreased and then increased, while the grassland gradually expanded [156]. Despite a decreasing trend in soil erosion in the PRB over the past 30 years [62], the Nanpan and Beipan River Basin and the northern NRB still face severe rocky desertification.

4.3. Water-Level Reduction and Saltwater Intrusion at Estuaries

Since the mid-1980s, the river channels in the PRD have shown an overall uneven trend of downcutting [73]. The cross-sections of riverbeds have been narrowing and deepening [74–76], with this trend intensifying from the 1990s to the early 21st century. Consequently, the water levels in the PRD rivers have noticeably decreased compared to the 1980s. For example, in 2005, the water level at the Sanshui station dropped by 2.45 m under similar runoff conditions compared to 1989 [11]. The issue of saltwater intrusion in the estuarine areas has gradually become a significant factor hindering economic development and threatening residents' livelihoods. Since 2009, saltwater intrusion has become a continuous phenomenon throughout the entire delta region during the dry season. In 2011–2012, the annual average extent of saltwater intrusion reached 36.3 km, twice that of 2001–2002. The duration of salinity exceeding thresholds has become longer, and the seasonality has happened earlier, indicating a deterioration of saltwater intrusion over the past decade. Specifically, between 2011 and 2012, the duration of salinity exceeding thresholds exceeded 700 h, whereas in 2003, it was less than 250 h. From 2005 to 2009, the intrusion of seawater occurred one month earlier than in previous years, from November to February, whereas previously, it occurred from December to March [157]. The decrease in water levels in upstream rivers is the primary cause of exacerbated saltwater intrusion in the estuary. In the delta region, there is a noticeable downward trend in water levels in the lower reaches of the West River and the North River, weakening the ability of the estuarine areas to resist seawater intrusion.

Modaomen and Lingding Bay (Figure 1c) are the two areas facing the most severe threats from saltwater intrusion in the PRE, and they are also regions with a higher prevalence of land reclamation and embankment projects. Between 1980 and 2018, the total reclaimed area in the PRE reached 769.2 km^2 [158]. Sand mining, dredging, and reclamation projects can narrow water channels and reduce water exchange capacity, not only hindering river runoff dispersion but also diminishing the effective wave height in the estuary, ultimately exacerbating seawater intrusion [159–161]. Another significant factor to consider is the rise in sea level. From 1959 to 2011, there was a noticeable upward trend in sea levels in the PRD, with a linear trend of 4.08 mm/yr [162]. The rising sea level enhances tidal range and ocean currents, causing tidal currents to arrive earlier and increasing the distance of saltwater intrusion. Moreover, the intrusion distance of seawater increases with the rising sea level and reduced river runoff. Previous studies indicate that during the dry season, with a sea level rise of 0.8 m, the intrusion distance of saltwater in Lingding Bay will increase approximately by 10 km, while in Modaomen, it will increase approximately by 3.7 km [163]. The saltwater intrusion in the PRE is highly sensitive to sea level rise [164]. Multiple factors contribute to the intrusion of seawater in the PRD region, and effective measures are urgently needed to counteract the upstream intrusion of saltwater in the estuarine areas.

4.4. Climate Change and Increased Drought Events in the Upper and Middle Reaches

The drought situation in the PRB has become more severe. Between 1960 and 2005, noticeable drought trends were observed in November, December, and January in the PRB [165]. The western and northeastern regions of the basin have lower drought risks, while the PRD faces higher drought risks [20], with longer average durations of droughts, especially during extreme drought conditions [166]. However, in the western regions of the basin, particularly in the Nanpan and Beipan Basin, the probability of drought occurrence is relatively higher. During the period from 1997 to 2010, short-term extreme hot and dry weather events in the PRB intensified, primarily occurring in the northeastern and western regions of the basin, with a significant increasing trend observed in the central-western and southeastern areas [167]. Since the 21st century, the number of events of moderate droughts and severe droughts and the duration and intensity of droughts have significantly increased in the PRB [168]. The most notable changes are seen in autumn droughts, with a trend of hydrological droughts shifting from summer to autumn [169].

5. Sustaining the Pearl River: Policy Overhaul and Strategic Adjustments

5.1. Basin Governance Framework

The establishment of a comprehensive management institution for the entire PRB has undergone a complex process [170]. Since its re-establishment in August 1979, the Pearl River Water Resources Commission (PRWRC) has played a crucial role in the integrated management of the PRB for over 40 years. However, China's Water Law, enacted in 2002, states that, "the central government implements a combined water resources management from river commissions and provincial governments". This provision has significantly limited the functions of the PRWRC as a basin management institution [171], and this limitation has persisted to the present day (Figure 9).

PRWRC is a branch organization established by the Ministry of Water Resources, responsible for the development, planning, and management of water resources in the PRB. However, functions such as the jurisdiction of hydropower, navigation, and water pollution control are under other departments subordinate to the State Council. Under these structures, although PRWRC can handle tasks like regional water project planning, water resource supervision, and administrative law enforcement [18], the actual control over the development and utilization of water energy resources are regulated by the electricity department; waterway management is overseen by the State Council's transportation department, and pollution prevention and control are led by the State Council's ecological environment department. This management divides responsibilities based on the functions of water resources, which significantly has undermined integrated water resources management. Each department prioritizes its own interests, neglecting the overall interests of the PRB [172–174].

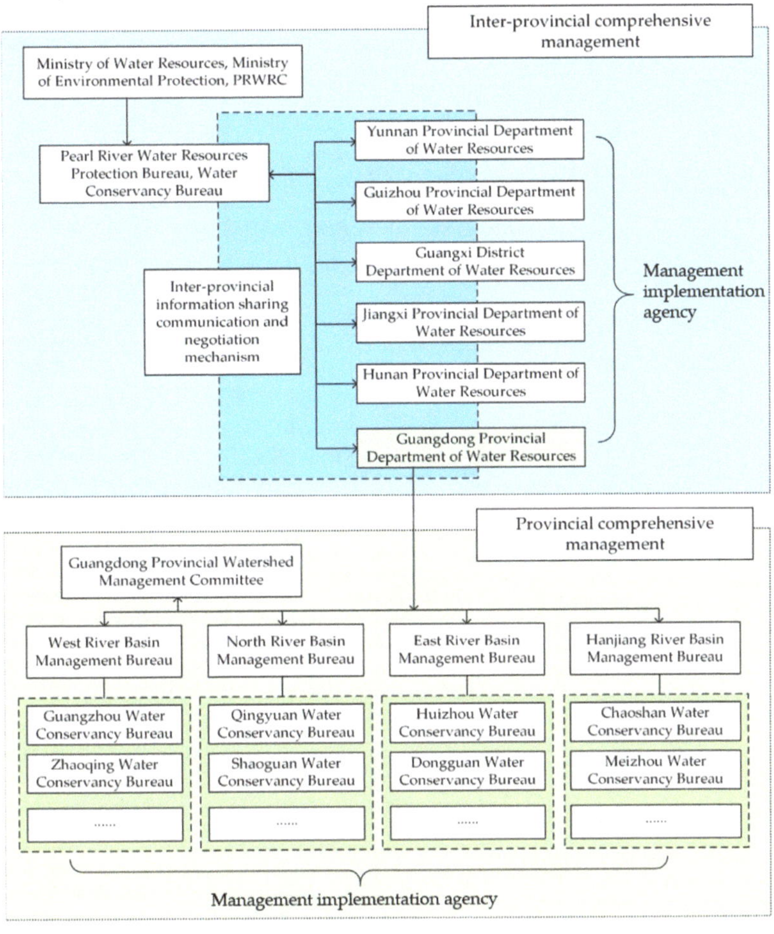

Figure 9. Framework of PRB Management Institutional (modified from Ref. [172]).

Therefore, it is recommended that the responsibilities and jurisdiction of PRWRC be legislated and clearly defined. This would establish a more comprehensive water resource management system, breaking away from the existing administrative fragmentation. The goal is to transform basin management into a model led by a single institution with collaboration from multiple departments [173–176]. This approach aims to strengthen unified basin management [177], fully leveraging the role of PRWRC as a comprehensive basin management institution and a mediator for trans-province water affairs negotiations.

5.2. Adapting to Future Climate Change

Compared to northern China, the PRB is an area abundant in precipitation, making it prone to flood hazards during the monsoon season. However, drought events also occur frequently during dry seasons. Both floods and droughts are significantly influenced by climate change, causing severe disruptions to local residents. Therefore, there is an urgent need to explore adaptive strategies for climate change. Currently, research on the impact factors of climate change in the PRB primarily focuses on the correlation between large-scale climate indices and climatic indicators such as rainfall and drought. However, there are limited studies on adaptation strategies, preparation efforts, and adaptive training for local residents. The natural ecosystems in the PRB exhibit relatively weak adaptive capacities to climate change. To address extreme climate events such as floods and droughts, decision-makers should optimize the layout of hydrological monitoring stations and enhance monitoring capabilities for key tributaries, small and medium-sized rivers, and small to medium-sized reservoirs. This includes improving monitoring capabilities for river and ecological flows and soil conservation. The use of new monitoring methods should be promoted, expanding the scope of real-time online monitoring and enhancing the capabilities of intelligent water security monitoring.

5.3. Estuary and Coastal Zone Development

With rising sea levels, the reduction in sediment load to the sea has significantly weakened the coastal resistance to erosion; this, in turn, has led to severe land loss in the delta region [178]. The decrease in sediment load has a serious impact on the morphological changes in estuarine and coastal areas. Not just the decrease in SSC but also the extensive reclamation of intertidal resources in estuarine areas for urbanization and industrialization has also widely occurred. From 1985 to 2015, the net loss of mangrove forests in the PRD region was approximately 11.73 km^2 [179]. Additionally, due to human activities such as sea reclamation, the coastline in the PRE has extended seaward in recent years [77–79]. Between 1976 and 2006, the average annual net extension was approximately 19.3 m, implying an elongation of river channels at the estuary. This extension was often achieved by blocking lateral channels that connect with deep water passages [180]. This will prevent floodwaters from being discharged into the sea in a short time, putting the estuary area at greater risk of flooding.

The significant reduction in sediment load is largely attributed to the sediment trapping in dams. Some studies have suggested taking into account sediment release measures during water impoundment and considering the creation of designated sediment bypass channels [119]. In estuarine areas, it is recommended to establish ecological reserves to prevent the fast degradation in wetlands, coupled with the implementation of ecological protection red lines in estuarine regions, use of policy regulation, ecological transformation and restoration of artificial shorelines, controlling such activities as sea reclamation and pollution discharge, aiming to maintain the ecological balance of in estuary regions [181–183].

5.4. Sustaining Water Resouces

The PRB experiences significant disparities in precipitation between the wet and dry seasons, with approximately 80% of the annual precipitation occurring during the wet season. This difference significantly affects the annual distribution of river flow, hindering the balanced allocation of water resources throughout the year and posing challenges to economic development and normal domestic activities.

The persistent issue of winter saltwater intrusion in the PRD region poses a serious threat to regional water security. PRWRC has addressed this by organizing water dispatches from key dams and sluicegates on various tributaries during the dry season. By releasing water from upstream reservoirs, water levels can be increased to cope with saltwater intrusion. By 2021, this practice has been implemented continuously for 16 rounds [184]. Water dispatch not only helps maintain a stable water supply downstream but also plays a crucial role in the ecological health of fish populations, agricultural irrigation, and shipping traffic in different river sections upstream and midstream. However, the legal and regulatory framework for water dispatches in the PRB is insufficient, especially concerning large-scale water transfers to ensure winter water supply to Macao and Zhuhai. Early initiatives heavily relied on administrative orders [173]. This situation has led to numerous conflicts and contradictions in the coordination of water dispatches, with economically disadvantaged upstream regions making sacrifices to ensure water stability for the developed downstream areas.

The "Regulations on Pearl River Water Dispatch" is a legal document addressing the allocation of water resources in the basin. It has been submitted for approval by the State Council multiple times over recent decades. However, formal and comprehensive legal provisions have not yet been released. To strengthen the legitimacy of basin water dispatches, it is advisable to expedite the promulgation and improvement of the "Regulations" [173]. Furthermore, enhancing information exchange between different departments and establishing an intelligent and informative platform for collaborative reservoir dispatch between upstream and downstream [185,186] can improve the efficiency of river flow dispatch. Introducing market mechanisms into river management, compensating upstream regions through water pricing to alleviate conflicts and contradictions in water dispatches [174,187], and using price leverage to promote water conservation and efficiency [176] can all facilitate the rational allocation and utilization of river water resources.

5.5. River Ecosystem Conservation

From the perspective of river biota and fish species, the biodiversity of the PRB exceeds that of the Yangtze River, securing its position as the top river system among the seven major rivers in China. The period from the 1960s to the 1980s witnessed a severe decline in fishery resources in the Pearl River system. The natural fish yield in Guangdong Province plummeted by 52% from 1966 to 1973, and in Guangxi Province, it dropped by 66.5% from 1962 to 1979. Several precious fish species, especially those inhabiting rivers and estuaries, were endangered [124]. Before the establishment of the People's Republic of China, large-scale water conservancy projects were not yet constructed in the PRB, and water quality pollution was scarce. However, with the massive construction of reservoirs starting in the 1960s, the ecological integrity of fish populations came under serious threat. In recent decades, improvements in water quality in the PRB have been achieved through measures such as controlling the discharge of pollutants from industrial and domestic wastewater, enhancing forest environmental protection, and improving solid waste management [9]. Nevertheless, the discharge of industrial and domestic wastewater has led to nutrient and heavy metal pollution in rivers [188], highlighting the persistent pollution issues in the PRB. Therefore, it is recommended to follow the example of the first basin-wide protection regulation implemented in 2021, the "Yangtze River Protection Law", and enact comprehensive regulations for the protection and utilization of water resources in the PRB [189]. This would strengthen and standardize ecological protection efforts across the entire basin.

Author Contributions: Conceptualization, H.O. and X.Y.; data curation, S.C. and W.F.; writing—original draft preparation, H.O.; writing—review and editing, S.C., W.F., J.Q., X.M., T.Z., X.Y. and L.P.; supervision, X.Y. All authors have read and agreed to the published version of the manuscript.

Funding: The National Key R&D Program of China (Grant No.: 2021YFE0117300) and Major Project of High-Resolution Earth Observation System (Grant No.: 30-Y60B01-9003-22/23).

Data Availability Statement: Data are contained within the article..

Acknowledgments: The authors are very grateful to scholars for their research on the Pearl River Basin. They also thank the editor and the anonymous reviewers for their professional and pertinent comments and suggestions.

Conflicts of Interest: The authors declare no conflicts of interest.

References

1. Syvitski, J.P.M.; Vörösmarty, C.J.; Kettner, A.J.; Green, P. Impact of Humans on the Flux of Terrestrial Sediment to the Global Coastal Ocean. *Science* **2005**, *308*, 376–380. [CrossRef] [PubMed]
2. Whipple, K.X. The Influence of Climate on the Tectonic Evolution of Mountain Belts. *Nat. Geosci.* **2009**, *2*, 97–104. [CrossRef]
3. Willett, S.D.; McCoy, S.W.; Beeson, H.W. Transience of the North American High Plains Landscape and Its Impact on Surface Water. *Nature* **2018**, *561*, 528–532. [CrossRef] [PubMed]

4. Ran, L.; Lu, X.; Fang, N.; Yang, X. Effective Soil Erosion Control Represents a Significant Net Carbon Sequestration. *Sci. Rep.* **2018**, *8*, 12018. [CrossRef] [PubMed]
5. Wu, J.; Cheng, S.-P.; He, L.-Y.; Wang, Y.-C.; Yue, Y.; Zeng, H.; Xu, N. Assessing Water Quality in the Pearl River for the Last Decade Based on Clustering: Characteristic, Evolution and Policy Implications. *Water Res.* **2023**, *244*, 120492. [CrossRef] [PubMed]
6. Ferguson, R.I.; Lewin, J.; Hardy, R.J. Fluvial Processes and Landforms. *Geol. Soc. Lond. Mem.* **2022**, *58*, 257–270. [CrossRef]
7. Yang, X.; Lu, X.; Ran, L. Sustaining China's Large Rivers: River Development Policy, Impacts, Institutional Issues and Strategies for Future Improvement. *Geoforum* **2016**, *69*, 1–4. [CrossRef]
8. Deng, S.; Chen, T.; Yang, N.; Qu, L.; Li, M.; Chen, D. Spatial and Temporal Distribution of Rainfall and Drought Characteristics across the Pearl River Basin. *Sci. Total Environ.* **2018**, *619–620*, 28–41. [CrossRef]
9. Ma, X.; Brookes, J.; Wang, X.; Han, Y.; Ma, J.; Li, G.; Chen, Q.; Zhou, S.; Qin, B. Water Quality Improvement and Existing Challenges in the Pearl River Basin, China. *J. Water Process Eng.* **2023**, *55*, 104184. [CrossRef]
10. Su, C.; Chen, X. Covariates for Nonstationary Modeling of Extreme Precipitation in the Pearl River Basin, China. *Atmos. Res.* **2019**, *229*, 224–239. [CrossRef]
11. Luo, X.-L.; Zeng, E.Y.; Ji, R.-Y.; Wang, C.-P. Effects of In-Channel Sand Excavation on the Hydrology of the Pearl River Delta, China. *J. Hydrol.* **2007**, *343*, 230–239. [CrossRef]
12. Chan, F.K.S.; Yang, L.E.; Scheffran, J.; Mitchell, G.; Adekola, O.; Griffiths, J.; Chen, Y.; Li, G.; Lu, X.; Qi, Y.; et al. Urban Flood Risks and Emerging Challenges in a Chinese Delta: The Case of the Pearl River Delta. *Environ. Sci. Policy* **2021**, *122*, 101–115. [CrossRef]
13. Lai, C.; Chen, X.; Wang, Z.; Wu, X.; Zhao, S.; Wu, X.; Bai, W. Spatio-Temporal Variation in Rainfall Erosivity during 1960–2012 in the Pearl River Basin, China. *CATENA* **2016**, *137*, 382–391. [CrossRef]
14. Pearl River Water Resources Commission. Pearl River Sediment Bulletin, Guangzhou, China, 2022. Available online: http://www.pearlwater.gov.cn/zwgkcs/lygb/nsgb/202311/P020231110315588395900.pdf (accessed on 1 February 2024).
15. Xiao, M.; Zhang, Q.; Singh, V.P.; Liu, L. Transitional Properties of Droughts and Related Impacts of Climate Indices in the Pearl River Basin, China. *J. Hydrol.* **2016**, *534*, 397–406. [CrossRef]
16. Zhang, Q.; Xu, C.-Y.; Chen, Y.D.; Jiang, J. Abrupt Behaviors of the Streamflow of the Pearl River Basin and Implications for Hydrological Alterations across the Pearl River Delta, China. *J. Hydrol.* **2009**, *377*, 274–283. [CrossRef]
17. Niu, J.; Chen, J.; Sivakumar, B. Teleconnection Analysis of Runoff and Soil Moisture over the Pearl River Basin in Southern China. *Hydrol. Earth Syst. Sci.* **2014**, *18*, 1475–1492. [CrossRef]
18. Cui, W.Z.; Chen, J.; Wu, Y.P.; Wu, Y.D. An Overview of Water Resources Management of the Pearl River. *Water Supply* **2007**, *7*, 101–113. [CrossRef]
19. Pearl River Water Resources Commission. Pearl River Bulletin of Soil and Water Conservation, Guangzhou, China, 2020. Available online: http://www.pearlwater.gov.cn/zwgkcs/lygb/stbc/202109/P020210914335796341337.pdf (accessed on 1 February 2024).
20. Zhang, Q.; Xiao, M.; Singh, V.P.; Li, J. Regionalization and Spatial Changing Properties of Droughts across the Pearl River Basin, China. *J. Hydrol.* **2012**, *472–473*, 355–366. [CrossRef]
21. Zhang, Q.; Gu, X.; Singh, V.P.; Shi, P.; Sun, P. More Frequent Flooding? Changes in Flood Frequency in the Pearl River Basin, China, since 1951 and over the Past 1000 Years. *Hydrol. Earth Syst. Sci.* **2018**, *22*, 2637–2653. [CrossRef]
22. Zhao, M.; Peng, J.; Liu, Y.; Li, T.; Wang, Y. Mapping Watershed-Level Ecosystem Service Bundles in the Pearl River Delta, China. *Ecolog. Econ.* **2018**, *152*, 106–117. [CrossRef]
23. Yang, T.; Shao, Q.; Hao, Z.-C.; Chen, X.; Zhang, Z.; Xu, C.-Y.; Sun, L. Regional Frequency Analysis and Spatio-Temporal Pattern Characterization of Rainfall Extremes in the Pearl River Basin, China. *J. Hydrol.* **2010**, *380*, 386–405. [CrossRef]
24. Chen, Y.D.; Yang, T.; Xu, C.-Y.; Zhang, Q.; Chen, X.; Hao, Z.-C. Hydrologic Alteration along the Middle and Upper East River (Dongjiang) Basin, South China: A Visually Enhanced Mining on the Results of RVA Method. *Stoch. Environ. Res. Risk Assess.* **2010**, *24*, 9–18. [CrossRef]
25. Chen, Y.D.; Zhang, Q.; Xu, C.-Y.; Lu, X.; Zhang, S. Multiscale Streamflow Variations of the Pearl River Basin and Possible Implications for the Water Resource Management within the Pearl River Delta, China. *Quat. Int.* **2010**, *226*, 44–53. [CrossRef]
26. Chen, Y.D.; Zhang, Q.; Chen, X.; Wang, P. Multiscale Variability of Streamflow Changes in the Pearl River Basin, China. *Stoch. Environ. Res. Risk Assess.* **2012**, *26*, 235–246. [CrossRef]
27. Huang, C.; Wang, Y.; Yang, Q.; Liu, F. Spatiotemporal variation of discharge and its influencing factors in the Pearl River Delta network area. *Period. Ocean Univ. China* **2022**, *52*, 97–106. (In Chinese) [CrossRef]
28. Zhang, S.; Lu, X.X.; Higgitt, D.L.; Chen, C.-T.A.; Han, J.; Sun, H. Recent Changes of Water Discharge and Sediment Load in the Zhujiang (Pearl River) Basin, China. *Glob. Planet. Chang.* **2008**, *60*, 365–380. [CrossRef]
29. Liu, F.; Yang, Q.; Chen, S.; Luo, Z.; Yuan, F.; Wang, R. Temporal and Spatial Variability of Sediment Flux into the Sea from the Three Largest Rivers in China. *J. Asian Earth Sci.* **2014**, *87*, 102–115. [CrossRef]
30. Wu, C.S.; Yang, S.L.; Lei, Y. Quantifying the Anthropogenic and Climatic Impacts on Water Discharge and Sediment Load in the Pearl River (Zhujiang), China (1954–2009). *J. Hydrol.* **2012**, *452–453*, 190–204. [CrossRef]
31. Wei, X.; Cai, S.; Ni, P.; Zhan, W. Impacts of Climate Change and Human Activities on the Water Discharge and Sediment Load of the Pearl River, Southern China. *Sci. Rep.* **2020**, *10*, 16743. [CrossRef]
32. Tang, J.; Duan, K.; Hao, L. Analysis on changes in water level-flow of the Pearl River Basin in recent years and its causes. *Pearl River.* **2021**, *42*, 1–7+28. (In Chinese) [CrossRef]

33. Dai, S.B.; Yang, S.L.; Cai, A.M. Impacts of Dams on the Sediment Flux of the Pearl River, Southern China. *CATENA* **2008**, *76*, 36–43. [CrossRef]
34. Zhang, W.; Wei, X.; Jinhai, Z.; Yuliang, Z.; Zhang, Y. Estimating Suspended Sediment Loads in the Pearl River Delta Region Using Sediment Rating Curves. *Cont. Shelf Res.* **2012**, *38*, 35–46. [CrossRef]
35. Yang, Y.; Wang, Y.; Cai, S.; Liu, F.; Yang, Q. Characteristics of interannual and intra-annual variation of sediment discharge in the Pearl River Basin over the past 60 years. *Res. Soil Water Conserv.* **2021**, *28*, 155–162. (In Chinese) [CrossRef]
36. Liu, F.; Hu, S.; Guo, X.; Luo, X.; Cai, H.; Yang, Q. Recent Changes in the Sediment Regime of the Pearl River (South China): Causes and Implications for the Pearl River Delta. *Hydrol. Process.* **2018**, *32*, 1771–1785. [CrossRef]
37. Wang, Y.; Peng, J.; Huang, C.; Wu, H.; Cai, S.; Liu, F. Abnormal changes diagnosis and its control factors of water and sediment discharge in the lower reaches of Pearl River Basin from 1960 to 2019. *Res. Soil Water Conserv.* **2022**, *29*, 21–31. (In Chinese) [CrossRef]
38. Cao, B.; Qiu, J.; Zhang, W.; Xie, X.; Lu, X.; Yang, X.; Li, H. Retrieval of Suspended Sediment Concentrations in the Pearl River Estuary Using Multi-Source Satellite Imagery. *Remote Sens.* **2022**, *14*, 3896. [CrossRef]
39. Guo, J.; Ma, C.; Bin, A.; Xu, X.; Huang, W.; Zhao, J. Assessing the Effects of the Hong Kong-Zhuhai-Macau Bridge on the Total Suspended Solids in the Pearl River Estuary Based on Landsat Time Series. *J. Geophys. Res. Oceans.* **2020**, *125*, e2020JC016202. [CrossRef]
40. Tan, C.; Huang, B.; Liu, K.; Chen, H.; Liu, F.; Qiu, J.; Yang, J. Using the Wavelet Transform to Detect Temporal Variations in Hydrological Processes in the Pearl River, China. *Quat. Int.* **2017**, *440*, 52–63. [CrossRef]
41. Wang, C.; Li, W.; Chen, S.; Li, D.; Wang, D.; Liu, J. The Spatial and Temporal Variation of Total Suspended Solid Concentration in Pearl River Estuary during 1987–2015 Based on Remote Sensing. *Sci. Total Environ.* **2018**, *618*, 1125–1138. [CrossRef]
42. Zhan, W.; Wu, J.; Wei, X.; Tang, S.; Zhan, H. Spatio-Temporal Variation of the Suspended Sediment Concentration in the Pearl River Estuary Observed by MODIS during 2003–2015. *Cont. Shelf Res.* **2019**, *172*, 22–32. [CrossRef]
43. Yunus, A.P.; Masago, Y.; Boulange, J.; Hijioka, Y. Natural and Anthropogenic Forces on Suspended Sediment Dynamics in Asian Estuaries. *Sci. Total Environ.* **2022**, *836*, 155569. [CrossRef] [PubMed]
44. Yin, S.; Gao, G.; Huang, A.; Li, D.; Ran, L.; Nawaz, M.; Xu, Y.J.; Fu, B. Streamflow and Sediment Load Changes from China's Large Rivers: Quantitative Contributions of Climate and Human Activity Factors. *Sci. Total Environ.* **2023**, *876*, 162758. [CrossRef] [PubMed]
45. Zhang, S.; Zhang, J.; Zhang, W.; Zhang, D.; Fu, J.; Zang, C. Spatiotemporal variability characteristics and driving forces of land use in the Pan-Pearl River Basin, China. *Chin. J. Appl. Ecol.* **2020**, *31*, 573–580. (In Chinese) [CrossRef]
46. Liu, H.; Jiang, H.; Hu, B. Eco-environmental Effects of Land Use Change in the Xijiang River Basin of Guangxi. *J. Nanning Norm. Univ.* **2020**, *37*, 104–111. (In Chinese) [CrossRef]
47. Li, X.; Zhou, J.; Jin, T.; Wang, J. LUCC spatial simulation of urban agglomeration in central Yunnan. *Ecol. Rural. Environ.* **2022**, *38*, 1318–1329. (In Chinese) [CrossRef]
48. Wang, L. Land Use Change and Its Eco-Environmental Effects of Urban Agglomeration in Central Yunnan. Master's Thesis, Yunnan University, Kunming, China, 2021. Available online: https://kns.cnki.net/KCMS/detail/detail.aspx?dbcode=CMFD&dbname=CMFD202202&filename=1021111261.nh&v= (accessed on 1 February 2024). (In Chinese)
49. Li, M. Study on Land Use Landscape Pattern Change and Ecosystem Service Value Accounting in Guizhou Province. Master's Thesis, Nanjing Forestry University, Nanjing, China, 2019. Available online: https://kns.cnki.net/KCMS/detail/detail.aspx?dbcode=CMFD&dbname=CMFD202001&filename=1020802391.nh&v= (accessed on 1 February 2024). (In Chinese)
50. Shi, W. Research on Conversion Trend and Spatial Pattern of Land-Use/Cover Change in Dongjiang Region Based on Geographic Information System. Master's Thesis, Nanjing University, Nanjing, China, 2012. Available online: https://kns.cnki.net/KCMS/detail/detail.aspx?dbcode=CMFD&dbname=CMFD2012&filename=1012376046.nh&v= (accessed on 1 February 2024). (In Chinese)
51. Deng, X.; Chen, Y. Land use change and its driving mechanism in Dongjiang River Basin from 1990 to 2018. *Bull. Soil Water Conserv.* **2020**, *40*, 236–242+258+331. (In Chinese) [CrossRef]
52. Peng, Z.; Gu, C.; Liu, Z.; Lin, W.; Zhou, P. Impact of land use change during 1989-2009 on eco-capacity in Dongjiang watershed. *Chin. J. Plant Ecol.* **2014**, *38*, 675–686. (In Chinese) [CrossRef]
53. Li, J.; Luo, X.; Zhu, S.; Zhang, W.; Xu, Z.; Huo, Z.; Tian, Z.; Huo, G. Effect of LUCC on hydrological response in Beijiang basin. *Yangtze River* **2020**, *51*, 89–94+129. (In Chinese) [CrossRef]
54. Yang, J.; Huang, X. 30 m Annual Land Cover and Its Dynamics in China from 1990 to 2019; Earth System Science Data, 2021. Available online: https://doi.org/10.5194/essd-2021-7 (accessed on 1 February 2024).
55. Hu, M.M.; Xia, B.C. Land-Use Variations in Regions with Rapid Economic Development—A Case Study in the Pearl River Delta. *J. Environ. Inform. Lett.* **2020**, *3*, 49–58. [CrossRef]
56. Ye, C.; Feng, Y. Ecological risk assessment for Pearl River Delta based on land use change. *Trans. Chin. Soc. Agric. Eng.* **2013**, *29*, 224–232+294. (In Chinese) [CrossRef]
57. Li, B.; Yang, Z.; Cai, Y.; Xie, Y.; Guo, H.; Wang, Y.; Zhang, P.; Li, B.; Jia, Q.; Huang, Y.; et al. Prediction and Valuation of Ecosystem Service Based on Land Use/Land Cover Change: A Case Study of the Pearl River Delta. *Ecol. Eng.* **2022**, *179*, 106612. [CrossRef]
58. Zhou, R.; Lin, M.; Gong, J.; Wu, Z. Spatiotemporal Heterogeneity and Influencing Mechanism of Ecosystem Services in the Pearl River Delta from the Perspective of LUCC. *J. Geog. Sci.* **2019**, *29*, 831–845. [CrossRef]

59. He, Y.; Wang, Y.; Yuan, Z.; Zheng, M.; Huang, B.; Liang, C. Land use change and its response to urbanization in the Pearl River Delta. *Ecol. Environ. Sci.* **2020**, *29*, 303–310. (In Chinese) [CrossRef]
60. Li, L.; Wang, Y.; Liu, C. Effects of Land Use Changes on Soil Erosion in a Fast Developing Area. *Int. J. Environ. Sci. Technol.* **2014**, *11*, 1549–1562. [CrossRef]
61. Chen, Y.; Yang, Q.; Liu, B.; Huang, C.; Wang, C.; Pang, G. Assessment of soil erosion intensity in Pearl River Basin based on CSLE model. *Sci. Soil Water Conserv.* **2021**, *19*, 86–93. (In Chinese) [CrossRef]
62. Mu, X.; Qiu, J.; Cao, B.; Cai, S.; Niu, K.; Yang, X. Mapping Soil Erosion Dynamics (1990–2020) in the Pearl River Basin. *Remote Sens.* **2022**, *14*, 5949. [CrossRef]
63. Chen, Y.D.; Zhang, Q.; Xu, C.-Y.; Yang, T.; Chen, X.; Jiang, T. Change-Point Alterations of Extreme Water Levels and Underlying Causes in the Pearl River Delta, China. *River Res. Appl.* **2009**, *25*, 1153–1168. [CrossRef]
64. Liu, F.; Yuan, L.; Yang, Q.; Ou, S.; Xie, L.; Cui, X. Hydrological Responses to the Combined Influence of Diverse Human Activities in the Pearl River Delta, China. *CATENA* **2014**, *113*, 41–55. [CrossRef]
65. Lu, X.X.; Zhang, S.R.; Xie, S.P.; Ma, P.K. Rapid Channel Incision of the Lower Pearl River (China) since the 1990s as a Consequence of Sediment Depletion. *Hydrol. Earth Syst. Sci.* **2007**, *11*, 1897–1906. [CrossRef]
66. Qin, B.; Tian, T.; Cheng, X.; Liu, L.; Zhang, Y. River evolution and reason analysis of the Beiiang River main stream in recent ten years. *Port Waterw. Eng.* **2022**, *1*, 131–135. (In Chinese) [CrossRef]
67. Zhang, Q.; Singh, V.P.; Peng, J.; Chen, Y.D.; Li, J. Spatial–Temporal Changes of Precipitation Structure across the Pearl River Basin, China. *J. Hydrol.* **2012**, *440–441*, 113–122. [CrossRef]
68. Zhang, Q.; Li, K.; Singh, V.P.; Chen, X.; Li, J. Changes in Stage–Flow Relation of the East River, the Pearl River Basin: Causes and Implications. *Hydrol. Res.* **2013**, *44*, 737–746. [CrossRef]
69. Tan, C.; Huang, B.S.; Yang, Q.S.; Qiu, J. Hydrodynamic Characteristics Change of the Pearl River Delta Channel Network in the Recent 50 Years. *Adv. Mater. Res.* **2012**, *573–574*, 415–420. [CrossRef]
70. Huang, C. The recent evolution and future trend of middle and lower reaches of Dongiiang River. *Guangdong Water Resour. Hydropower* **2022**, *12*, 37–42. Available online: https://kns.cnki.net/KCMS/detail/detail.aspx?dbcode=CJFD&dbname=CJFDLAST2023&filename=GDSD202212007&v= (accessed on 1 February 2024). (In Chinese)
71. Liu, W.; Zhong, K.; Liu, K. An analysis on the evolution of the river course at lower reaches of the Beiiang by remote sensing technique. *Trop. Geogr.* **2006**, *2*, 123–128. Available online: https://kns.cnki.net/KCMS/detail/detail.aspx?dbcode=CJFD&dbname=CJFD2006&filename=RDDD200602005&v= (accessed on 1 February 2024). (In Chinese)
72. Zhong, K.; Liu, W.; Huang, J. A riverway evolution analysis based on remote sensing technique: A case study of the lower reaches of the Beijiang River. *Remote Sens. Nat. Resour.* **2006**, *3*, 69–73. Available online: https://kns.cnki.net/KCMS/detail/detail.aspx?dbcode=CJFD&dbname=CJFD2006&filename=GTYG200603017&v= (accessed on 1 February 2024). (In Chinese)
73. Yuan, F.; He, Y.; Wu, M.; Liu, B. Fluvial processes in the Pearl River Delta in recent decades. *J. Sediment Res.* **2018**, *43*, 40–46. (In Chinese) [CrossRef]
74. Zhen, G. Research of the Human Activity Impact on Hydrodynamic Environment in Pearl River Delta. Ph.D. Thesis, Wuhan University, Wuhan, China, 2005. Available online: https://kns.cnki.net/KCMS/detail/detail.aspx?dbcode=CDFD&dbname=CDFD9908&filename=2006035294.nh&v= (accessed on 1 February 2024). (In Chinese)
75. Qian, Y. Impacts and management measures of disordered sand mining in the Pearl River Delta rivers. *Pearl River* **2004**, *2*, 44–46+58. Available online: https://kns.cnki.net/KCMS/detail/detail.aspx?dbcode=CJFD&dbname=CJFD2004&filename=RMZJ200402016&v= (accessed on 1 February 2024). (In Chinese)
76. Liu, B.; Peng, S.; Liao, Y.; Long, W. The Causes and Impacts of Water Resources Crises in the Pearl River Delta. *J. Clean. Prod.* **2018**, *177*, 413–425. [CrossRef]
77. Wang, X.-Z.; Zhang, H.-G.; Fu, B.; Shi, A. Analysis on the Coastline Change and Erosion-Accretion Evolution of the Pearl River Estuary, China, Based on Remote-Sensing Images and Nautical Charts. *J. Appl. Remote Sens.* **2013**, *7*, 073519. [CrossRef]
78. Zhang, W.; Xu, Y.; Hoitink, A.J.F.; Sassi, M.G.; Zheng, J.; Chen, X.; Zhang, C. Morphological Change in the Pearl River Delta, China. *Mar. Geol.* **2015**, *363*, 202–219. [CrossRef]
79. Li, X.; Damen, M.C.J. Coastline Change Detection with Satellite Remote Sensing for Environmental Management of the Pearl River Estuary, China. *J. Mar. Syst.* **2010**, *82*, S54–S61. [CrossRef]
80. Zhao, D. Morphological Evolution of the Pearl River Delta in the Past 165 Years and Its Response to Human Activities. Ph.D. Thesis, Zhejiang University, Hangzhou, China, 2017. Available online: https://kns.cnki.net/KCMS/detail/detail.aspx?dbcode=CDFD&dbname=CDFDLAST2018&filename=1018000274.nh&v= (accessed on 1 February 2024). (In Chinese)
81. Wang, B.; Zhu, J.; Wu, H.; Yu, F.; Song, X. Dynamics of Saltwater Intrusion in the Modaomen Waterway of the Pearl River Estuary. *Sci. China Earth Sci.* **2012**, *55*, 1901–1918. [CrossRef]
82. Wu, C.; Ji, C.; Shi, B.; Wang, Y.; Gao, J.; Yang, Y.; Mu, J. The Impact of Climate Change and Human Activities on Streamflow and Sediment Load in the Pearl River Basin. *Int. J. Sediment Res.* **2019**, *34*, 307–321. [CrossRef]
83. Gu, X.; Zhang, Q.; Wang, Z. Evaluation on stationarity assumption of annual maximum peak flows during 1951-2010 in the Pearl River Basin. *J. Nat. Resour.* **2015**, *30*, 824–835. (In Chinese) [CrossRef]
84. Shi, X. Achievements in flood control in the Pearl River Basin. *Pearl River.* **1990**, *5*, 2–6. Available online: https://kns.cnki.net/KCMS/detail/detail.aspx?dbcode=CJFD&dbname=CJFD9093&filename=RMZJ199005000&v= (accessed on 1 February 2024). (In Chinese)

85. Zhang, Y.; Tian, P.; Yang, L.; Zhao, G.; Mu, X.; Wang, B.; Du, P.; Gao, P.; Sun, W. Relationship between Sediment Load and Climate Extremes in the Major Chinese Rivers. *J. Hydrol.* **2023**, *617*, 128962. [CrossRef]
86. Qiu, J.; Cao, B.; Park, E.; Yang, X.; Zhang, W.; Tarolli, P. Flood Monitoring in Rural Areas of the Pearl River Basin (China) Using Sentinel-1 SAR. *Remote Sens.* **2021**, *13*, 1384. [CrossRef]
87. Wang, Z.; Chen, X.; Huang, G. Spatio-temporal change characteristics of mean temperature in the Pearl River Basin during 1961–2000. *Trop. Geogr.* **2007**, *4*, 289–293+322. (In Chinese) [CrossRef]
88. Shi, J.; Wang, Z.; Chen, X. Change trend of average temperature in the Pearl River Basin and its spatial distribution characteristics. *Pearl River* **2007**, *5*, 4–6+12. Available online: https://kns.cnki.net/KCMS/detail/detail.aspx?dbcode=CJFD&dbname=CJFD2007&filename=RMZJ200705002&v= (accessed on 1 February 2024). (In Chinese)
89. Liu, L.; Jiang, T.; Yuan, F. Observed (1961–2007) and projected (2011–2060) climate change in the Pearl River Basin. *Clim. Chang. Res.* **2009**, *5*, 209–214. Available online: https://kns.cnki.net/KCMS/detail/detail.aspx?dbcode=CJFD&dbname=CJFD2009&filename=QHBH200904008&v= (accessed on 1 February 2024). (In Chinese)
90. Xie, Y.; Li, J.; Chen, W.; Luo, S. Spatio-temporal variation of average temperature over the Pearl River basin during 1959–2013. *Acta Sci. Nat. Univ. Sunyatseni* **2016**, *55*, 30–38. (In Chinese) [CrossRef]
91. Tian, Q.; Prange, M.; Merkel, U. Precipitation and Temperature Changes in the Major Chinese River Basins during 1957–2013 and Links to Sea Surface Temperature. *J. Hydrol.* **2016**, *536*, 208–221. [CrossRef]
92. Fischer, T.; Gemmer, M.; Lüliu, L.; Buda, S. Temperature and Precipitation Trends and Dryness/Wetness Pattern in the Zhujiang River Basin, South China, 1961–2007. *Quat. Int.* **2011**, *244*, 138–148. [CrossRef]
93. Fischer, T.; Gemmer, M.; Liu, L.; Su, B. Trends in Monthly Temperature and Precipitation Extremes in the Zhujiang River Basin, South China (1961–2007). *Adv. Clim. Chang. Res.* **2010**, *1*, 63–70. [CrossRef]
94. Zhang, F.; Lei, X.; Jiang, Y.; Bai, J. Analysis on character of temperature variation in upstream of Pearl River basin during 55 years. *J. Water Resour. Water Eng.* **2012**, *23*, 20–25. Available online: https://kns.cnki.net/KCMS/detail/detail.aspx?dbcode=CJFD&dbname=CJFD2012&filename=XBSZ201202006&v= (accessed on 1 February 2024). (In Chinese)
95. Huang, Q.; Chen, Z. Regional study on the trends of extreme temperature and precipitation events in the Pearl River Basin. *Adv. Earth Sci.* **2014**, *29*, 956–967. (In Chinese) [CrossRef]
96. Liu, Q.; Wu, X.; Chen, X.; Yang, B. Temporal and spatial variation characteristics of extreme temperature in the Pearl River Basin during 1960-2012. *J. Nat. Resour.* **2015**, *30*, 1356–1366. (In Chinese) [CrossRef]
97. Hu, X.; Cao, Y.; Wang, J. Analysis on spatial and temporal variation characteristics of extreme climate events in the Pear River Basin within Yunnan Province from 1964 to 2013. *Pearl River* **2019**, *40*, 95–102+109. (In Chinese) [CrossRef]
98. Xu, F.; Zhang, W.; Zhao, L.; Wang, J.; Gong, Q. Spatio-temporal variability in extreme temperature from 1960 to 2018 in the Pearl River Basin, China. *Mt. Res.* **2022**, *40*, 343–354. (In Chinese) [CrossRef]
99. Zhang, N.; Sun, Z.; Zeng, G. Change of extreme temperatures in China during 1955–2005. *Trans. Atmos. Sci.* **2008**, *1*, 123–128. (In Chinese) [CrossRef]
100. Qin, B.; Wu, X.; Zhu, T.; Lai, C. Research on spatial-temporal variation characteristics and influencing factors of annual maximum daily temperature in the Pearl River Basin under the background of global warming. *Pearl River* **2020**, *41*, 1–6+61. (In Chinese) [CrossRef]
101. Peng, S. 1-Km Monthly Mean Temperature Dataset for China (1901–2022). National Tibetan Plateau Data Center 2020. Available online: https://data.tpdc.ac.cn/en/data/71ab4677-b66c-4fd1-a004-b2a541c4d5bf/ (accessed on 1 February 2024).
102. Lu, W.; Liu, B.; Chen, J.; Chen, X. Variation trend of precipitation in the pear river basin in recent 50 years. *J. Nat. Resour.* **2014**, *29*, 80–90. (In Chinese) [CrossRef]
103. Lu, W.; Liu, B.; Chen, X.; Wu, L. Characteristics of precipitation period in Pear River Basin. *J. China Hydrol.* **2013**, *33*, 82–86. Available online: https://kns.cnki.net/KCMS/detail/detail.aspx?dbcode=CJFD&dbname=CJFD2013&filename=SWZZ201302019&v= (accessed on 1 February 2024). (In Chinese)
104. Peng, J.; Zhang, Q.; Liu, C. Changing properties of precipitation regimes over the Pearl River Basin. *Pearl River* **2012**, *33*, 13–17. (In Chinese) [CrossRef]
105. Liu, B.; Chen, J.; Lu, W.; Chen, X.; Lian, Y. Spatiotemporal Characteristics of Precipitation Changes in the Pearl River Basin, China. *Theor. Appl. Climatol.* **2016**, *123*, 537–550. [CrossRef]
106. Zhao, Y.; Zou, X.; Cao, L.; Xu, X. Changes in Precipitation Extremes over the Pearl River Basin, Southern China, during 1960–2012. *Quat. Int.* **2014**, *333*, 26–39. [CrossRef]
107. Zhang, Q.; Xu, C.-Y.; Zhang, Z. Observed Changes of Drought/Wetness Episodes in the Pearl River Basin, China, Using the Standardized Precipitation Index and Aridity Index. *Theor. Appl. Climatol.* **2009**, *98*, 89–99. [CrossRef]
108. Wang, Z.; Chen, X.; Zhang, L.; Li, Y. Spatio-temporal change characteristics of precipitation in the Pearl River Basin in recent 40 years. *J. China Hydrol.* **2006**, *6*, 71–75. Available online: https://kns.cnki.net/KCMS/detail/detail.aspx?dbcode=CJFD&dbname=CJFD2006&filename=SWZZ200606021&v= (accessed on 1 February 2024). (In Chinese)
109. Cai, S.; Niu, K.; Mu, X.; Yang, X.; Pirotti, F. Spatiotemporal Changes in Extreme Precipitation in China's Pearl River Basin during 1951–2015. *Water* **2023**, *15*, 2634. [CrossRef]
110. Wu, X.; Chen, X.; Tang, Y.; Wang, Z.; Lai, C. Spatiotemporal variations and the causes of non-stationary extreme precipitation in the Pearl River basin. *J. Hydraul. Eng.* **2015**, *46*, 1055–1063. (In Chinese) [CrossRef]

111. Tang, Y.; Chen, X. Multi-scale spatio-temporal characteristics and influence of precipitation variation in Zhujiang River Basin during the last 50 years. *Sci. Geogr. Sinica.* **2015**, *35*, 476–482. (In Chinese) [CrossRef]
112. Zhang, Q.; Xu, C.-Y.; Becker, S.; Zhang, Z.X.; Chen, Y.D.; Coulibaly, M. Trends and Abrupt Changes of Precipitation Maxima in the Pearl River Basin, China. *Atmos. Sci. Lett.* **2009**, *10*, 132–144. [CrossRef]
113. Weng, Q. A discussion on ecological problems in Guangxi. *Acad. Forum.* **2001**, *2*, 80–83+1. (In Chinese) [CrossRef]
114. Xu, M.; Wang, D.; Zhao, Y. A review of the development history of water pollution prevention and control in China. *Chin. J. Plant Ecol.* **2012**, *1*, 63–67. (In Chinese) [CrossRef]
115. Liao, Y.; Fan, J. Overview of the Pearl River Basin and opinions on development and governance. *Pearl River* **1980**, *1*, 16–38. Available online: https://kns.cnki.net/KCMS/detail/detail.aspx?dbcode=CJFD&dbname=CJFD7984&filename=RMZJ198001005&v= (accessed on 1 February 2024). (In Chinese)
116. Liu, G.; Jin, Q.; Li, J.; Li, L.; He, C.; Huang, Y.; Yao, Y. Policy Factors Impact Analysis Based on Remote Sensing Data and the CLUE-S Model in the Lijiang River Basin, China. *CATENA* **2017**, *158*, 286–297. [CrossRef]
117. Huang, W.; Wang, H. Drought and Intensified Agriculture Enhanced Vegetation Growth in the Central Pearl River Basin of China. *Agric. Water Manag.* **2021**, *256*, 107077. [CrossRef]
118. Yang, M.; Luo, C.; Wang, Y. Review and prospect of water conservancy planning in the Pearl River Basin. *Pearl River* **2009**, *30*, 10–12. (In Chinese) [CrossRef]
119. Chen, W.; Zeng, Z. A brief discussion on the sediment problem in reservoirs in the Pearl River Basin. *Pearl River* **1997**, *5*, 26–28. Available online: https://kns.cnki.net/KCMS/detail/detail.aspx?dbcode=CJFD&dbname=CJFD9697&filename=RMZJ199705008&v= (accessed on 1 February 2024). (In Chinese)
120. Fan, J. Discussion on major comprehensive planning issues in the Pearl River Basin. *Pearl River* **2001**, *4*, 1–4. Available online: https://kns.cnki.net/KCMS/detail/detail.aspx?dbcode=CJFD&dbname=CJFD2001&filename=RMZJ200104000&v= (accessed on 1 February 2024). (In Chinese)
121. Li, Z. Review and prospect of water resources protection work in the Pearl River Basin. *Pearl River* **1999**, *5*, 12–14. Available online: https://kns.cnki.net/KCMS/detail/detail.aspx?dbcode=CJFD&dbname=CJFD9899&filename=RMZJ199905006&v= (accessed on 1 February 2024). (In Chinese)
122. Cai, S.; Lu, S. Some thoughts on comprehensively implementing the "Water Law" and promoting legal water management in the Pearl River. *Pearl River* **2013**, *34*, 1–4. (In Chinese) [CrossRef]
123. Long, Y.; Song, L.; Shu, Y.; Li, B.; Peijnenburg, W.; Zheng, C. Evaluating the Spatial and Temporal Distribution of Emerging Contaminants in the Pearl River Basin for Regulating Purposes. *Ecotoxicol. Environ. Saf.* **2023**, *257*, 114918. [CrossRef]
124. Cui, S.; Wang, X.; Deng, J. A tentative discussion on river ecological problems and countermeasures in the Pearl River system. *Water Resour. Dev. Res.* **2005**, *9*, 7–11. (In Chinese) [CrossRef]
125. Yue, Z. Strengthen comprehensive management of river basins, improve water management and management capabilities, and promote the construction of a green Pearl River. *China Water Resour.* **2014**, *24*, 73–74. Available online: https://kns.cnki.net/KCMS/detail/detail.aspx?dbcode=CJFD&dbname=CJFDLAST2015&filename=SLZG201424022&v= (accessed on 1 February 2024). (In Chinese)
126. Shen, K.; Jin, G. The policy effects of local governments' environmental governance in China-A study based on the evolution of the "River-Director" system. *Soc. Sci. China* **2018**, *5*, 92–115+206. Available online: https://kns.cnki.net/kcms2/article/abstract?v=6e6DrGrvy4NjC2LY_x-hlq9dwxO0-_s-J1ugbcaHxHydxzWjEn805HsWBkq3l_gsbnC7xaShVTezDl4qFm_rTAh0eSJDEMx1e1geIsXczRYqBftzr3kcOsq3i6g757Vp&uniplatform=NZKPT&language=gb (accessed on 1 February 2024). (In Chinese)
127. Zhang, Q.; Wang, Y.; Singh, V.P.; Gu, X.; Kong, D.; Xiao, M. Impacts of ENSO and ENSO Modoki+A Regimes on Seasonal Precipitation Variations and Possible Underlying Causes in the Huai River Basin, China. *J. Hydrol.* **2016**, *533*, 308–319. [CrossRef]
128. Liu, Z.; Menzel, L.; Dong, C.; Fang, R. Temporal Dynamics and Spatial Patterns of Drought and the Relation to ENSO: A Case Study in Northwest China. *Int. J. Climatol.* **2016**, *36*, 2886–2898. [CrossRef]
129. Vicente-Serrano, S.M.; Aguilar, E.; Martínez, R.; Martín-Hernández, N.; Azorin-Molina, C.; Sanchez-Lorenzo, A.; Kenawy, A.E.; Tomás-Burguera, M.; Moran-Tejeda, E.; López-Moreno, J.I.; et al. The Complex Influence of ENSO on Droughts in Ecuador. *Clim. Dyn.* **2017**, *48*, 405–427. [CrossRef]
130. Ouyang, R.; Liu, W.; Fu, G.; Liu, C.; Hu, L.; Wang, H. Linkages between ENSO/PDO Signals and Precipitation, Streamflow in China during the Last 100 Years. *Hydrol. Earth Syst. Sci.* **2014**, *18*, 3651–3661. [CrossRef]
131. Qiu, J.; Wang, Y.; Xiao, J. Spatiotemporal Distribution of Droughts in the Xijiang River Basin, China and Its Responses to Global Climatic Events. *Water* **2017**, *9*, 265. [CrossRef]
132. Zhang, Q.; Li, J.; Singh, V.P.; Xu, C.-Y.; Deng, J. Influence of ENSO on Precipitation in the East River Basin, South China: ENSO and Precipitation in South China. *J. Geophys. Res. Atmos.* **2013**, *118*, 2207–2219. [CrossRef]
133. Huang, C.; Zhang, Q.; Singh, V.P.; Gu, X.; Shi, P. Spatio-Temporal Variation of Dryness/Wetness across the Pearl River Basin, China, and Relation to Climate Indices: Dryness/Wetness across the Pearl River Basin. *Int. J. Climatol.* **2017**, *37*, 318–332. [CrossRef]
134. Huang, S.; Wang, L.; Wang, H.; Huang, Q.; Leng, G.; Fang, W.; Zhang, Y. Spatio-Temporal Characteristics of Drought Structure across China Using an Integrated Drought Index. *Agric. Water Manag.* **2019**, *218*, 182–192. [CrossRef]
135. Huang, C.; Zhang, Q.; Xiao, M. Influences of ENSO, NAO, IOD and PDO on precipitation regimes in the Pearl River basin. *Acta Sci. Nat. Univ. Sunyatseni.* **2016**, *55*, 134–142. (In Chinese) [CrossRef]

136. Huang, C.; Zhang, Q.; Chen, X.; Xiao, M. Spatio-temporal variations of dryness/wetness and influence from climate factors in Pearl River Basin. *J. China Hydrol.* **2017**, *37*, 12–20+59. Available online: https://kns.cnki.net/KCMS/detail/detail.aspx?dbcode=CJFD&dbname=CJFDLAST2017&filename=SWZZ201705003&v= (accessed on 1 February 2024). (In Chinese)
137. Heng, C.; Yoon, S.-K.; Kim, J.-S.; Xiong, L. Inter-Seasonal Precipitation Variability over Southern China Associated with Commingling Effect of Indian Ocean Dipole and El Niño. *Water* **2019**, *11*, 2023. [CrossRef]
138. Gu, X.; Zhang, Q.; Singh, V.P.; Shi, P. Hydrological Response to Large-Scale Climate Variability across the Pearl River Basin, China: Spatiotemporal Patterns and Sensitivity. *Glob. Planet. Chang.* **2017**, *149*, 1–13. [CrossRef]
139. Cui, L.; Chen, X.; An, J.; Yao, C.; Su, Y.; Zhu, C.; Li, Y. Spatiotemporal Variation Characteristics of Droughts and Their Connection to Climate Variability and Human Activity in the Pearl River Basin, South China. *Water* **2023**, *15*, 1720. [CrossRef]
140. Yan, X.; Zhang, M. Numerical simulation of the Indian ocean dipole influence on climate variations over east Asian monsoon region during equator East Pacific SSTA. *J. Trop. Meteorol.* **2004**, *4*, 375–382. (In Chinese) [CrossRef]
141. Chan, J.C.L. PDO, ENSO and the Early Summer Monsoon Rainfall over South China. *Geophys. Res. Lett.* **2005**, *32*, L08810. [CrossRef]
142. Yuan, L. Discussion of Relationship between Tropical Pacific Sea Surface Temperature Anomaly and Winter Rainfall over South China. Master's Thesis, Nanjing University of Information Science and Technology, Nanjing, China, 2013. Available online: https://kns.cnki.net/KCMS/detail/detail.aspx?dbcode=CMFD&dbname=CMFD201401&filename=1013340651.nh&v= (accessed on 1 February 2024). (In Chinese)
143. Song, X.; Zhang, J.; Zou, X.; Zhang, C.; AghaKouchak, A.; Kong, F. Changes in Precipitation Extremes in the Beijing Metropolitan Area during 1960–2012. *Atmos. Res.* **2019**, *222*, 134–153. [CrossRef]
144. Wang, X.; Hou, X.; Zhao, Y. Changes in Consecutive Dry/Wet Days and Their Relationships with Local and Remote Climate Drivers in the Coastal Area of China. *Atmos. Res.* **2021**, *247*, 105138. [CrossRef]
145. Wang, H.; Chen, H. Climate Control for Southeastern China Moisture and Precipitation: Indian or East Asian Monsoon?: Climate Control for Moisture in China. *J. Geophys. Res. Atmos.* **2012**, *117*, D12109. [CrossRef]
146. Tan, M. Circulation Effect: Response of Precipitation δ18O to the ENSO Cycle in Monsoon Regions of China. *Clim. Dyn.* **2014**, *42*, 1067–1077. [CrossRef]
147. Deng, S.; Yang, N.; Li, M.; Cheng, L.; Chen, Z.; Chen, Y.; Chen, T.; Liu, X. Rainfall Seasonality Changes and Its Possible Teleconnections with Global Climate Events in China. *Clim. Dyn.* **2019**, *53*, 3529–3546. [CrossRef]
148. Jia, L.; Zhou, S. The effect of Indian Ocean SST anomaly on Indian monsoon and summer precipitation over Tibetan Plateau. *Plateau Meteorol.* **2003**, *S1*, 132–137. Available online: https://kns.cnki.net/KCMS/detail/detail.aspx?dbcode=CJFD&dbname=CJFD2003&filename=GYQX2003S1017&v= (accessed on 1 February 2024). (In Chinese)
149. Luo, L.; Jiang, T.; Sun, H.; Jing, C.; Su, B. Variation of droughts and floods and their connection with atmospheric circulation in the Pearl River Basin. *J. Arid Land Resour. Environ.* **2017**, *31*, 142–147. (In Chinese) [CrossRef]
150. Wu, Z.-Y.; Lu, G.-H.; Liu, Z.-Y.; Wang, J.-X.; Xiao, H. Trends of Extreme Flood Events in the Pearl River Basin during 1951–2010. *Adv. Clim. Chang. Res.* **2013**, *4*, 110–116. [CrossRef]
151. Zhang, Q.; Gu, X.; Singh, V.P.; Xiao, M.; Xu, C.-Y. Flood Frequency under the Influence of Trends in the Pearl River Basin, China: Changing Patterns, Causes and Implications: Flood Frequency under Influences of Trends. *Hydrol. Process.* **2015**, *29*, 1406–1417. [CrossRef]
152. Gu, X.; Zhang, Q. Non-stationary flood risk analysis in Pearl River Basin, considering the impact of hydrological trends. *Geogr. Res.* **2014**, *33*, 1680–1693. (In Chinese) [CrossRef]
153. Zhang, Q.; Gu, X.; Singh, V.P.; Xiao, M.; Xu, C.-Y. Stationarity of Annual Flood Peaks during 1951–2010 in the Pearl River Basin, China. *J. Hydrol.* **2014**, *519*, 3263–3274. [CrossRef]
154. Wu, X.; Huang, W.; Ning, X.; Song, J.; Wu, Z.; Dan, X.; Wang, H.; Lu, L. Dynamic change of land rocky desertification and river sediment content in Pearl River Basin. *Bull. Soil Water Conserv.* **2021**, *41*, 22–30. (In Chinese) [CrossRef]
155. Wang, J.; Kang, Q.; Yang, D. Current status of soil erosion and rocky desertification in the upper reaches of the Pearl River, their causes and prevention strategies. *Subtrop. Soil Water Conserv.* **2014**, *26*, 38–41. Available online: https://kns.cnki.net/KCMS/detail/detail.aspx?dbcode=CJFD&dbname=CJFD2014&filename=FJSB201403012&v= (accessed on 1 February 2024). (In Chinese)
156. Dou, X.; Peng, Q.; Zhang, W.; Huang, W.; Yi, Q.; Li, M. Impacts of LUCC and climate change on runoff in the Nanpan River Basin based on Scenario analysis. *J. Catastrophology* **2020**, *35*, 84–89. (In Chinese) [CrossRef]
157. Liu, B.; Peng, S.; Liao, Y.; Wang, H. The Characteristics and Causes of Increasingly Severe Saltwater Intrusion in Pearl River Estuary. *Estuar. Coast. Shelf Sci.* **2019**, *220*, 54–63. [CrossRef]
158. Wang, X.; Yan, F.; Su, F. Changes in Coastline and Coastal Reclamation in the Three Most Developed Areas of China, 1980–2018. *Ocean Coast. Manag.* **2021**, *204*, 105542. [CrossRef]
159. Shen, Y.; Zhang, H.; Tang, J. Hydrodynamics and Water Quality Impacts of Large-Scale Reclamation Projects in the Pearl River Estuary. *Ocean Eng.* **2022**, *257*, 111432. [CrossRef]
160. Cao, Y.; Zhang, W.; Zhu, Y.; Ji, X.; Xu, Y.; Wu, Y.; Hoitink, A.J.F. Impact of Trends in River Discharge and Ocean Tides on Water Level Dynamics in the Pearl River Delta. *Coast. Eng.* **2020**, *157*, 103634. [CrossRef]
161. Zhang, H.-X.; Shen, Y.-M.; Tang, J. Wave and Storm Surge Evolutions in the Pearl River Estuary with Large-Scale Land Reclamation Impacts. *Ocean Eng.* **2023**, *273*, 113977. [CrossRef]

162. He, L.; Li, G.; Li, K.; Shu, Y. Estimation of Regional Sea Level Change in the Pearl River Delta from Tide Gauge and Satellite Altimetry Data. *Estuar. Coast. Shelf Sci.* **2014**, *141*, 69–77. [CrossRef]
163. Chen, Y.; Zuo, J.; Zou, H.; Zhang, M.; Zhang, K. Responses of Estuarine Salinity and Transport Processes to Sea Level Rise in the Zhujiang (Pearl River) Estuary. *Acta Oceanol. Sin.* **2016**, *35*, 38–48. [CrossRef]
164. Yin, K.; Xu, S.; Huang, W.; Xie, Y. Effects of Sea Level Rise and Typhoon Intensity on Storm Surge and Waves in Pearl River Estuary. *Ocean Eng.* **2017**, *136*, 80–93. [CrossRef]
165. Zhang, Q.; Xiao, M.; Singh, V.P.; Chen, X. Copula-Based Risk Evaluation of Droughts across the Pearl River Basin, China. *Theor. Appl. Climatol.* **2013**, *111*, 119–131. [CrossRef]
166. Xiao, M.; Zhang, Q.; Kong, D.; Zhang, Z. Research on drought behaviors in the Pearl River Basin based on markov chain. *J. N. China Univ. Water Resour. Electr. Power (Nat. Sci. Ed.)* **2016**, *37*, 15–23. (In Chinese) [CrossRef]
167. Zhang, H.; Wu, C.; Hu, B.X. Recent Intensification of Short-term Concurrent Hot and Dry Extremes over the Pearl River Basin, China. *Int. J. Climatol.* **2019**, *39*, 4924–4937. [CrossRef]
168. Chen, Z. Empirical diagnostic analysis on spatial and temporal variations of droughts in the Pearl River basin, China. *Acta Sci. Nat. Univ. Sunyatseni.* **2020**, *59*, 33–42. (In Chinese) [CrossRef]
169. Wu, Z.; Bai, B.; He, H.; Li, Y.; Xu, H. Temporal and spatial characteristics of hydrological drought in the Pearl River Basin from 1981 to 2020. *J. Hohai Univ. (Nat. Sci.)* **2023**, *51*, 1–9. (In Chinese) [CrossRef]
170. Liu, Z. Review of the establishment of the Pearl River Water Resources Commission of the Ministry. *Pearl River* **1999**, *5*, 4–5. Available online: https://kns.cnki.net/KCMS/detail/detail.aspx?dbcode=CJFD&dbname=CJFD9899&filename=RMZJ199905002&v= (accessed on 1 February 2024). (In Chinese)
171. Shen, D. The 2002 Water Law: Its Impacts on River Basin Management in China. *Water Policy* **2004**, *6*, 345–364. [CrossRef]
172. Song, L. Study on the Resources Management System in Pearl River Basin. Master's Thesis, South China University of Technology, Guangzhou, China, 2009. Available online: https://kns.cnki.net/KCMS/detail/detail.aspx?dbcode=CMFD&dbname=CMFD2011&filename=2010048300.nh&v= (accessed on 1 February 2024). (In Chinese).
173. Zhang, J. Study and comparison on reforming models in China's major river-basin management-system. *J. Chongqing Univ. (Soc. Sci. Ed.)* **2014**, *20*, 18–22. (In Chinese) [CrossRef]
174. Xing, H. Thoughts on the water resources management system and mechanism of the Pearl River Basin. *Pearl River* **2008**, *1*, 3–6. Available online: https://kns.cnki.net/KCMS/detail/detail.aspx?dbcode=CJFD&dbname=CJFD2008&filename=RMZJ200801006&v= (accessed on 1 February 2024). (In Chinese)
175. Chen, Q. Study on the Rule of Law in the Cooperative Governance of the Pearl River Basin. Master's Thesis, Guangdong University of Foreign Studies, Guangzhou, China, 2021. Available online: https://kns.cnki.net/KCMS/detail/detail.aspx?dbcode=CMFD&dbname=CMFD202201&filename=1021763253.nh&v= (accessed on 1 February 2024). (In Chinese)
176. Liu, J.; Zhou, Q. Discussing countermeasures and measures for water resources management in the Pearl River based on foreign experience. *Water Resour. Plan. Design.* **2006**, *3*, 32–35. Available online: https://kns.cnki.net/KCMS/detail/detail.aspx?dbcode=CJFD&dbname=CJFD2006&filename=SLGH200603009&v= (accessed on 1 February 2024). (In Chinese)
177. Qian, Y. The Water Supply and Its Development in the Pearl River. Master's Thesis, Hohai University, Nanjing, China, 2005. Available online: https://kns.cnki.net/KCMS/detail/detail.aspx?dbcode=CMFD&dbname=CMFD0506&filename=2006013253.nh&v= (accessed on 1 February 2024). (In Chinese)
178. Edmonds, D.A.; Toby, S.C.; Siverd, C.G.; Twilley, R.; Bentley, S.J.; Hagen, S.; Xu, K. Land Loss Due to Human-Altered Sediment Budget in the Mississippi River Delta. *Nat. Sustain.* **2023**, *6*, 644–651. [CrossRef]
179. Ai, B.; Ma, C.; Zhao, J.; Zhang, R. The Impact of Rapid Urban Expansion on Coastal Mangroves: A Case Study in Guangdong Province, China. *Front. Earth Sci.* **2020**, *14*, 37–49. [CrossRef]
180. Weng, Q. A Historical Perspective of River Basin Management in the Pearl River Delta of China. *J. Environ. Manag.* **2007**, *85*, 1048–1062. [CrossRef] [PubMed]
181. Chen, W.; Liu, P.; Chen, J. Thoughts on the regulation and protection for the Pearl River estuary. *China Water Resour.* **2020**, *20*, 36–39. Available online: https://kns.cnki.net/KCMS/detail/detail.aspx?dbcode=CJFD&dbname=CJFDLAST2020&filename=SLZG202020025&v= (accessed on 1 February 2024). (In Chinese).
182. Cui, S. Discussion on the urban water sources in the Pearl River estuary regions: Problems and solution. *China Water Resour.* **2010**, *1*, 32–35. Available online: https://kns.cnki.net/KCMS/detail/detail.aspx?dbcode=CJFD&dbname=CJFD2010&filename=SLZG201001015&v= (accessed on 1 February 2024). (In Chinese).
183. Dominicis, M.D.; Wolf, J.; Hespen, R.V.; Zheng, P.; Hu, Z. Mangrove Forests Can Be an Effective Coastal Defence in the Pearl River Delta, China. *Commun. Earth Environ.* **2023**, *4*, 13. [CrossRef]
184. Li, J.; Li, X.; Li, C. Practice and reflection of Pearl River water dispatching in dry season. *China Flood Drought Manag.* **2021**, *31*, 68–72. (In Chinese) [CrossRef]
185. Tang, Q.; Chen, L.; Cao, P. A brief discussion on the construction ideas of water management remote sensing monitoring system in the Pearl River Basin. *Pearl River* **2013**, *34*, 3–5. (In Chinese) [CrossRef]
186. Gan, H.; Wu, H. Study on the construction of the digital twin Pearl River Basin. *China Flood Drought Manag.* **2022**, *32*, 36–39. (In Chinese) [CrossRef]
187. Xing, H. Research on The legal mechanism of valley water resources administration. *Law Sci. Mag.* **2007**, *3*, 103–105. (In Chinese) [CrossRef]

188. Wang, F.; Hao, R.J. Environmental Pollution in Pearl River Delta, China: Status and Potential Effects. *J. Environ. Inform. Lett.* **2020**, *3*, 110–123. [CrossRef]
189. Weng, Z.; Li, L.; Liu, Y. Construction of water culture and water policy and regulation in the Guangdong-Hong Kong-Macao Greater Bay Area. *China Water Resour.* **2023**, *2*, 62–66. Available online: https://kns.cnki.net/KCMS/detail/detail.aspx?dbcode=CJFD&dbname=CJFDAUTO&filename=SLZG202302016&v= (accessed on 1 February 2024). (In Chinese)

water

MDPI

Article

The Impact of the Three Gorges Reservoir Operations on Hydraulic Characteristics in the Backwater Region: A Comprehensive 2D Modeling Study

Yaqian Xu [1,2,*], Shengde Yu [3], Defu Liu [1,2,*], Jun Ma [2] and Mingying Chuo [2]

1 State Key Laboratory of Water Resources Engineering and Management, Wuhan University, Wuhan 430072, China
2 Key Laboratory of Health Intelligent Perception and Ecological Restoration of River and Lake, Ministry of Education, Hubei University of Technology, Wuhan 430068, China; junma@hbut.edu.cn (J.M.); mchuo@unc.edu (M.C.)
3 Department of Earth and Environmental Sciences, University of Waterloo, 200 University Avenue West, Waterloo, ON N2L 3G1, Canada; s228yu@uwaterloo.ca
* Correspondence: artryxu@whu.edu.cn (Y.X.); dfliu@189.cn (D.L.)

Abstract: The Three Gorges Reservoir (TGR), a landmark of human engineering, has significantly altered the hydrodynamics and ecology of its surrounding environment. Our research explores the hydrodynamic and ecological changes in the TGR, focusing on their implications for reservoir-induced water quality and water resource issues. We designed a 2D hydrodynamic and water quality model and implemented 15 operational scenarios with an advanced dynamic storage capacity method for the TGR during flood season, drawdown and impoundment periods. Our simulations well reproduced and predicted water levels, discharge rates, and thermal conditions of the TGR, providing critical insights. The dynamic storage capacity method significantly improved the precision of water level simulations. This approach achieved modeling errors below 0.2 m when compared to real measurements from seven stations. We performed a detailed analysis of the sensitive, sub-sensitive, and insensitive areas during three reservoir operation periods. The drawdown period showed the most extensive impact range (468 km river channel), while the impoundment period had the least impact range (76 km river channel). Furthermore, we quantified the delay of temperature waves during these periods, observing a maximum delay of approximately 120 km and a minimum delay of less than 10 km, which underscores the variability in hydrodynamic responses under different operational scenarios. Our findings reveal the complex sensitivities of the TGR to varied operational modes, aiding in the development of eutrophication and water resources control strategies. Our modeling application provides different operational scenarios and insights for ecological management strategies in large dam systems globally, informing future water resource management and policy-making, ensuring sustainable and effective management of large reservoir systems.

Keywords: large dam system; operational scenarios; sensitivity zoning; TGR 2D hydrodynamic model; water resource management

Citation: Xu, Y.; Yu, S.; Liu, D.; Ma, J.; Chuo, M. The Impact of the Three Gorges Reservoir Operations on Hydraulic Characteristics in the Backwater Region: A Comprehensive 2D Modeling Study. *Water* **2024**, *16*, 2045. https://doi.org/10.3390/w16142045

Academic Editors: Guilin Han and Xiaolong Liu

Received: 27 June 2024
Revised: 16 July 2024
Accepted: 17 July 2024
Published: 19 July 2024

1. Introduction

Across millennia, aquatic ecosystems have evolved with their hydrothermal environments, forming complex and dynamic interactions [1,2]. The evolution of aquatic ecosystems is not static; they are constantly changing and adapting to environmental pressures and human activities, such as climate variations, land use changes, and water resource management, etc. The recent surge in dam constructions has significantly altered these ecological systems [3–5], bringing both benefits in hydropower and water management, but also altering the hydrothermal dynamics of riverine ecosystems [6,7]. These changes present potential risks to aquatic ecosystems [8–10]. For example, dam construction and management can contribute to greenhouse gas emissions, while eutrophication may occur in reservoirs resulting in poor water quality [11–13].

The Three Gorges Reservoir (TGR) stands as a prime example of such infrastructural advancements and their potential environmental impacts. Following the construction of the TGR, dramatic changes have occurred in its aquatic habitat, including an expanded water surface, reduced flow rates, and increased sediment deposition. This transformation has led to a new equilibrium in the ecosystems, creating a hybrid river–reservoir environment [14–16].

Although numerous studies have explored the long-term ecological and hydrothermal consequences of reservoir operations [17–19], there remains a gap in understanding the short-term impacts of water level fluctuations, particularly in large systems like the TGR. We hypothesize that daily water level fluctuations, caused by short-term scheduling in the TGR, result in significant changes in water temperature and levels. These changes affect the hydrodynamics and ecological health of the reservoir's different regions. Understanding these variations in the TGR could provide insights for managing similar large dam systems globally.

Focusing on the TGR, the current study aims to understand the complex relationship between dam operations, water level fluctuations, and hydrothermal dynamics. Our modeling application, focused on the Three Gorges Dam (TGD), provides different operational scenarios and insights for ecological management strategies. This approach ensures sustainable and effective management of large reservoir systems similar to the TGD, informing future water resource management and policy-making.

2. Methods

2.1. Study Area and Data Collection

Located in the Yangtze River, China, the TGR serves as a prime example of a standard channel reservoir. With its impressive capacity of 39.3 billion cubic meters (3.93×10^{10} m^3), the TGR ranks as the world's foremost hydropower station. Since its first filling in June 2003, which increased the water level to 135 m, followed by a further increase to 175 m in 2010, the TGR has witnessed regular fluctuations in water levels between 145 and 175 m [15].

This study focuses on the backwater region of the TGR, extending from the Cuntan hydrologic station in Chongqing City to the Three Gorges Dam (TGD) in Zigui, Yichang City, covering a span of approximately 604 km (Figure 1). There are 6 larger tributaries considered in the subsequent model: Wujiang (WJ), Longhe (LH), Xiaojiang (XJ), Modaoxi (MDX), Daninghe (DNH), and Xiangxi River (XXR).

2.2. Model Development

2.2.1. Model Grid Generalization

CE-QUAL-W2 is a two-dimensional (longitudinal-vertical) laterally averaged hydrodynamic and water quality model developed by the U.S. Army Corps of Engineers' Waterways Experiment Station. It is suitable for numerical simulation of long and narrow rivers, estuaries, and reservoirs. This model has been successfully used in many peer-reviewed studies on hydrodynamics and water quality [20]. The model has been applied to rivers [21], lakes [22,23], reservoirs [24,25], estuaries [26,27], and their combinations, including entire river basins with multiple reservoirs and river segments. Additionally, CE-QUAL-W2 also performs well in modeling laboratory tanks [28], showcasing its reliability and feasibility in both natural and controlled settings.

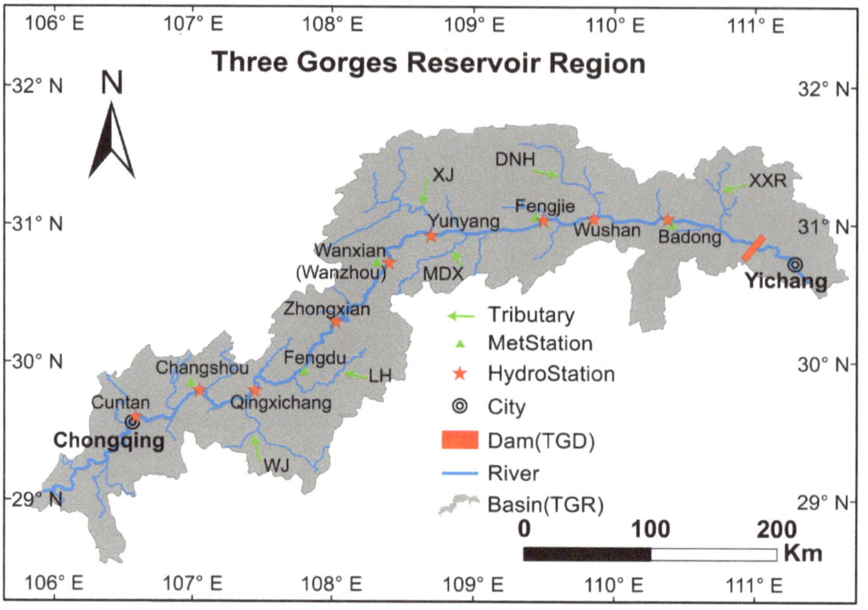

Figure 1. The spatial representation of the Three Gorges Reservoir (TGR) (highlighting the backwater area, extending from the Cuntan hydrologic station in Chongqing City to the Three Gorges Dam in Zigui, Yichang City. Key sections including Cuntan, Changshou, Wanxian, Yunyang, Fengjie, Wushan, Badong, and the TGD are displayed in green or red shapes).

The computational grid system of the TGR model used a rectangular grid. This grid system (Figure 2) was divided into 7 branches (waterbodies), with 8 key cross-sections including Cuntan, Changshou, Wanxian, Yunyang, Fengjie, Wushan, Badong, and the TGD (Table 1). The grid was defined by a longitudinal length of 2000 m, vertical layer spacing of 1 m, and a width for each layer derived from the average width of the section, calculated using data from over 100 large cross-sections. The TGR model consisted of 316 segments and 214 layers. It specified two distinct slopes, one for the permanent backwater region, set at zero, and another for the fluctuating backwater area, set at a non-zero value. In the modeling process, the mainstream of the TGR was considered with detailed bathymetry, while tributaries were treated as point source loadings and were not included in the bathymetric data.

Table 1. Bathymetry parameters for all waterbodies.

Branch ID	Upstream /HydroStation	Downstream /HydroStation	Number of Segments	Branch Length (km)	Tributary	MetStation
BR_1	Cuntan	Changshou	39	78	—	Changshou
BR_2	Changshou	Wanxian	124	248	WJ & LH	Fengdu
BR_3	Wanxian	Yunyang	19	38	XJ	Wanzhou
BR_4	Yunyang	Fengjie	37	74	MDX	Fengjie
BR_5	Fengjie	Wushan	24	48	DNH	Fengjie
BR_6	Wushan	Badong	22	44	—	Badong
BR_7	Badong	TGD	37	76	XXR	Badong

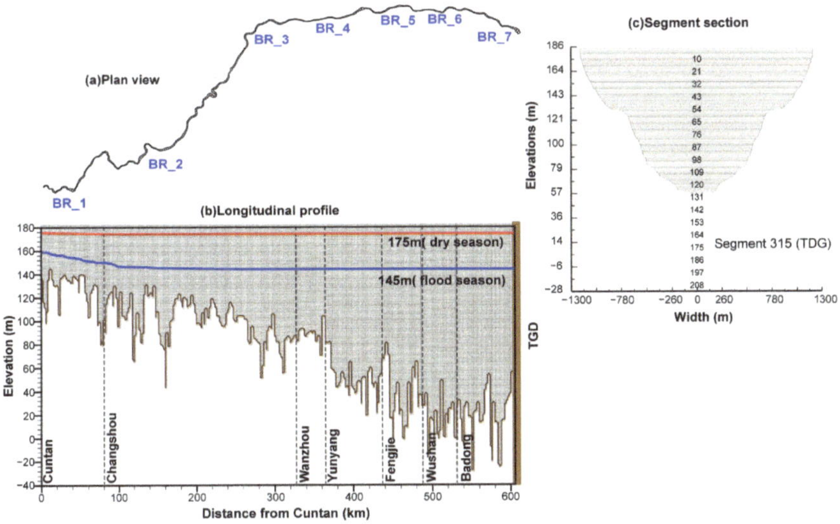

Figure 2. Grids of TGR model: (**a**) plan view, (**b**) longitudinal profile, and (**c**) segment section for TGD. There were 316 segments and 214 layers in total. Length = 2000 m. Layer height = 1 m.

A more detailed description of the data required for modeling in this study is provided in the Supplementary Materials (Figures S1–S5 and Tables S1 and S2).

2.2.2. Assessment of Model Performance

To assess the model's effectiveness, two primary metrics were utilized, the mean error (ME) and the absolute mean error (AME). The ME offers insights into the model's bias, whereas the AME provides an average error measure, reflecting the model's overall performance. These metrics are widely recognized in the scientific community for evaluating the congruence between models and observed data [20,24,26]. The calculation formula is as follows:

$$ME = \frac{\sum_{i=1}^{n}(X_{obs,i} - X_{model,i})}{n} \tag{1}$$

$$AME = \frac{\sum_{i=1}^{n}|X_{obs,i} - X_{model,i}|}{n} \tag{2}$$

where n is the number of observations, $X_{obs,i}$ is the i_{th} observation, and $X_{model,i}$ is the i_{th} simulated value.

This model's starting date was 1 May 2014 (Julian day = 121) and ending date was 30 April 2015 (Julian day = 485). The calibration and validation of the model demonstrated that the simulated water elevation was quite close to the observed elevation, with an $|ME|$ less than 0.2 and an AME less than 0.37. Additionally, the $|ME|$ of the surface water temperature in key sections was less than 0.35 °C (Figures S4 and S5).

2.2.3. Reservoir Operation Scenarios

As an annual regulation reservoir, the TGR has four operation periods during a hydrological year: a dry season (winter), a drawdown period (spring), a flood season (summer), and an impoundment period (autumn). During the dry season, the TGR maintains a high water level of 175 m from November to April of the following year. The drawdown period extends from mid-May to June. In the flood season, the water level is lowered to 145 m from June to August. Lastly, the impoundment period occurs from September to November, preparing the reservoir for the upcoming hydrological cycle.

According to the tide-type ecological operation method [14,29] and the TGR regulation rule (China Three Gorges Corporation), five specific regulation scenarios have been implemented for each operation period. These regulation scenarios are significant in managing the reservoir's hydrological dynamics.

The water level process under the tide-type ecological operation method has been parameterized into six types: initial water level (IWL), daily fluctuation in the first fluctuation period (DF1), daily fluctuation in the second fluctuation period (DF2), duration of the first fluctuation period (DLDF1), duration of the second fluctuation period (DLDF2), days with an unchanged water level (DWLU), and duration for the scheduling period (DSP). These parameters for the typical regulation scenarios are detailed in Table 2, providing a comprehensive view of each scheme's characteristics. Additionally, Figure 3 graphically depicts the water level process, illustrating the fluctuations and patterns across different periods within the TGR operation, each water level process lasting 15 days.

Table 2. Parameters of reservoir regulation scenarios.

Scenario Number	IWL (m)	DF1 (m/Day)	DLDF1 (Day)	DWLU (Day)	DF2 (m/Day)	DLDF2 (Day)	DSP (Day)
S00	145	0	0	15	0	0	15
S01	145	1	10	5	−1	0	15
S02	145	1	7	7	−1	1	15
S03	145	1	4	4	−1	4	15
S04	145	1	1	1	−1	1	15
S10	165	0.2	4	0	−0.6	11	15
S11	165	−0.6	15	0	0	0	15
S12	165	−0.6	11	0	0.2	4	15
S13	165	−0.6	4	0	0.1	11	15
S14	165	−0.3	15	0	0	0	15
S20	150	0	15	0	0	0	15
S21	150	1	2	2	1	2	15
S22	150	1	15	0	0	0	15
S23	150	3	3	10	3	2	15
S24	150	3	3	0	0.5	12	15

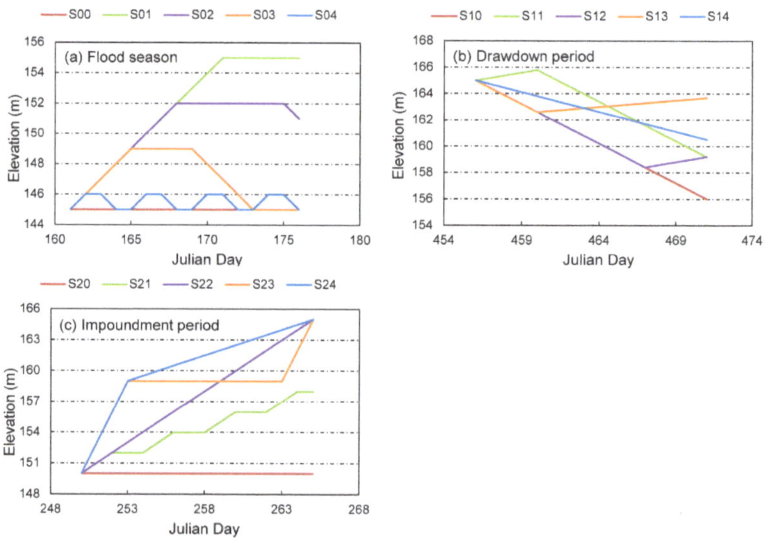

Figure 3. Water level processes for different regulation schemes: (**a**) flood season, (**b**) drawdown period, and (**c**) impoundment period with different initial water levels (145 m, 165 m, and 150 m).

2.3. Defining Sensitivity Zones in Channel Reservoirs

Channel reservoirs are characterized by a pronounced longitudinal gradient change pattern, with deeper waters near the dam head and shallower waters in the tail and tributary backwater areas. These unique bathymetric morphologies and flow characteristics give rise to hydrodynamic properties that differ from typical rivers and lakes, where flow velocity and other parameters exhibit a combination of both riverine and lake features. From the dam's head to the reservoir's end, a gradient change is apparent, leading to the categorization of the reservoir into river-type, transition-type, and lake-type zones [30]. Due to varying discharge rates and distances from the dam, different sections of the backwater area respond differently to water level fluctuations at the dam. For instance, a daily rise of 1.5 m in the water level at the TGD, sustained over a week, could potentially cause the water level in the Wushan section (Da'ning River) to increase by up to 1.4 m per day. However, the Yunyang section (Xiaojiang River) and Wanzhou section only reach a maximum of 1.3 m per day, lagging by approximately one day. The Qingxichang section only achieves a maximum change of 0.5 m per day, with almost no significant change occurring in the Cuntan section.

To investigate the extent of the backwater area impacted by the TGD's water level fluctuations in different seasons, 'sensitivity zoning' was employed. This method separated the channel reservoir into different sensitivity zones based on the correlation analysis of various hydrodynamic characteristics across different sections (Figure 4). A correlation coefficient ($|R|$) value of $\geq$0.9 denotes a 'Sensitive Area'; a value between 0.5 and <0.9 signifies a 'Sub-sensitive Area'; and a value of <0.5 indicates an 'Insensitive Area'.

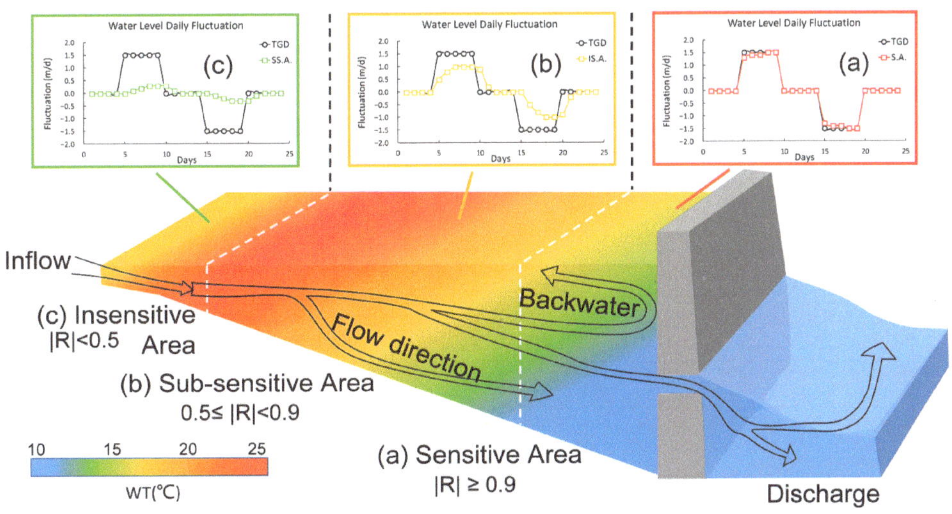

Figure 4. Longitudinal zoning of sensitivity in channel reservoirs and their response to tide-type operations at TGD. (**a**–**c**) Daily water level fluctuations in different zones. V.S. tide-type operation at TGD. S.A., SS.A., and IS.A. represent "sensitive area", "sub-sensitive area" and "insensitive area".

3. Results

3.1. Correlation Analysis of Water Level Daily Fluctuations in Different Scenarios

To understand the impact of reservoir operations on water level fluctuations, a correlation analysis was conducted on key sections of the TGR, with the correlation coefficients presented in Table 3. The analysis revealed that the sensitivity of areas within the reservoir varies across different operation periods (Figure 5).

Table 3. Correlation coefficients (|R|) of typical sites during different periods. ($p < 0.01$).

Site Name	Flood Season	Drawdown Period	Impoundment Period
Cuntan	0.1895	0.6439	0.3967
Changshou	0.3482	0.8334	0.2894
Qingxichang	0.7086	0.8984	0.3730
Zhongxian	0.8179	0.9444	0.5771
Wanxian	0.8686	0.9747	0.6663
Yunyang	0.8918	0.9785	0.6927
Fengjie	0.9381	0.9894	0.7397
Wushan	0.9682	0.9947	0.7583
Badong	0.9964	0.9987	0.9099

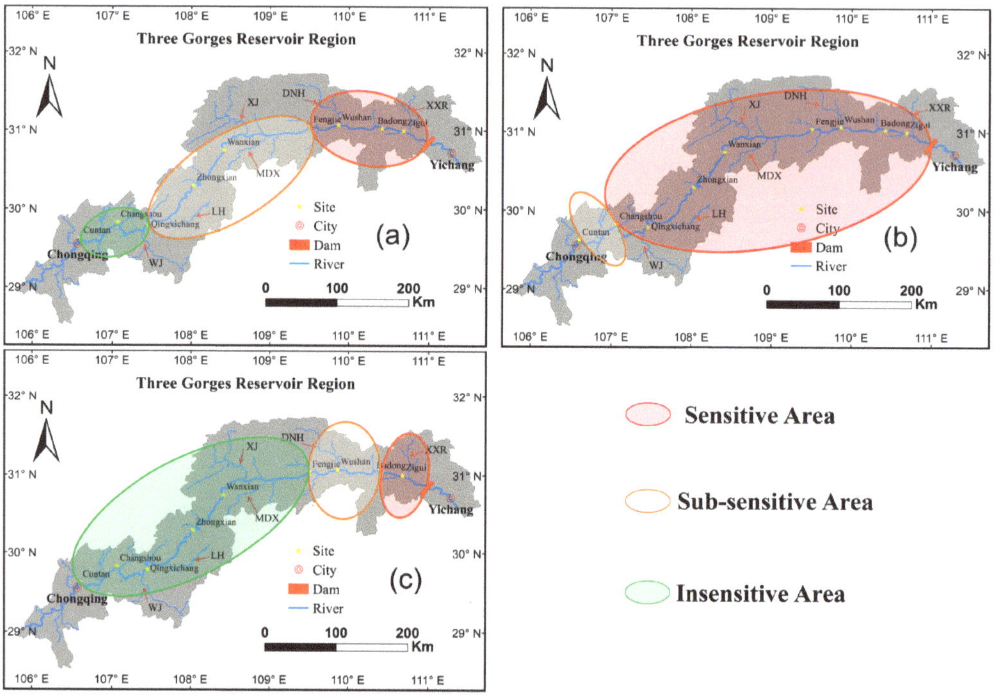

Figure 5. Zoning of sensitivity in TGR during different operation periods: (**a**) flood season, (**b**) drawdown period, and (**c**) impoundment period.

In the flood season, the section from Fengjie to the TGD was identified as a sensitive area, spanning a river length of 168 km, which constituted 27.8% of the total river length. The area from Qingxichang to Fengjie was classified as a sub-sensitive area, covering 300 km (49.7% of the total river length). Finally, the segment from Cuntan to Qingxichang was designated as an insensitive area, with a river length of 136 km, accounting for 22.5% of the total river length. This scenario reflects the influence of operational strategies on water level dynamics.

During the drawdown period, the sensitive area extended from Qingxichang to the TGD, covering 468 km (77.5% of the total river length), marking the largest sensitive zone. The sub-sensitive area during this period ranged from Changshou to Qingxichang, covering a distance of 136 km (22.5% of the total river length). These variations highlight the changing impact of water level management across the reservoir.

In the impoundment period, the sensitive area was from Badong to the TGD, covering 76 km (12.6% of the total river length). The sub-sensitive area extended from Zhongxian to Badong, covering 288 km (47.7% of the total river length). Additionally, the river channel from Cuntan to Zhongxian was identified as the insensitive area, spanning 240 km (39.7% of the total river length). This distribution was primarily due to the lower discharge rates during this period. Contrary to common expectations, the sensitive area in the impoundment period was smaller than in the flood season, mainly because the upstream section is more influenced by inflow and less influenced by the backwater effect of the reservoir. However, if the discharge exceeds inflow, the extent of the impact of water level scheduling increases, highlighting the complexity of managing reservoir operations.

This correlation analysis underscores the importance of tailoring reservoir management strategies to the specific dynamics of each operational period, ensuring an optimal balance between water resource utilization and environmental conservation.

3.2. Surface Water Temperature Variability under Different Scenarios

As depicted in Figure 6, surface water temperatures in the TGR display varying patterns under different operational scenarios, with each scenario revealing its unique characteristics.

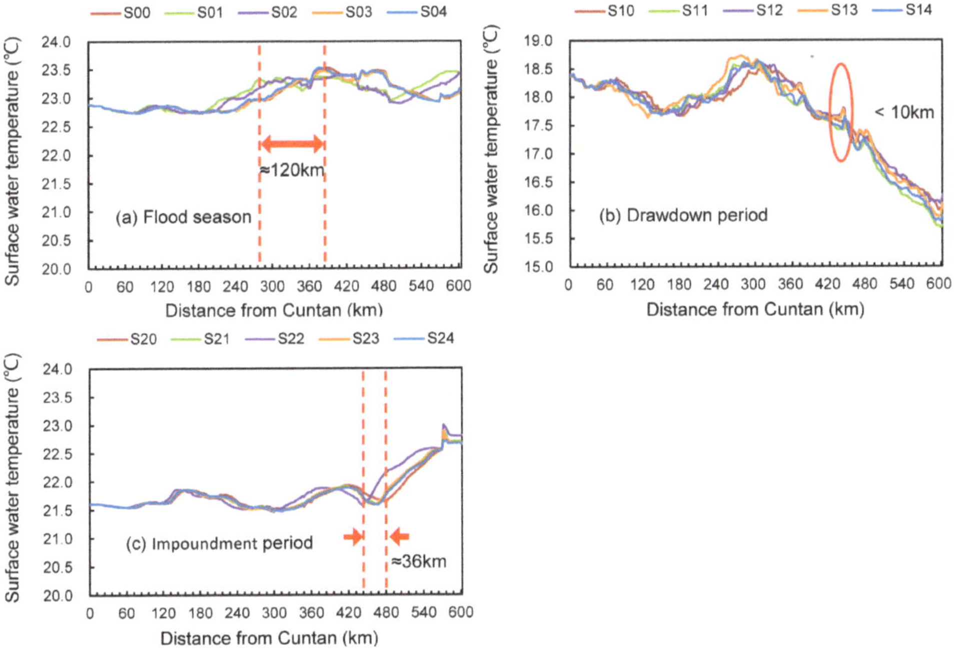

Figure 6. Surface water temperature in different scenarios: (**a**) flood season, (**b**) drawdown period, and (**c**) impoundment period.

(a) Flood Season Analysis:

The flood season revealed similar temperature change patterns under five regulation conditions, with comparable highest, lowest, and average surface temperatures. The spatial distribution of temperatures, including the location of peak temperatures and the high-temperature zone, was influenced by water level processes, but not significantly by water level fluctuations during the scheduling period.

In S01 to S04, water levels in the TGR exceeded the 145 m benchmark observed in the baseline S00. This indicates higher water storage in these scenarios compared to baseline. Temperature Waves: A notable feature in these scenarios (S01–S04) was the delayed propagation of surface water temperature waves compared to the baseline S00. Interestingly, a direct relationship was observed; higher water levels corresponded to a slower transfer speed of the temperature wave. When the water level in front of the dam rose by 10 m (comparing S01 with S00), the propagation of the temperature wave from upstream to downstream was delayed over a range of approximately 120 km.

(b) Drawdown Period Analysis:

The spatial distribution of surface water temperature in the mainstream during the drawdown period remained consistent across different regulation conditions, with stability upstream and a downward trend midstream and downstream. The highest and average surface temperatures varied minimally (<0.2 °C), and the lowest temperature difference was small (<0.4 °C). The temperature in front of the dam was more influenced by water level fluctuations than by final water level height. For instance, scenarios 10 and 12, both ending at 159.2 m, exhibited a 0.6 °C temperature difference in front of the dam due to differing water level trajectories.

Across all five regulation schemes (S10 to S14), a consistent observation was the minimal fluctuation in the reservoir's water volume. The consistency in water volume led to uniformity in surface water temperature waves across these scenarios. The spatial difference in the phase of temperature waves was essentially negligible, not exceeding 10 km, because of the minimal differences in water volume during the drawdown period.

(c) Impoundment Period Analysis:

The water temperature trends observed during the impoundment period showed similarities to those in the flood season. During the impoundment period, when the water level in front of the dam rises (using S20 as a benchmark), the delay in the propagation of the water temperature wave along the river at the end of the regulation period ranged from 15 to 36 km. The maximum spatial delay occurred in scenario S22, as this scenario maintained a daily fluctuation of 1 m/d for continuous storage over 15 days, whereas other scenarios experienced a slowdown in the rate of water storage.

3.3. Variation of Vertical Water Temperature in Different Scenarios

Variations in water temperature within the TGR were noticeable, both vertically and longitudinally (Figure 7), and were significantly influenced by different operational scenarios.

During the flood season, under five conditions (S00–S04), the TGR exhibited consistent vertical temperature conditions. In the upstream sections, the water temperature was almost uniformly mixed, showing negligible vertical stratification. However, a dispersion phenomenon occurred in the vertical temperature profile, resulting in temporary stratification in some parts of the river, such as the middle section of the reservoir. This is attributed to the discharge of water from the mid-upper layer of the reservoir, leading to faster surface water flow compared to the bottom layer. Each condition presented specific stratification profiles, with onset points ranging from 455 km to 462 km and peak stratification occurring between 550 km and 560 km. The differences in water temperature between the surface and bottom layers varied, with maximum differences reaching up to 11.98 °C, although variations near the dam were generally minimal, usually under 1 °C.

During the drawdown period, similar conditions were observed under five different conditions (S10–S14), characterized by upstream temperature homogeneity and downstream stratification. Stratification onset points shifted from around 280 km to 310 km, with peaks up to 560 km, and temperature differences ranged from 4.16 °C to 5.97 °C. The temperature differential near the dam was more significant, reaching up to 2.8 °C. The trend of outflow exceeding inflow during this period implies that operational scheduling only alters the length of the stratified section without significantly impacting the vertical temperature structure.

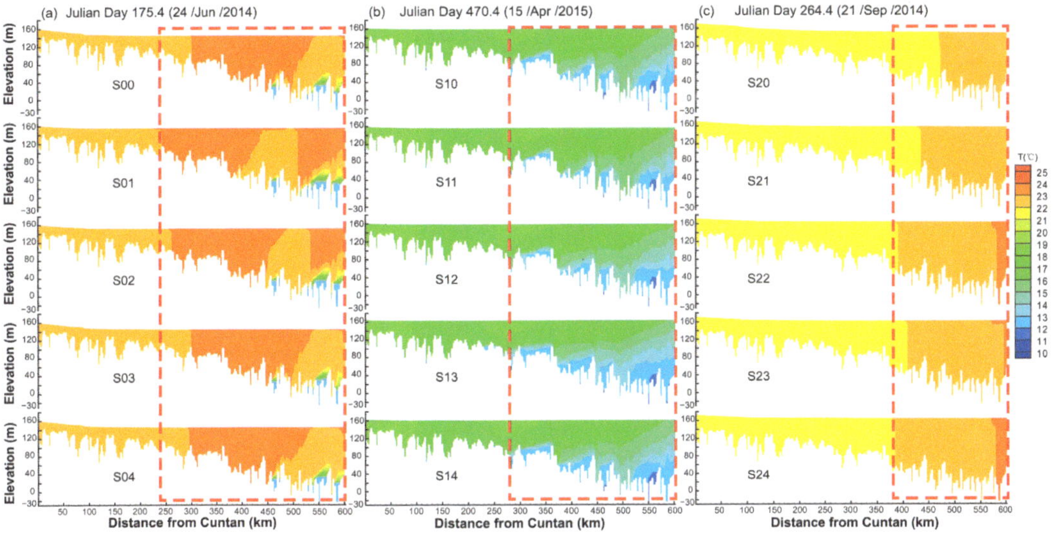

Figure 7. Water temperature longitudinal profile after different scenarios: (**a**) flood season, Julian Day = 175.4, (**b**) drawdown period, Julian Day = 470.4, and (**c**) impoundment period, Julian Day = 264.4.

During the impoundment period, stratification was minimal. However, longitudinal temperature distribution was evident, influenced by water level variations. In conditions S22 to S24, higher water levels led to clearer stratification layers, marked by localized high surface temperatures extending from the CJXX to the dam front (approximately 32 km river length). For both the flood season and impoundment periods, our results showed that if the water level remained constant, temperature variations occurred longitudinally along the river. During rising water conditions, an "S"-shaped thermal stratification formed near the dam, caused by the strengthening backwater effects (S01, S02, and S22, S23, S24).

4. Discussion

4.1. Modeling Backwater Fluctuations: Implications, Challenges and Perspectives

The backwater region presents an interesting characteristic; it is like a natural river at low water levels and transforms into a reservoir-like state at higher water levels. This characteristic poses unique challenges for reservoir simulation and water quality management. Typically, such reservoir simulations assume a uniform water level [20]; however, natural rivers exhibit a gradient on the water surface [31]. This condition becomes particularly crucial in the case of water quality and resource management for the TGR.

Ignoring the variations of the fluctuating backwater region during simulations can lead to significant errors [32,33]. This is particularly evident when considering the reservoir's capacity. Simulated estimates (or static capacity) may not accurately reflect the actual capacity (or dynamic capacity), especially at lower water levels. This discrepancy can be attributed to factors such as the elevation at the riverbed's base at the end of the TGR. During low water level periods, the error in estimating water levels becomes non-negligible. In some scenarios, this can even result in a complete disruption of the flow, underscoring the need for more detailed modeling approaches that account for the variation of the backwater region (Figure 8).

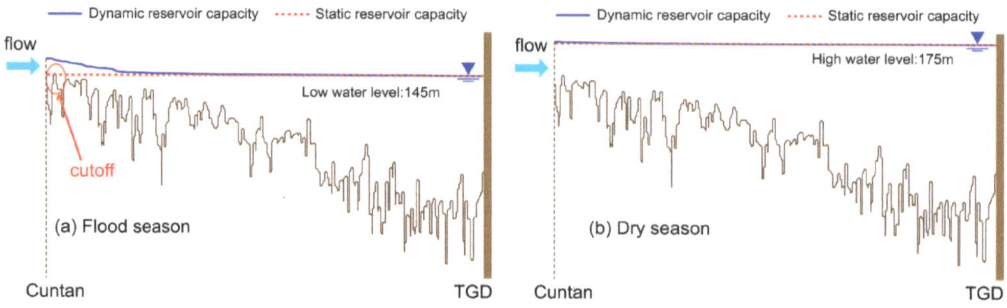

Figure 8. Dynamic and static reservoir capacity of TGR under different water levels: (**a**) flood season and (**b**) dry season.

Our results highlight the complexities of modeling dynamic water systems like the TGR and underscore the importance of integrating natural river characteristics into simulation models for more accurate and effective reservoir water resource and water quality management. However, the reliability of these results depends on the suitability of the modeling data. Over the course of long-term reservoir operation, sedimentation will occur and alter the bottom topography, thereby impacting the hydrodynamic characteristics of the reservoir area. It is reasonable for this TGR model to match the simulation period of the model with the bathymetric data surveyed in 2013, spanning from 1 May 2014 to 30 April 2015. The topographic data will need to be updated accordingly if there are any updates made to the study period.

4.2. Role of Reservoir Operations on Tributaries in Different Sensibility Zones

The sensitivity zones of different tributary estuaries vary across seasons, as does their susceptibility to fluctuations in water levels upstream of the dam. Firstly, the XXR falls within a highly responsive area during all three periods and exhibits extreme sensitivity to water level fluctuations in the TGR. This forms the basis for ecological operation and algae control measures in the XXR. Secondly, the representative tributaries in the middle and at the end of the reservoir are only sensitive during the drawdown period, such as XJ and LH, respectively. Therefore, if environmental issues like algal blooms arise in spring within these rivers, emergency regulations can be implemented through water level adjustments. During the flood season and impoundment period, however, this approach may not be effective; other alternative methods for environmental governance are required.

How does reservoir scheduling impact the water environment conditions of different tributaries? There are two aspects to consider. Firstly, water level fluctuations affect various estuaries located within distinct sensitive areas with varying potential energy [34]. Secondly, water temperature waves accelerate or decelerate propagation processes along the main stream due to operations at the TGR resulting in increased or decreased temperatures at river mouths, thereby influencing water exchange patterns among major tributaries.

XXR experiences thermal stratification for the majority of the year [35]. Lakes often display static stratification [36], while rivers, due to their high flow velocity, typically prevent stratification. However, reservoirs and especially their tributaries can manifest characteristics of both lakes and rivers. This unique combination can lead to the formation of density currents [37,38]. The depth of the intrusion layer is dictated by the temperature differential between the intrusion flow and the existing stratification in the bay [26,39–41].

Notably, the depth of this intrusion layer is influenced by water level fluctuations. For instance, density currents shifted from the middle to the surface layer during rising water levels [39]. Such shifts are crucial since only these transfers can effectively mitigate algal blooms in XXR. In this study, we conducted a sensitivity analysis on the reservoir area impacted by the operations of the TGR during different periods, providing valuable support for enhancing and refining the TGR regulation rule. For instance, the flood limit water level (145 m) could be raised to help control blooms in XXR at appropriate times during the flood season. By reservoir scheduling to induce these shifts, it provides an innovative yet straightforward strategy for algal bloom management. This approach essentially converges on the concept of ecology operation.

With the development of cascade hydropower projects and more attention to the ecological environment, the ability of reservoirs to effectively balance eco-environmental and dynamoelectric benefits through operational strategies is a matter of significant concern for all stakeholders involved [42]. There have been relevant studies on the impact of reservoir operations on aquatic ecosystems of tributaries [3,43] and the enhancement of tributary water environments through these operations [44].

5. Conclusions

This current study presents a thorough examination of the water level and temperature of the TGR on the Yangtze River. We analyzed the effects of reservoir operations on hydraulic characteristics, with a focus on the TGR's backwater region. We conclude that:Large reservoir modeling research needs to consider dynamic storage capacity to achieve higher accuracy and to better reflect real conditions.

(1) During different operational periods of the reservoir, the range of areas affected by changing water levels varies. The sensitive zone is largest during the drawdown period (468 km river channel), making it more feasible to achieve ecological benefits through operational regulation. Conversely, during the impoundment period, the sensitive zone is the smallest (76 km river channel), posing challenges for enhancing ecological and environmental benefits through operational regulations. During the flood season, the sensitive zone is moderate in size (168 km river channel), but ecological and environmental benefits need to be considered with priority of ensuring flood control objectives.

(2) The fluctuation in reservoir water levels alters the propagation speed of surface temperature waves in the river direction by increasing or decreasing the water volume. The propagation of surface water temperature waves from upstream to downstream could be delayed over a range of approximately 120 km in the flood season and could be delayed by about 36 km in the impoundment period. However, the spatial difference in the phase of temperature waves did not exceed 10 km during the drawdown period.

(3) The study of surface and vertical water temperature variabilities revealed different patterns across the flood season, drawdown, and impoundment periods. In most cases, the vertical water temperature in the TGR was well-mixed, but occasionally, short-term stratification occurred due to the flow velocity among the surface, middle, and bottom layers.

In summary, our research offers a comprehensive understanding of the TGR's hydrodynamic and thermal characteristics during short-term scheduling. These insights provide significant scientific evidence for developing effective reservoir management strategies, beneficial for the TGR and other large-scale reservoirs worldwide.

Supplementary Materials: The following supporting information can be downloaded at https://www.mdpi.com/article/10.3390/w16142045/s1, Figure S1: Inflows and outflows; Figure S2: Air temperature and dew point temperature in typical weather stations of TGR; Figure S3: Wind rose diagrams in typical weather stations of TGR; Figure S4: Comparisons of the simulated and observed water level; Figure S5: Comparisons of the simulated and observed surface water temperature; Table S1: Location and data information; Table S2: Location and data information. References [45,46] were cited in the Supplementary Materials.

Author Contributions: Y.X.: writing—original draft, conceptualization, software, validation, formal analysis, writing—review and editing. S.Y.: writing—original draft, writing—review and editing. D.L.: supervision, writing—review and editing. J.M.: software. M.C.: writing—review and editing. All authors have read and agreed to the published version of the manuscript.

Funding: This work was supported by the National Natural Science Foundation of China (Grant Nos. U2040220, 52179065), the Natural Science Foundation of Hubei Province, China (No.2024AFD369), the Open Project Funding of Key Laboratory of Intelligent Health Perception and Ecological Restoration of Rivers and Lakes, Ministry of Education, Hubei University of Technology (HGKFZ01), the Innovation Demonstration Base of Ecological Environment Geotechnical and Ecological Restoration of Rivers and Lakes (2020EJB004), and the Major Science and Technology Program of The Ministry of Water Resources (SKS-2022077).

Data Availability Statement: The data presented in this study are available on request from the corresponding author.

Acknowledgments: The writers acknowledge engineers in China Three-gorge Corporation for their help with data collection. The writers also acknowledge the assistance of anonymous reviewers.

Conflicts of Interest: The authors declare no conflicts of interest.

References

1. Benjankar, R.; Tonina, D.; McKean, J.A.; Sohrabi, M.M.; Chen, Q.; Vidergar, D. Dam Operations May Improve Aquatic Habitat and Offset Negative Effects of Climate Change. *J. Environ. Manag.* **2018**, *213*, 126–134. [CrossRef] [PubMed]
2. Smucker, N.J.; Beaulieu, J.J.; Nietch, C.T.; Young, J.L. Increasingly Severe Cyanobacterial Blooms and Deep Water Hypoxia Coincide with Warming Water Temperatures in Reservoirs. *Glob. Chang. Biol.* **2021**, *27*, 2507–2519. [CrossRef] [PubMed]
3. Zarfl, C.; Lumsdon, A.E.; Berlekamp, J.; Tydecks, L.; Tockner, K. A Global Boom in Hydropower Dam Construction. *Aquat. Sci.* **2015**, *77*, 161–170. [CrossRef]
4. Grill, G.; Lehner, B.; Thieme, M.; Geenen, B.; Tickner, D.; Antonelli, F.; Babu, S.; Borrelli, P.; Cheng, L.; Crochetiere, H.; et al. Mapping the World's Free-Flowing Rivers. *Nature* **2019**, *569*, 215–221. [CrossRef] [PubMed]
5. Maavara, T.; Chen, Q.; Van Meter, K.; Brown, L.E.; Zhang, J.; Ni, J.; Zarfl, C. River Dam Impacts on Biogeochemical Cycling. *Nat. Rev. Earth Environ.* **2020**, *1*, 103–116. [CrossRef]
6. Wang, Y.; Zhang, N.; Wang, D.; Wu, J.; Zhang, X. Investigating the Impacts of Cascade Hydropower Development on the Natural Flow Regime in the Yangtze River, China. *Sci. Total Environ.* **2018**, *624*, 1187–1194. [CrossRef] [PubMed]
7. Zhang, P.; Yang, Z.; Cai, L.; Qiao, Y.; Chen, X.; Chang, J. Effects of Upstream and Downstream Dam Operation on the Spawning Habitat Suitability of Coreius Guichenoti in the Middle Reach of the Jinsha River. *Ecol. Eng.* **2018**, *120*, 198–208. [CrossRef]
8. Wu, H.; Chen, J.; Xu, J.; Zeng, G.; Sang, L.; Liu, Q.; Yin, Z.; Dai, J.; Yin, D.; Liang, J.; et al. Effects of Dam Construction on Biodiversity: A Review. *J. Clean. Prod.* **2019**, *221*, 480–489. [CrossRef]
9. Song, J.-H.; Ma, L.; Her, Y.; Li, Y. Immediate Influences of a Large Dam Construction on Local Storm Event Patterns and Weather Variables: A Case Study of the Three Gorges Project. *Weather* **2020**, *75*, 99–103. [CrossRef]
10. Barbarossa, V.; Schmitt, R.J.P.; Huijbregts, M.A.J.; Zarfl, C.; King, H.; Schipper, A.M. Impacts of Current and Future Large Dams on the Geographic Range Connectivity of Freshwater Fish Worldwide. *Proc. Natl. Acad. Sci. USA* **2020**, *117*, 3648–3655. [CrossRef] [PubMed]
11. Yu, Z.; Wang, L. Factors Influencing Thermal Structure in a Tributary Bay of Three Gorges Reservoir. *J. Hydrodyn.* **2011**, *23*, 407–415. [CrossRef]
12. Winemiller, K.O.; McIntyre, P.B.; Castello, L.; Fluet-Chouinard, E.; Giarrizzo, T.; Nam, S.; Baird, I.G.; Darwall, W.; Lujan, N.K.; Harrison, I.; et al. Balancing Hydropower and Biodiversity in the Amazon, Congo, and Mekong. *Science* **2016**, *351*, 128–129. [CrossRef] [PubMed]
13. Yang, H.; Guo, W.; Wang, H. Anthropogenic Intensification of the Eco-Hydrothermal Regime Transition in Regulated Rivers: The Cumulative Effect of Cascade Reservoirs. *J. Environ. Manag.* **2024**, *349*, 119478. [CrossRef] [PubMed]
14. Liu, D.; Yang, Z.; Ji, D.; Ma, J.; Cui, Y.; Song, L. A Review on the Mechanism and Its Controlling Methods of the Algal Blooms in the Tributaries of Three Gorges Reservoir. *J. Hydraul. Eng.* **2016**, *47*, 443–454. (In Chinese) [CrossRef]
15. Wang, L.; Li, H.; Yang, Z.; Zhang, J.; Deng, C.; Yang, F.; Zheng, B. Four Effects of Water Environment Evolution in Early Period (2003–2012) after Impoundment of the Three Gorges Reservoir. *Res. Environ. Sci.* **2020**, *33*, 1109–1118. (In Chinese) [CrossRef]
16. Cai, Q.; Hu, Z. Studies on Eutrophication Problem and Control Strategy in the Three Gorges Reservoir. *Acta Hydrobiol. Sin.* **2006**, *30*, 7–11. (In Chinese) [CrossRef]
17. Blaen, P.J.; Hannah, D.M.; Brown, L.E.; Milner, A.M. Water Temperature Dynamics in High Arctic River Basins. *Hydrol. Process.* **2013**, *27*, 2958–2972. [CrossRef]
18. Sun, Y.; Tang, D.; Dai, H.; Liu, P.; Sun, Y.; Cui, Q.; Zhang, F.; Ding, Y.; Wei, Y.; Zhang, J.; et al. A Real-Time Operation of the Three Gorges Reservoir with Flood Risk Analysis. *Water Sci. Technol.-Water Supply* **2016**, *16*, 551–562. [CrossRef]
19. Xu, Y.; Guo, X.; Dan, Y.; Yang, Z.; Ma, J.; Liu, D.; Xu, Y. Impact of Cascade Reservoirs on Continuity of River Water Temperature: A Temperature Trend Hypothesis in River. *Hydrol. Process.* **2021**, *35*, e13994. [CrossRef]
20. Cole, T.M.; Wells, S.A. *CE-QUAL-W2: A Two-Dimensional, Later-Ally Averaged, Hydrodynamic and Water Quality Model, Version 4.0 User Manual*; Department of Civil and Environmental Engineering, Portland State University: Portland, OR, USA, 2016.

21. Al-Murib, M.D.; Wells, S.A. Hydrodynamic and Total Dissolved Solids Model of the Tigris River Using CE-QUAL-W2. *Environ. Process.* **2019**, *6*, 619–641. [CrossRef]
22. Berger, C.; Wells, S.A.; Annear, R.L. *Laurance Lake Temperature Model*; Technical Report EWR-01-04; Water Quality Research Group, Department of Civil and Environmental Engineering, Portland State University: Portland, OR, USA, 2005.
23. Wells, S.A.; Berger, C.J. Modeling the Response of Dissolved Oxygen to Phosphorus Loading in Lake Spokane. *Lake Reserv. Manag.* **2016**, *32*, 270–279. [CrossRef]
24. Long, L.; Ji, D.; Liu, D.; Yang, Z.; Lorke, A. Effect of Cascading Reservoirs on the Flow Variation and Thermal Regime in the Lower Reaches of the Jinsha River. *Water* **2019**, *11*, 1008. [CrossRef]
25. He, W.; Wang, H.; Zhang, J.; Xu, H.; Xiao, Y. Diurnal Variation Characteristics of Thermal Structure in a Deep Reservoir and the Effects of Selective Withdrawal. *J. Environ. Manag.* **2023**, *333*, 117459. [CrossRef] [PubMed]
26. Ma, J.; Liu, D.; Wells, S.A.; Tang, H.; Ji, D.; Yang, Z. Modeling Density Currents in a Typical Tributary of the Three Gorges Reservoir, China. *Ecol. Model.* **2015**, *296*, 113–125. [CrossRef]
27. Li, Y.; Sun, J.; Lin, B.; Liu, Z. Thermal-Hydrodynamic Circulations and Water Fluxes in a Tributary Bay of the Three Gorges Reservoir. *J. Hydrol.* **2020**, *585*, 124319. [CrossRef]
28. Xu, Y.; Ma, J.; Yang, Z. Calibration of Kinetic Parameters and Numerical Simulation of Growth of Microcystis Aeruginosa Cultured in Laboratory. *J. Ecol. Rural Environ.* **2018**, *34*, 267–275. (In Chinese)
29. Yang, Z.; Liu, D.; Ji, D.; Ma, J. Feasibility Study on Tide-Type Ecological Operation Method to Control Algal Bloom in Tributaries of Three Gorges Reservoir. *Water Resour. Power* **2015**, *33*, 48–50+109. (In Chinese)
30. Zhang, Y.; Zheng, B.; Fu, G.; Liu, H. On the Assessment Methodology and Standards for Nutrition Status in Channel Type Reservoirs Based on Zoning of Eutrophication Sensitivity. *Acta Sci. Circumstantiae* **2006**, *26*, 1016–1021. (In Chinese) [CrossRef]
31. Zhang, S.; Jing, Z.; Yi, Y.; Wu, Y.; Zhao, Y. The Dynamic Capacity Calculation Method and the Flood Control Ability of the Three Gorges Reservoir. *J. Hydrol.* **2017**, *555*, 361–370. [CrossRef]
32. Jing, Z.; An, W.; Zhang, S.; Xia, Z. Flood Control Ability of River-Type Reservoirs Using Stochastic Flood Simulation and Dynamic Capacity Flood Regulation. *J. Clean. Prod.* **2020**, *257*, 120809. [CrossRef]
33. Zhang, R.; Zhang, S.; Wen, X.; Jing, Z. Refined Scheduling Based on Dynamic Capacity Model for Short-Term Hydropower Generation. *Water Resour. Manag.* **2023**, *37*, 21–35. [CrossRef]
34. Ji, D.; Zhou, Z.; Yang, Z.; Long, L.; Wang, Y. Study on the Gravity Wave Induced by the Diurnal Variation of Outflow Discharge in the Three Gorges Reservoir. *J. Hydraul. Eng.* **2021**, *52*, 111–119. (In Chinese) [CrossRef]
35. Long, L.; Xu, H.; Yan, M.; Yang, X.; Zhao, X.; Ji, D.; Cui, Y.; Yang, Z.; Liu, D.; Yang, Z. Variable Drivers of Surface Turbulence under Condition of Stratified Density Currents in a Tributary Bay of Three Gorges Reservoir. *J. Hydrol.* **2023**, *623*, 129849. [CrossRef]
36. Boehrer, B.; Schultze, M. Stratification of Lakes. *Rev. Geophys.* **2008**, *46*, RG2005. [CrossRef]
37. Jin, J.; Wells, S.A.; Liu, D.; Yang, G.; Zhu, S.; Ma, J.; Yang, Z. Effects of Water Level Fluctuation on Thermal Stratification in a Typical Tributary Bay of Three Gorges Reservoir, China. *PeerJ* **2019**, *7*, e6925. [CrossRef]
38. Noori, R.; Tian, F.; Ni, G.; Bhattarai, R.; Hooshyaripor, F.; Klöve, B. ThSSim: A Novel Tool for Simulation of Reservoir Thermal Stratification. *Sci. Rep.* **2019**, *9*, 18524. [CrossRef] [PubMed]
39. Yang, Z.J.; Liu, D.F.; Ji, D.B.; Xiao, S.B. Influence of the Impounding Process of the Three Gorges Reservoir up to Water Level 172.5 m on Water Eutrophication in the Xiangxi Bay. *Sci. China Technol. Sci.* **2010**, *53*, 1114–1125. [CrossRef]
40. Chuo, M.; Ma, J.; Liu, D.; Yang, Z. Effects of the Impounding Process during the Flood Season on Algal Blooms in Xiangxi Bay in the Three Gorges Reservoir, China. *Ecol. Model.* **2019**, *392*, 236–249. [CrossRef]
41. Xu, Y.; Xu, P.; Yang, Z.; Liu, D.; Ma, J. Dynamic Replenishment Process of Nutrients in Tributary of Channel Reservoir. *Environ. Sci.* **2018**, *39*, 765–773. (In Chinese) [CrossRef]
42. Xu, C.; Zhu, D.; Guo, W.; Ouyang, S.; Li, L.; Bu, H.; Wang, L.; Zuo, J.; Chen, J. Multi-Objective Ecological Long-Term Operation of Cascade Reservoirs Considering Hydrological Regime Alteration. *Water* **2024**, *16*, 1849. [CrossRef]
43. Zhi, X.; Chen, L.; Chen, S.; Yu, J.; Jiang, J.; Xu, Y.; Li, L.; Meng, X.; Shen, Z. Heterogeneity and Influencing Factors of Algal Blooms in the Reservoir-Impacted Tributary: Evidence from Remote Sensing and Physical-Based Model. *J. Hydrol.* **2024**, *634*, 131058. [CrossRef]
44. Zhang, H.; Chang, J.; Gao, C.; Wu, H.; Wang, Y.; Lei, K.; Long, R.; Zhang, L. Cascade Hydropower Plants Operation Considering Comprehensive Ecological Water Demands. *Energy Convers. Manag.* **2019**, *180*, 119–133. [CrossRef]
45. MET4 and MET4A Calculation of Dew Point. Archived from the Original on 26 May 2012. Available online: https://archive.ph/20120526034637/http://www.paroscientific.com/dewpoint.htm (accessed on 7 October 2014).
46. Barenbrug, A.W.T. *Psychrometry and Psychrometric Charts*, 3rd ed.; Cape and Transvaal Printers Ltd.: Cape Town, South Africa, 1974.

MDPI AG
Grosspeteranlage 5
4052 Basel
Switzerland
Tel.: +41 61 683 77 34

Water Editorial Office
E-mail: water@mdpi.com
www.mdpi.com/journal/water